NATO ASI Series

Advanced Science Institutes Series

A series presenting the results of activities sponsored by the NATO Science Committee, which aims at the dissemination of advanced scientific and technological knowledge, with a view to strengthening links between scientific communities.

The Series is published by an international board of publishers in conjunction with the NATO Scientific Affairs Division

A Life Sciences B Physics	Plenum Publishing Corporation London and New York
C Mathematical and Physical Sciences D Behavioural and Social Sciences E Applied Sciences	Kluwer Academic Publishers Dordrecht, Boston and London
F Computer and Systems Sciences G Ecological Sciences H Cell Biology I Global Environmental Change	Springer-Verlag Berlin Heidelberg New York London Paris Tokyo Hong Kong Barcelona Budapest

PARTNERSHIP SUB-SERIES

1. Disarmament Technologies	Kluwer Academic Publishers
2. Environment	Springer-Verlag/Kluwer Acad. Publishers
3. High Technology	Kluwer Academic Publishers
4. Science and Technology Policy	Kluwer Academic Publishers
5. Computer Networking	Kluwer Academic Publishers

The Partnership Sub-Series incorporates activities undertaken in collaboration with NATO's Cooperation Partners, the countries of the CIS and Central and Eastern Europe, in Priority Areas of concern to those countries.

NATO-PCO DATABASE

The electronic index to the NATO ASI Series provides full bibliographical references (with keywords and/or abstracts) to about 50000 contributions from international scientists published in all sections of the NATO ASI Series. Access to the NATO-PCO DATABASE is possible via a CD-ROM "NATO Science & Technology Disk" with user-friendly retrieval software in English, French and German (© WTV GmbH and DATAWARE Technologies Inc. 1992).

The CD-ROM can be ordered through any member of the Board of Publishers or through NATO-PCO, Overijse, Belgium.

Series I: Global Environmental Change, Vol. 56

Springer
Berlin
Heidelberg
New York
Barcelona
Hong Kong
London
Milan
Paris
Singapore
Tokyo

Ice Physics and the Natural Environment

Edited by

John S. Wettlaufer
Applied Physics Laboratory and
Department of Physics
University of Washington
Seattle, Washington 98105-5640, USA

J. Gregory Dash
Department of Physics
University of Washington
Seattle, Washington 98105-1560, USA

Norbert Untersteiner
Department of Atmospheric Sciences
University of Washington
Seattle, Washington 98105-1640, USA

With 100 Figures and 7 Tables

Springer
Published in cooperation with NATO Scientific Affairs Division

Proceedings of the NATO Advanced Research Workshop "Ice Physics in the Natural and Endangered Environment", held at Maratea, Italy, September 1997

Cataloging-in-Publication Data applied for

Die Deutsche Bibliothek - CIP-Einheitsaufnahme

Ice physics and the natural environment : with 7 tables ; [proceedings of the NATO Advanced Research Workshop "Ice Physics in the Natural and Endangered Environment" held at Maratea, Italy, September 1997] / ed. by John S. Wettlaufer ... Publ. in cooperation with NATO Scientific Affairs Division. - Berlin ; Heidelberg ; New York ; Barcelona ; Hong Kong ; London ; Milan ; Paris ; Singapore ; Tokyo : Springer, 1999
(NATO ASI series : Ser. I, Global environmental change ; Vol. 56)
ISBN 3-540-65155-1

ISBN-13: 978-3-642-64226-5 e-ISBN-13: 978-3-642-60030-2
DOI: 10.1007/978-3-642-60030-2

Softcover reprint of the hardcover 1st edition 1999

Typesetting: Camera ready by the authors/editors
Printed on acid-free paper
SPIN: 10571794 31/3137 - 5 4 3 2 1 0

ACKNOWLEDGEMENTS

The idea of an ice physics conference came about in March 1996, during a visit to Seattle by Professor John Nye of the University of Bristol. Professor Nye has had a renowned career applying physics to a wide range of topics ranging from optics to crystallography and glaciology, and his work has touched researchers in many departments of the University of Washington. One of us (Untersteiner) worked with Nye previously, and the other two had recently published a review on the surface melting of ice and its environmental consequences. We were pleased the review was well received, as a number of people told us they appreciated its broad reach. Stimulated by Professor Nye's visit, and a wonderful evening surrounding Krystyna Untersteiner's cooking, we began to think about a meeting on an extension of this general approach; that is, to explore ice phenomena, ranging from microphysics to its largest size scales and manifestations. In the next few days, we sketched the outlines of a meeting designed as a NATO Advanced Study Institute, with which Dash and Untersteiner had some experience years before as directors. We invited three eminent scientists to join us on an Advisory Board: Professor Nye, Professor Krystof Birkenmajer from Kraków, Poland, and Prof. Fiorenzo Ugolini of Florence, Italy. We then began to survey the interest of colleagues and the process of constructing a proposal to the NATO Scientific and Environmental Affairs Division. The response was positive and we rushed to put the proposal together to meet the NATO deadline. We were informed in mid-December 1996 that the proposal would be supported and the organizational effort began to accelerate. In the eight months that followed, we announced the ASI, reviewed applications, organized the lecture program, designed the conference materials, developed the financial protocol and communicated with all those involved. None of this would have been possible without the hard work of Peggy Hartman, Suzan Huney, Dian Petersen and Agnes Sieger of the Applied Physics Laboratory. Ms. Hartman was involved in the difficult job of putting our rushed proposal together. Ms. Petersen built up the organizational edifice from the beginning of the effort, from, among many other things, maintaining the electronic communications with the participants, to designing the conference materials and putting together mailings. Ms. Huney administered the financial aspects of the ASI, and was an indispensable on-site budget and organizational coordinator. Ms. Sieger provided her expertise in the difficult task of technical editing and typesetting for the volume. We are indebted to their hard work and to the administration of the Applied Physics Laboratory, Robert Spindel and John Harlett, Director and Deputy Director, for grants in kind of staff time and materials.

Our sincere thanks go to the authors, lecturers and participants who constituted the Institute. Without the efforts of the authors to meet the deadlines, this volume would have been a never-ending project. The rather striking interdisciplinary nature of the Institute forced us to request that the lecturers provide both an up-to-date assessment of their fields for their disciplinary colleagues, and a broadly accessible overview for the nonspecialist. Although in principle this may be viewed as an unfair and unrealistic demand, the lecturers rose to the occasion and infused the Institute with a spirit of cooperation as well as an equally important and lively sense of scrutiny. We think this shines through in the present volume. We would like to thank the referees, many of whom were mailed manuscripts from the Institute venue, for their invaluable contribution to the success of the project. It is difficult to ask a busy colleague to review manuscripts of this nature, yet everyone was both receptive and timely, making our editorial work dramatically less burdensome.

Finally, we thank the NATO Scientific and Environmental Affairs Division for the grant that supported the travel and subsistence expenses for the lecturers, the subsidies for the participants, and other expenses of the Institute. We owe sincere thanks to Dr. L. Veiga da Cunha, Director of the NATO ASI Program. Without Dr. Veiga da Cunha's prompt, courteous and simple administrative style, our work would have been less pleasurable and certainly less efficient. We also thank the Tourist Office of the Regione Basilicata for providing additional funds that were used for Institute expenses, and for arranging special entertainment at the end of the first week. The National Science Foundation and the Office of Naval Research also provided funds to assist in the support of travel and subsistence. In particular, we would like to thank Dr. J. DeCorpo from the ONR for his timely response to our request for support. The choice of the location of the Institute in Acquafredda di Maratea, Italy (about 150 km south of Naples) is primarily due to the fact that one of us (Untersteiner) had Directed an ASI there in 1981, and had such a high opinion of the geography of the region and the staff of the Hotel Villa Del Mare, directed by Mr. A. Guzzardi. Our experience was no different and although the days were long, we were fortunate to enjoy superb hospitality and professional service from the hotel staff, and we would particularly like to thank Ms. Maria Armiento, Ms. Velia Filizzola and Mr. Guzzardi.

J.S. Wettlaufer, J.G. Dash and N. Untersteiner
Seattle, July 1998

Contents

Part 2 Lecture Notes

Part 3 Articles

PARTICIPANTS

Yevgeny Aksenov
University of Cambridge
Scott Polar Research Institute
Lensfield Road
Cambridge CB2 1ER
U.K.

Todd Arbetter
University of Colorado
Dept. of Aerospace Engineering Science
Campus Box 429
Boulder, CO 80309-0429
USA

Nicole Audiffren
LAMP
94, Av. des Landais
Aubere 635177
France

Neil Bacon
University of Washington
Department of Physics
Box 351560
Seattle, WA 98195
USA

Ian Baker
Dartmouth College
Thayer School of Engineering
Hanover, NH 03755
USA

Marcia Baker
University of Washington
Geophysics Program
Box 351650
Seattle, WA 98198
USA

Enrique Ricardo Batista
University of Washington
Department of Physics
Box 351560
Seattle, WA 98195-1560
USA

Krzysztof Birkenmajer
Polish Academy of Sciences
Institute of Geological Sciences
Senacka 3
Krakow 31-002
Poland

Ido Braslavsky
Weizmann Institute of Science
Dept. of Physics of Complex Systems
Rehovot 76100
Israel

Kurt Cuffey
University of Washington
Dept. Geological Sciences
Box 351310
Seattle, WA 98195-1310
USA

Benjamin Demirdjian
CRMC2-CNRS
Campus de luminy
Case 913
13288 Marseille Cedex 9
France

John Dempsey
Clarkson University
Department of Civil
& Environmental Engineering
Potsdam, NY 13699-5710
USA

Igor Dmitrenko
Kiel University
Center for Marine Geological Investigation
Kiel D-24148
Germany

Bo Elberling
University of Copenhagen
Institute of Geography
Øster Voldgade 10
Copenhagen DK-1350
Denmark

Sam Fain
University of Washington
Physics Department
Box 351560
Seattle, WA 98195-1560
USA

Daniel Feltham
University of Dundee
Department of Mathematics
Dundee DD14HN
U.K.

Alexander Finaev
Institute of Geology
Lab. of Glaciology & Meteorology
Dushanbe 734025
Tajikistan

Masami Fukuda
Hokkaido University
Institute of Low Temperature Science
Sapporo 060
Japan

Brigitte Furevik
University of Bergen
Nansen Environ. & Remote Sens. Ctr.
Edv. Greigsvei 3A
Solheimsviken N-5037
Norway

Yoshinori Furukawa
Hokkaido University
Inst. of Low Temperature Science
Sapporo 060
Japan

Sebastian Gerland
Norwegian Polar Institute
Storgata 25A
Tromso N-9008
Norway

Mats Granskog
Univ. of Helsinki
Department of Geophysics
P. Box 4
Helsinki FIN-00014
Finland

Mayo Greenberg
Huygens Laboratorium
Postbus 9504
Leiden 2300 RA
Netherlands

Kaisa Helena Hietala
University of Helsinki
Department of Geophysics
Helsinki FIN-00014
Finland

Van Hodgkin
University of Washington
Department of Physics
Box 351560
Seattle, WA 98195-1560
USA

Chris Jeffery
University of British Columbia
Department of Geography
British Columbia V6T 1Z2
CANADA

Sigfus Johnsen
Niels Bohr Institute
Dept. of Gephysics
Copenhagen DK-2200
Denmark

Tammy Johnson
Univ. of California Santa Barbara
ICESS
Santa Barbara, CA 93106-4060
USA

Peter Johnston
Universita Degli Studi di Milano
Via Emanueli, 15
Milano 20126
Italy

Alexander Kolesnikov
UMIST
Department of Physics
P.O. Box 88
Manchester M60 1QD
UK

Konrad Kossacki
Warsaw University
Institute of Geophysics
Poland

Geert-Jan Kroes
Leiden Institute of Chemistry
Gorlaeus Laboratories
RA Leiden 2300
NETHERLANDS

Ronald Kwok
Jet Propulsion Laboratory
4800 Oak Grove Drive
Pasadena, CA 91109
USA

Jacek Leliwa-Kopystynski
University of Warsaw
Institute of Geophysics
Pasteura 7
Warszawa 02093
Poland

Jichen Li
UMIST
Department of Physics
P.O. Box 88
Manchester M60 1QD
U.K.

Ole Lønne
UNIS
POB 156
Longyearbyen N-9170
Norway

Minoru Maruyama
Osaka City University
Department of Physics
Sumiyoshi-ku
Osaka 558
Japan

Brian Mason
University of Washington
Department of Physics
Box 351560
Seattle, WA 98195
USA

Felix Ng
University of Oxford
Mathematical Institute
24-29 St. Giles
Oxford OX1 3LB
U.K.

John Nye
University of Bristol
HH Wills Physics Laboratory
Bristol BS81TL
U.K.

Anders Omstedt
SMHI
Nörrkoping S-60176
Sweden

David Oxtoby
University of Chicago
James Franck Institute
Chicago, IL 60637
USA

Anja Paelli
University of Oulu
Department of Geophysics
Oulu 90571
Finland

W.S.B. Paterson
University of British Columbia
Vancouver, BC V6T 1Z4
Canada

Thomas Peter
Max-Planck-Institute fur Chemie
P.O. Box 3060
Mainz
Germany

Victor Petrenko
Dartmouth College
Thayer School of Engineering
Hanover, NH 03755
USA

Allan Phelps
University of Alaska
Dept. of Biology
3211 Providence Drive
Anchorage, AK 99508
USA

Lydia Popova
ul. Krupskoy, 8-1-187
Moscow 117311
Russia

Olga Popovitcheva
Institute of Nuclear Physics
Moscow State University
119899 INP
Moscow
Russia

Alan Rempel
University of Cambridge
Dept. of Applied Math. & Theoretical Phys.
Silver Street
Cambridge CB3 9EW
U.K.

Lane Seeley
University of Washington
Department of Physics
Box 351560
Seattle, WA 98195
USA

Gerald Seidler
University of Washington
Department of Physics
Box 351560
Seattle, WA 98195-1560
USA

Shaun Sellers
University of East Anglia
School of Mathematics UEA
NR4 7TJ Norwich
UK

Anna Sinisalo
University of Oulu
Department of Geophysics
Oulu 90571
Finland

Ronald Sletten
University of Washington
Quaternary Research Center
Box 351360
Seattle, WA 98195-1360
USA

Lars Hendrik Smedsrud
University of Bergen
Geophysical Institute
Alle gaten 70
Bergen 5007
Norway

Sergey Sokratov
Russian Academy of Science
Institute of Geography
Staromonetniy
Moscow 1099017
Russia

Jostein A. Solhaug
Nansen Environmental &
Remote Sensing Center
Edvard Griegsvei 3A
Solheimsviken N-5027
Norway

Ryszard Staroszczyk
University of East Anglia
School of Mathematics
Norwich NR4 773
UK

Nadja Steiner
Institut fur Meereskunde Kiel
Dusternbrookerweg 20
Kiel 24105
Germany

Jean Suzanne
CRMC2-CNRS
Department de Physique
Case 901
13288 Marseille Cedex 9
France

Alan Thorndike
University of Puget Sound
Department of Physics
138 Thompson Hall
Tacoma, WA 98416
USA

F.C. Ugolini
Universita di Firenze
Dept. Sci. del Suolo e Nutrizione della
Pianta
P. le Cascine
Firenze 15-50144
Italy

Juha Petteri Uotila
Finnish Institute of Marine Research
P.O. Box 33
Helsinki FIN-00931
Finland

Peter Wadhams
University of Cambridge
Scott Polar Research Institute
Lensfield Road
Cambridge CB2 1ER
U.K.

John Wettlaufer
University of Washington
Applied Physics Laboratory
Box 355640
Seattle, WA 98105-6698
USA

Larry Wilen
Ohio University
Dept. of Physics and Astronomy
252B Clippinger Labs
Athens, OH 45701
USA

Peter Williams
Carleton University
Geotechnical Science Laboratories
Ottawa, Ont. K1S 5B6
Canada

Introduction

The Advanced Study Institute *Ice Physics in the Natural and Endangered Environment* was held at Acquafredda di Maratea, Italy, from September 7 to 19, 1997. The ASI was designed to study the broad range of ice science and technology, and it brought together an appropriately interdisciplinary group of lecturers and students to study the many facets of the subject. The talks and poster presentations explored how basic molecular physics of ice have important environmental consequences, and, conversely, how natural phenomena present new questions for fundamental study. The following summary of lectures discusses these linkages, in order that overall unity of the subject and this volume can be perceived. Not all of the lecturers and participants were able to contribute a written piece, but their active involvement was crucial to the success of the Institute and thereby influenced the content of the volume.

We began the Institute by retracing the history of the search for a microscopic understanding of melting. Our motivation was straightforward. Nearly every phenomenon involving ice in the environment is influenced by the change of phase from solid to liquid or vice-versa. Hence, a sufficiently deep physical picture of the melting transition enriches our appreciation of a vast array of geophysical and technical problems. We found that the theory of melting possesses a long and distinguished history in the study of condensed matter, and in particular the interplay between dimensionality and phase behavior. Crucial in the history is the distinction between positional and topological long-range order; considering only one is not sufficient to understand melting. The search originally focused on a microscopic picture of continuous solid breakdown from within the bulk of the crystal, with intermolecular interactions treated as harmonic. Lower dimensional systems seemed to offer the most suitable subjects for studies of continuous melting, but we learned that the best realizations of two-dimensional matter—adsorbed monolayer films—melt just as abruptly as three-dimensional matter. Progress was made by considering the role of excitations such as vacancies, interstititals, and dislocations, the latter observed using synchrotron x-ray topography by Ian Baker in ice. But for all the interest continuous melting has captured, and for all the new physics that has grown out of it, it has never been observed definitively in ordinary solids. Such effort has thus been important for science but has left melting in an ambiguous state. By pointing a finger at the surface of solids, the situation has been greatly clarified. All real crystals are finite and their surfaces anticipate their destiny, melting from the outside in.

This process, called surface melting, is basic to melting in all materials. It is observed in ice at its interfaces with the vapor and other materials and results in the existence of mobile interfacial liquid at temperatures well below the normal freezing point. It was seen throughout the Institute to be crucial in many varied settings from frost heave to crystal growth.

Initiated by David Oxtoby's first lecture we turned our attention to a complementary question: How does solid form in the first place? In other words, within a bulk fluid or vapor below the normal freezing point, how is the solid nucleated? In melting, the anharmonicity of the outer atoms of the surface initiates the breakdown of the solid phase. In nucleation, although the bulk solid phase is the stable thermodynamic one, the surface separating it from the liquid phase must be *created.* The surface is the antagonist of the story, and Oxtoby showed us that there is more to this boundary than meets the eye. Classical nucleation theory fails to account for the structural variation associated with moving from one phase to the other. Plumbing the depths of the interface and properly accounting for the thermodynamic history of the material are crucial for the understanding of nucleation in general. Major roadblocks in our understanding of the nucleation of ice impede our ability to properly treat the microscopics of water so as to incorporate the tenets of nonclassical nucleation theory. Oxtoby made convincing arguments by analogy. For example, in an "argon-like" material (well described by a hard-sphere Lennard-Jones potential) the liquid anticipates the solid through an ordering process which lowers the free energy barrier to nucleation. (In particular, a metastable bcc ordering occurs at the interface between a liquid and an fcc solid.) By analogy, and reference to experimental data, Oxtoby brought to discussion the question of whether cubic ice is nucleated first and hexagonal ice either grows from the cubic seed or appears through a structural phase transition. He showed us that simulations demonstrate the importance of electric fields in ice nucleation and illuminated our ignorance of the solid/liquid interfacial energy as a function of temperature, a quantity so important in nucleation and crystal growth. The experimental studies of ice nucleation being performed by Lane Seeley, Jerry Seidler and Greg Dash might shed light on some of these issues. We learned about heterogeneous nucleation in general and were impressed by its importance in atmospheric science both in the context of cloud physics, by Marcia Baker, and in the context of atmospheric chemistry, by Thomas Peter.

Nucleation and surface melting take us from microscopic to macroscopic when we consider crystal growth. The length scale of importance in the nucleation and growth of ice single crystals and polycrystals ranges from molecular (Angstroms) to meters, and hence a proper description requires an arsenal ranging from statistical mechanics to transport theory. On one end of the scale just after nucleation, when a crystal is smaller than any characteristic scale in the diffusion field of the source material, its evolution depends solely on the molecular kinetic processes limiting the rate at which molecules attach themselves to the solid, the so-called "interface controlled" limit. As such a crystal becomes larger, it begins to reach further into the diffusion field, and the growth itself modifies the field, crossing into the "diffusion limited" regime. Each regime, and the boundaries between them, presents us with a hierarchy of challenging and interesting problems. However, crystal growth is a history dependent process, and all of these processes conspire to create the object that motivates many of the ASI participants: the snowflake. Many aspects of these phenomena were addressed by Braslavsky, Dash, Furukawa, Hodgkin, Maruyama, Wilen, and myself.

Real ice surfaces are now being scrutinized using the tools of surface science which we learned from Sam Fain and Victor Petrenko, who are both employing Atomic Force Microscopy in their laboratories. The Institute was also exposed to several types of "model" ice surfaces and their consequences. Enrique Batista and Yoshinori Furukawa displayed computer simulations of ice surfaces and their diffusional and structural properties. In contrast, Jean Suzanne showed us how similar to ice the structure of water adsorbed on MgO is, and offers this "ice surface" as a potential model for chemisorption studies of atmospheric importance. When water molecules are deposited ballistically on a very cold substrate one may observe amorphous ices as studied by Alexander Kolesnikov and Ji-Chen Li using neutron spectroscopy and as shown by Mayo Greenberg to pervade extraterrestrial environments.

John Nye took us inside an ice polycrystal to observe liquid water present at temperatures below the normal melting point due solely to the curvature and impurity depressions of the equilibrium temperature. The water resides at the tri-grain junctions in microscopic veins (10–100 μm) and at nodes where four grains come together. The impurity depression is simply that which we find in normal solution theory, but the solution resides in the veins and nodes because of the strong impurity segregation properties of ice. The consequence is that, if the temperature is sufficiently high, ice polycrystals communicate liquid throughout their volume. Plunging down between the faces of two grains the veins appear to vanish, but does the water itself disappear? Some participants pointed out that, on a microscopic scale, ice grain boundaries themselves can interfacially melt, a process facilitated by the presence of impurities, so that still more liquid (or quasi-liquid) will be present throughout the polycrystal. The effect might provide a chemical reservoir in polycrystalline polar stratospheric cloud particles (PSC's) allowing the surface to exchange material with the interior and thus influence chemical reaction rates. Surely many laboratory experiments using vapor deposited ice films experience this phenomenon. It had previously been stressed that these contributions to liquid or quasi-liquid in polycrystalline ice might play an important role in facilitating the mobility of important chemicals in ice cores, a kind of effective diffusivity enhancing the smearing of isochrones. This diffusivity would thus depend on the crystallinity of the ice, and quite clearly the use of laboratory values on single crystals might misguide analyzers of ice core chemistry. The lectures of Sigfus Johnson forced the participants to revisit the issue and set John Nye into action; we expect to see a new publication on the topic soon.

This perspective on water in ice also touched on the concept of what we mean by the triple point. How do we think about the temperature defining equilibrium between ice/water/vapor? Such a query has a long history in the thermodynamics of phase equilibrium wherein bulk coexistence concerns a free energy state excluding the contribution of interfacial energy. Practically, however, we must construct an ice bath consisting of polycrystalline ice and water, which leads us to consider the question "should the equilibrium temperature be taken as the true triple point?" Or, given the practical need to use polycrystalline-crushed-ice, should we define a "practical triple point" and specify the degree of comminution?

The phenomenon also defines the boundary between "temperate" and "cold" ice in polythermal glaciers. The former is loosely defined by glaciologists as ice at a temperature "near" 0°C, while the latter is "cold" and "dry." Both types of ice are present in polythermal glaciers, and the boundary between them moves in response to changes in the environment, but it moves slowly, on the order of 10 meters in 100 years and slows down with time.

The latter study began to bridge the gap from microscopic to macroscopic, taking the participants up in scale. Nye made a leap in his second lecture on water in ice by educating us on how a particular Icelandic glacier "leaps" periodically in a process called a Jökulhlaup. A complicated sequence of events leads to the release of a tremendous surge of water out from under a glacier, driving disastrous floods in the region. Nye's physically based picture gets the gist of the event, and Felix Ng has showed us some recent numerical treatments of this impressive piece of glaciology.

Basic questions concerning interfacial adhesion and the role of surface melting meet common ground in the important problem of the icing of structures. Most notable are ships, which under icing conditions lose their gravitational stability, aircraft, which lose their aerodynamic integrity, and man-made structures, which buckle under ice loading. Victor Petrenko showed us many examples of disasters caused by icing of these structures and focused our attention on one aspect of the problem that is basic and unsolved—ice adhesion. Can we design structures that thwart adhesion?

Some of the most important environmental threats are above our heads, in the atmosphere. Marcia Baker and Thomas Peter took the Institute on a tour of the atmospheric phenomena that sit at the heart of climate change. They exposed us to an impressive range of space and time scales from the microphysics of cloud formation and glaciation to the heterogeneous chemistry taking place on polar stratospheric cloud particles. Baker describes a cloud as an extremely complex dynamic multiphase structure, but one serving as a laboratory for the study of nucleation and crystal growth. A cloud is an amalgam of chemistry and physics influencing the global radiation balance and thereby the earth's climate. We learned that practitioners study ice crystal growth in clouds both by flying through them with well-instrumented aircraft or, as Neil Bacon and Brian Swanson are doing, suspending single ice particles in an electromagnetic trap. When ice particles collide inside a cloud, Brian Mason and Greg Dash's laboratory collisions tell us that charge will be exchanged. Perhaps this is at the heart of the striking and dangerous phenomenon of thunderstorm electrification? Important clues again implicate the microscopics of surfaces.

Moving up into the stratosphere and onto the surface of PSC's, which act as substrates for chemical reactions, Thomas Peter discussed the reaction $HCl + ClONO_2 \rightarrow Cl_2 + HNO_3$ regarded by atmospheric chemists to be the most important heterogeneous chlorine activation reaction. The reaction can convert a rather inert species into highly reactive ozone-destroying species within hours under the right PSC formation conditions. His analysis of recent experiments reveals a discrepancy in the reaction rates of up to two orders of magnitude! Hence, we were exposed to the present and serious doubts concerning our understanding of the most crucial link in the chain reaction leading to stratospheric ozone destruction. The drama will continue to be played out on model PSC surfaces in laboratories, but it raises questions as to whether each laboratory has its "own" PSC surface and how in the future experimental preparations can best uniformly mimic the surfaces made by nature.

One can consider the glaciation in clouds from another perspective, also of importance in climate change. Snow falls in accordance with meteorological conditions and it buries clues concerning the deposition conditions of past climates. This is the basis of paleoclimate reconstruction through the analysis of ice cores retrieved from the

Greenland and Antarctic ice sheets. Simply stated perhaps but certainly not trivially carried out. The ice doesn't necessarily cooperate; it flows under its own weight and hence the history–depth relationship requires an understanding of ice rheology and the physics of ice flow. Stan Paterson provided us with a primer on the latter field, and Kurt Cuffey presented current ideas concerning the rheology. Sigfus Johnsen showed us the time history of climate captured in ice cores, how to read it and, perhaps the most difficult, how to retrieve it—an immense logistical enterprise. An outstanding question concerns the mechanism by which the so-called Heinrich events occurred. These "events" describe layers of sediment observed in the North Atlantic sediment cores containing the geochemical signature of material originating in Hudson Strait, Greenland, and Iceland. There is evidence that an armada of icebergs made during the retreat of the Laurentide ice sheet may have carried the debris composing the Heinrich layers into the ocean where melting released them. How did the debris get rafted? Can such an ice sheet surge? How much debris is required? Some of the present theories were called into question, but no new ones rose up to replace them.

Where there is ice there is rock, and Krzysztof Birkenmajer showed us that a great deal of information concerning the former is recorded in the latter. We were given a guided tour of how the great global glaciations shaped high latitude continents and land forms. The information is varied and difficult to obtain, and has taken him into both polar regions many times. Reading the climatic clues contained in rock, the reshaping of local land forms and continental masses involves tying the sometimes disparate techniques of glaciology and geology together. The result is a unique approach of viewing the earth's history.

Our attention was drawn to the role of ice in the history of the solar system and the galaxy during Mayo Greenberg's lectures. Interstellar dust grains only a tenth of a micron in size are covered by water ice and scattered throughout the Milky Way. The striking fact is that this dominates the galactic water mass. Comets follow in the mass hierarchy, and Konrad Kossacki showed us how the distribution of ice within these bodies influences their mechanical and thermal properties. Following comets in the extra terrestrial ice mass scale is the ice found on some planets and moons, and Jacek Leliwa-Kopystynski showed us that water ice comprises half of the mass of icy satellites. The conditions experienced by interstellar dust are recreated in the laboratory, allowing ice formation and concomitant chemical reactions to be revealed. Ground- and space-based observations of comets demonstrate that they form as an aggregation of interstellar dust, and hence from their early genesis consist of icy masses. More speculative was the issue of whether life on earth arose from prebiotic molecules under volatiles on interstellar grains which were deposited on our planet. The icy moon of Jupiter, Europa, was also scrutinized, but its shell will have to await more definitive observation before judgment can be rendered.

Looking back at the earth from space was Alan Thorndike with an eye toward a minimal description of ice and climate. By area, floating sea ice represents the most dramatic seasonal variation in global ice cover. A minimal description of the balance of energy at the surface of the ocean leads one to the conclusion that the presence of ice is required. For some it is sufficient to know that we observe ice in the polar seas, but to others addressing the question of whether an ice-free polar ocean can exist requires a fundamental understanding of the conditions necessary for the existence of the ice-covered state. With this in hand, Thorndike took the view that one can now model how the low latitudes and the polar regions interact. The important physical features are that the polar regions are white and hence reflect radiation, and there is a transport of heat between low and high latitudes. The temperatures of these regions adjust within the constraint that the system is in radiative equilibrium. The concepts form a minimal model of climate. With such a model the issue of whether an ice-free or ice-covered Arctic Ocean describes a stable equilibrium can be addressed. A model like this can be used as a testing ground for global climate simulations run on enormous computers. The advantage is that a minimal model provides relatively simple solutions in terms of the basic properties of the materials involved: the ice, the ocean, and the atmosphere.

Every day one of the polar oceans is freezing and, as the hemispherical winters evolve, the implications play an increasingly dramatic role in the lives of those inhabiting high-latitude regions. In terms of population density, the effects are dominant in the northern hemisphere, where seas, lakes, and estuaries freeze and modify local conditions. A minimal model won't tell us whether a storm near Helsinki will result in snowfall over Stockholm. For short-term forecasting and regional predictions in ice-infested areas, one needs to put as much physics into the air/sea/ice model as is dictated by the particular environment of interest and the observations available. Anders Omstedt introduced the Institute to this practical "maximalism" beginning from the simplest situation of a shallow lake and developing the necessary tools for dynamic/thermodynamic predictions in freezing seas. We were cautioned about the proper way to assimilate meteorological and oceanographic data in real-time predictions and exposed to the varied problems associated with ice formation in waters near populations. Most notably, shipping is influenced severely, but less well known is the role that freezing lakes play in the modification of regional climate. Areas such as Scandinavia, with a tremendous areal density of lakes, lend a strong impetus to understand these processes.

The motion of ice in frozen seas has an important influence on large-scale heat and mass balance, as seen in Ron Kwok's analysis of sea ice motion which blends position information taken from buoys and satellite data. As the ice moves into warmer waters it melts, thereby removing heat from the ocean and leaving the surface waters less saline. The large-scale effect of this process can have significant oceanographic consequences as was shown by Peter Wadhams. As one piece of sea ice encounters another, the weaker one may be deformed irreversibly. J. Dempsey showed how this process scales using concepts of fracture mechanics. Many participants are occupied with various aspects of modeling sea ice growth (D. Feltham), observing its motion, modeling its large-scale mechanical and geometric properties, and making other field observations on snow and glacial ice (S. Gerland, K. Hietala, T. Johnsen, A. Paelli, A. Sinisalo, N. Steiner, P. Uotila, L. Smedsrud).

Anyone who has traveled to a marginal ice zone, where sea ice meets the open ocean, is struck by a particular contrast with the central ice pack: the tremendous abundance of animal life. The food chain, at the crudest level, is obvious. There is more light in the water column and biological blooms are frequent. Fish feed on these.

Birds and seals feed on fish, and polar bears feed on seals. Thick ice hinders the process but, as Ole Lønne showed us, it constitutes an important link in the chain. Lønne, who goes into the Arctic but then ventures *under* the sea ice, studies this link, which takes the form of three species of shrimp-like animals called sympagic macro-fauna. These creatures cannot swim very well and hence cling to the inner walls of brine channels within sea ice. The wind driven motion of sea ice moves a colony long distances while its population evolves locally on its ice floe home. However, when the ice meets the open ocean, rapid ablation occurs and the macro-fauna lose their shelter and are thus suspended throughout the water column, where fish and birds have ready access to them. Herein lies a very interesting case wherein biological and physical approaches may blend to solve a puzzle. Most of the perennial ice produced in the Arctic Ocean is exported through Fram Strait and eventually melts. Lønne has observed the macro-fauna population densities on perennial ice. Because ice is continually lost, there is a sink of macro-fauna in the marginal seas. How is the population maintained and transported? Where is the source? How long do they live? The Institute brought together experts on sea ice motion who together with Lønne can address these and other important biological questions relying on an understanding of ice physical science.

This discussion of such a pristine substance might have swayed a less educated group into believing that all ice is clean. Many of the participants are experts on just how dirty important natural environmental and technical systems can be. Fiorenzo Ugolini first emphasized the very large fraction of polar regions that consist either of ice on land or partially, perennially and permanently frozen soils. Soil science relies on tools and techniques from geochemistry, geology, botany, geography, and those that its practitioners must invent themselves. In the polar regions one must fold ice into the picture and the complete tapestry is a rich one. We learned that polar soils can be desert-like because of the low precipitation, and have the added exposure to intense cold and seasons where the sun either never shines or never sets. When humidity is adequate a tundra can form, but the paucity of liquid water has made agriculture in the northern regions extremely challenging. The successes and failures were traced by Ugolini with the main ingredients of the former being the choice of intrinsically robust crops, choice of location and, interestingly, the sociological factors surrounding the process. The expense of food and other supplies in remote polar regions is due mainly to high transportation costs. This must be weighed against the relatively technical process of local production of produce and other consumables. Were it substantially simpler to purchase food than to grow it nearby, what would you do? Refrigeration is never a problem!

The natural thermal fluctuations in saturated soils result in the deformation in the direction of the heat flux vector of the material known as "frost heave." We endow the term with quotations for the simple fact that the phenomenon can be explained by no single mechanism. There are two very important manifestations of frost heave. Perhaps the most aesthetically pleasing and curious is the "cryoturbation" of land forms into impressive features including sorted stone circles and ice wedge polygons. The motion of the soil on this scale has severely deleterious effects on man-made structures such as pipelines. Although there is a contrast between observing naturally occurring geomorphological features and engineering pipelines in order to avoid the associated effects of differential ground motion, understanding both requires a concerted focus on the basic mechanisms of frost heave.

What is certain, as was stressed in the presentations of Masami Fukuda, Peter Williams, Alan Rempe, and myself, is that the mobility and concomitant pressure of water at temperatures below the bulk freezing point is important. I stressed that there are a number of physical mechanisms responsible for this water: interfacial melting, curvature, proximity of soil grains, and impurities. Each has unique temperature dependencies and hence each contributes to the volume fraction of water with varied strengths depending on the temperature. Each of these has been studied in specific experiments using techniques of modern condensed matter physics. Alan Rempel displayed the consequences of these in a model porous medium. Bo Elberling is studying gas transport in partially saturated, partially frozen soils.

Williams and Fukuda introduced us to the topic of frost heave from an engineering perspective, within which we are presented with two very different challenges. There are two extensive oil pipelines in the northern hemisphere, in Alaska and the Arctic of the Former Soviet Union. Essential to environmental safety during their operation and to the efficiency of transport is engineering that maintains the mechanical integrity of the pipeline in the presence of differential ground motion that can be caused by frost heave. The costs in terms of environmental damage and the price we pay for transportation are immense, and the phenomena driving frost heave are the most important scientific problem coupling the two. This was one of Peter Williams' main messages. Masami Fukuda introduced us to the second challenge: artificial ground freezing. A frozen soil has a vastly superior mechanical integrity than its unfrozen counterpart, making improvements in freezing technology paramount to engineering safety and structural stability. We were treated to a history of the field and shown the large-scale ground freezing projects in Japan that he has been involved in. The main usage is in stabilization of water-saturated soils during pipeline or tunnel construction or other excavations. The same techniques provide promise in the area of waste containment because in addition to mechanical superiority, the very low diffusivity and small phase segregation of impurities in ice spotlight frozen soils as important potential barriers for nuclear and nonnuclear materials. This is the main focus of Greg Dash's second chapter.

The last point, as we learned from Lydia Popova, should not be taken lightly. Dumping of military and nonmilitary radioactive wastes, the testing of nuclear weapons, the use of nuclear weapons as tools for excavation, radioactive fallout from military and nonmilitary sources, and oil spills have all left their signatures in the Arctic of the Former Soviet Union. Approximately sixty percent of this region is either perennially or seasonally ice covered, which provided the Institute with a clear link between a vast array of scientific fields, ranging from solute trapping in growing contaminated sea water to the mobility of contaminants in permafrost, and the question of long-term environmental danger. Popova gave us an appreciation for the difficulty in assessing the present day state of damage done by these and other activities, simply identifying the sites and the extent of the varied environmental threats is a monumental task requiring a concerted effort by both government and industry. The cost of identifying existing sites and preventing future ones is substantial. The cost of not doing so is so immense it escapes assessment. Future identification, monitoring, remediation, and containment strategies will build on scientific and engineering directed at both fundamental and applied aspects. These problems occurring in northern regions will rely heavily on knowledge gained in fields where the physics and chemistry of ice underlie the environmental consequences.

At the conclusion of the ASI, most participants felt that the meeting was successful. Because we came from many different disciplines, we were novices in many of the sessions, and the lectures had to be more descriptive and less detailed than they are in more specialized conferences. Some of us came away with a deep respect for the difficulties and subtle aspects of another discipline. In I.I. Rabi's 1955 Morris Loeb Lecture he stated his belief that "the value content of science or literary scholarship lies not in the subject matter alone, or even in greater part. It lies chiefly in the spirit and living tradition in which these disciplines are pursued. The spirit is almost always conditioned by the subject." It is hoped that this collection will convey the spirit of the Institute and hold the interest of novices and professional ice scientists.

J. S. Wettlaufer

History of the Search for a Theory of Melting

JG Dash

Department of Physics, University of Washington, Seattle WA 98195-1560

Abstract. An understanding of the microscopic melting of crystals has eluded researchers for more than a Century. The struggle has stimulated the development of quantum mechanics, concepts of long range order, the role of dimensionality in condensed matter, and more sensitive experimental techniques to test the theories. After years of probing the mechanism within the bulk material, we have discovered that the answer has been lying on the surface.

1 Introduction

Of all the phenomena exhibited by condensed matter, phase changes are the among the most dramatic, and of these melting and freezing are especially striking. The transformation between ice and water has been familiar to most of the world's population ever since people migrated out of the tropics. Consequently, we now know a great deal about the melting of ice and of many other solids under various conditions (see e.g. Turnbull (1956), Ubbelhode (1965), and Bilgram (1987)). However, melting is not yet a closed subject. Important aspects of the general phenomenon and of the melting of ice in particular have been discovered quite recently and are still being explored.

2 Abrupt and Continuous Melting

The question "How does a solid turn into a liquid?" can be answered on three levels.

The first level is dismissive: it points out that the melting of crystalline matter is abrupt, a first order phase change. Liquid and solid can coexist in equilibrium over a range of temperatures and pressures; it argues that since there are no intermediate equilibrium states it is hopeless to try to follow the progress of the transition.

The second level, which has engaged many investigators for long years, speculates that the phase change might be continuous or nearly so, and that given sufficient resolution it should be possible to trace a gradual precursor of melting in the solid almost all the way through to the transition. Such *one-phase melting* or *continuous melting* ideas have taken encouragement from several facts. Melting involves a relatively small increase in volume, typically about ten percent.

NATO ASI Series, Vol. I 56
Ice Physics and the Natural Environment
Edited by John S. Wettlaufer, J. Gregory Dash and Norbert Untersteiner

The heat of fusion is much smaller than the heat of vaporization, therefore the cohesive energy between molecules is not greatly different in the two phases. The specific heat does not change much across the transition, hence the molecular motions are similar. Taken together, these changes suggest that the transition, or at least its precursor, can be treated as a relatively mild perturbation. It is the history of this approach that occupies the major portion of this paper.

We will only touch on the third level at the end. For a fuller explanation, the reader is referred to the chapters by David Oxtoby (1998) and John Wettlaufer (1998) later in this volume.

The idea of continuous melting was stimulated by the success of van der Waals' (1873) theory of the liquid-vapor critical point. In the basic model, a substance is described in terms of two parameters: one for the strength of a long range attraction between molecules, and the other for molecular size. With these fundamental interactions, vapor-liquid equilibrium can be treated by the methods of classical statistical mechanics. The two-phase system can be tracked into the neighborhood of the critical point, where the transition becomes continuous, and where the intermediate regime can be studied in detail. Soon after van der Waals presented his model, it was suggested that solid-liquid transitions might be treated in the same manner, hence one could study the melting process at a solid-liquid critical point. One of Planck's earliest papers dealt with the possible coordinates of a critical point between ice and water (Planck 1882). Experimentalists studied a number of materials, and extended their liquid-solid phase boundaries to high pressures and temperatures, in the hope of finding critical end points. The quest, which was one of the motivations for high pressure research (Bridgman 1931), led to the discovery of many new phases of solids (including those of ice), but no solid-liquid endpoints have been found to date, and none are expected.

3 A Melting Criterion

In 1910 Frederick Lindemann proposed a simple idea: solids melt when the amplitude of atomic thermal vibrations exceeds a fraction of the interatomic spacing. His quantitative model made use of Einstein's explanation of the low temperature specific heats of solids, which proposed that the atoms vibrate as quantized harmonic oscillators. Einstein's theory had been published only three years earlier, and its adoption by Lindemann appears to be the first application of quantum theory to condensed matter after Einstein's own paper.

The theory is quite simple. The average thermal vibrational energy is related to the temperature by the equipartition relation, which gives:

$$m(2\pi\nu_E)^2\delta x^2 = kT, \tag{1}$$

where m is the atomic mass, ν_E is the vibration frequency, δx is the root-mean-square thermal average amplitude of vibration; k is Boltzmann's constant and T

is the absolute temperature. The frequency is related to Θ_E, the characteristic 'Einstein temperature' of the low temperature specific heat, by

$$h\nu_E = k\Theta_E, \tag{2}$$

where h is Planck's constant. Lindemann's basic assumption is that melting occurs when δx reaches a critical fraction of the nearest neighbor spacing. The equations combine into a simple expression relating the melting point T_m to properties of the bulk solid. Somewhat modernized from the original formulation, the melting rule is:

$$\Theta_D = c[T_m/MV^{2/3}]^{1/2}, \tag{3}$$

where Θ_D is the Debye temperature of the specific heat, M is the molecular weight, V is the molar volume, and c is a constant. Its empirical value for many simple solids indicates that melting occurs when the rms displacement reaches about 1/8 of the atomic spacing. But the success of Eq.3 should not obscure the fact that the theory has some serious defects. It is based on harmonic forces, which never give way, whereas melting must involve bond breaking. Another fault is that the model describes melting as an individual atomic property, whereas a phase transition is a cooperative process. Nevertheless, the practical success of the model lent support to the idea that melting begins gradually, at temperatures below the melting point.

4 Excitations in Solids

There are several different modes of thermal excitation of a solid, and each type contributes to a gradual disordering of the crystal. They include, in addition to atomic vibrations, dislocations, point defects such as vacancies and interstitials, and in the case of molecular crystals, orientational defects. Each has a characteristic excitation energy governing an exponential increase with temperature, which can appear to be precursors of the transition.

Experiments designed to detect continuous melting have had to contend with many difficulties. Dissolved impurities generally cause the transition to be broadened, due to progressive shifting of the melting point as the impurity concentrations in liquid and solid change along the phase boundary. Size effects also broaden the transition, due to melting point depression caused by surface curvature. The combination of accelerating changes with rising temperature, plus finite widths, of the transition can suggest the continuous melting predicted by some theories. But in pure and well ordered large crystals, there are no great changes of thermal, structural, and mechanical properties as T is raised all the way to T_m, where the changes are dramatic and abrupt, within experimental resolution.

The most important excitations with regard to melting are dislocations, which cause local disruptions of the crystal structure. The theory of melting by dislocations was proposed by Mott and Gurney (1938) and elaborated by many others, with developments continuing to the modern era (Cotterill 1979). The excitation energy of a dislocation is composed of a core energy and the energy of long range strain, which depends on the elasticity modulus. Since the presence of dislocations decreases the elastic modulus, the excitation energy is progressively reduced as the density of dislocations increases. Therefore their creation accelerates as the temperature increases, and if the feedback is sufficiently strong it can cause an avalanche of dislocations. The material becomes filled with dislocations; the total excitation energy is the latent heat, and their mobility accounts for the liquid's fluidity. The theory is controversial, one of the principal points of resistance being the description of the liquid as a crystal filled with dislocations. However, while dislocation-mediated melting of bulk matter is still debated, its application to 2D matter in the 1970's attracted strong interest and wide acceptance. It was stimulated by several currents in theory and experiment, which began many years earlier.

5 Long Range Order and Dimensionality

A prediction of continuous melting in lower dimensional systems originated with Rudolf Peierls(1935). Twenty five years after the publication of Lindemann's idea, Peierls treated the collectivity of atomic vibrations in models of one-dimensional (1D) and three-dimensional (3D) crystals. A simple model of a 1D crystal is a regular chain of identical masses connected by identical elastic bonds, with separation d. Fig. 1 shows the array, perfectly regular and motionless at zero temperature. The location of each atom is exact, the n-th atom being located at nd, no matter how distant. But at finite temperature, the atoms have thermal motions, which causes them to vibrate about their equilibrium positions. To quantify the deviations, consider the separation between two atoms some distance apart; define their instantaneous displacements from the ideal lattice positions by $u_n = x_n - nd$ and $u_m = x_m - md$. The deviation of their separation $\delta_{nm} = (u_n - u_m)$, and a gauge of the average deviation is given by its mean squared thermal average. The result of the calculation is:

$$< \delta_{nm}^2 >= (n-m)^2 d^2 kT/mv^2, \tag{4}$$

where v is the speed of elastic waves. Eq.(4) predicts that δ_{nm}^2 grows monotonically with separation, as long as T is finite. Thus the precision of location is limited in extent: *long range order* is destroyed by thermal vibrations as soon as T rises above absolute zero.

It was satisfying that the corresponding 3D model was found to be qualitatively different: the deviation does not increase with distance. In 3D the atomic positions are governed by linkages in three directions, so that the effects of the uncoordinated vibrations of neighbors tend to cancel. Peierls' calculation was based on harmonic forces; if one adopts Lindemann's limit for the deviation of nearest neighbor separations, the analysis leads to a similar melting rule.

The intermediate case of two dimensions is particularly interesting. The theory predicts that long range order in 2D is destroyed above $T = 0$, but much more gradually than in 1D, by varying as the logarithm of nd.

The destruction of long range order at finite temperature led Landau (1937) to conclude that one- and two-dimensional crystals could not exist. Frenkel (1946) scoffed at the conjecture, pointing out that long periodic chain-like molecules and large regular sheets of molecules are stable to quite elevated temperatures. Nevertheless, the theory stimulated interest in lower dimensional systems for studies of continuous melting. The first problem was how one might realize 2D solids and if obtained, how to probe them.

6 Realizations of Two-Dimensional Matter

Laboratory approximations of 2D matter had been studied in the Nineteenth Century, as monomolecular oil films on water. In these films the rod-like amphiphilic molecules are bound to the water-air interface, but are free to move about on the surface. Some of the films exhibit changes from low density to condensed states, analogous to the condensation of ordinary gases to liquids. Structural and orientational transitions have been observed, but no clear examples of solid-liquid phase changes have been seen.

In the 1950's research began on a different type of monolayer film: rare gas molecules adsorbed on uniform solid substrates at low temperatures. Adsorption was not then a new technique, for it had been in regular use to estimate surface areas and particle sizes in powders. What was new was the development of more homogeneous substrates that could better approximate the uniform planar surfaces of simple theoretical models (Dash 1975). These systems could be studied by methods that are sensitive to phase changes, such as calorimetry. With this technique, Goodstein (1965) found that the heat capacity of adsorbed monolayers of He^3 and He^4 had the temperature dependence predicted for 2D solids at low temperature, and when heated they displayed pronounced specific heat maxima, suggesting continuous phase transitions. The finding surprised many people including Peierls (1968, 1979), who had expected that a 2D solid would disorder so gradually that there would be no thermal signal. It was then realized that another form of long range order, the breakdown of which is more relevant to melting, could persist in 2D matter above absolute zero.

Mermin (1968) analyzed distortions of the harmonic 2D lattice in terms of *directional* or *topological order*, by the directions of bonds between an atom and its neighbors. Imagine a complete closed path laid out along the bonds at zero temperature, enclosing an area of the lattice and returning to the starting point. (Fig. 2A) As temperature rises the bond directions wander, so that the path becomes irregular. The system is then analyzed to relate the temperature and size of loop at which the path fails to return to the starting point. Mermin showed that the loop becomes distorted as temperature rises but the path remains closed.

How can long range positional order disappear as soon as it is warmed above absolute zero, yet perfect long range topological order persist to high temperature? In physical terms we are considering two different qualities: crystal structure and elasticity. As temperature rises the crystal structure becomes progressively disordered, a change which can be detected by the decreased intensity of x ray diffraction lines. However, as the structure is stretched and distorted, the material persists as a solid as long as the lattice is not ruptured. Melting requires bond breaking: in effect, dislocations. With the stimulus of the new experiments and theory, ideas of dislocation-mediated melting and lower dimensional order were brought together.

7 Dislocation-Mediated Melting in Adsorbed Films

Michael Kosterlitz and David Thouless (1972, 1973) proposed a fundamental theory of the destruction of two-dimensional topological order, with applications to crystals, superfluids, and ferromagnets. In the case of crystals, the phase transition occurs via dislocations, as proposed for the melting of bulk matter, although the theory is much more tractable in two dimensions. Richard Feynman proposed a model of dislocation-mediated melting at about the same time as the original KT paper, and although their approaches are somewhat different, they achieve essentially the same results. Feynman's work is obscure because it was never published on its own, and is reported only in a conference paper by colleagues (Elgin 1972). The circumstances of its origin lend it a special appeal, for it was written during three days in a topless bar (Goodstein 1972).

The breakdown of crystalline order involves the creation of free dislocations from bound pairs (Fig. 2B). A free dislocation involves a 'core', which has a diameter on the order of a lattice length, and a lattice strain, which falls off as the logarithm of distance. A close-bound pair of dislocations corresponds to a short length of extra atoms or a row of missing atoms. The energy of the pair is the sum of strain and core energies, and has the form

$$E = Lln(r/r_o) + const., \tag{5}$$

where L is the elastic modulus, r is the pair separation, and r_o is a length on the order of a lattice parameter. At low temperatures there are few thermally excited pairs. As T increases their number grows, their average separation increases, and they develop increased translational and rotational mobility. When the mean separation length $< r >$ reaches the size of the array the lattice becomes laced with paths that permit the pieces to slide. With many such pairs in different orientations the system can be *stirred*; it has become a liquid. The thermal average is readily calculated:

$$< r >= const. \int rexp[-Llnr/kT]rdr. \tag{6}$$

The melting point is reached when $< r >$ diverges. The integrand can be recast into the form

$$r^2exp(-Llnr)/kT = exp[lnr^{(2-L/kT)}]. \tag{7}$$

Thus, the melting occurs when the exponent becomes positive, hence

$$T = T_m = L/2k. \tag{8}$$

Feynman estimated the magnitude of L in terms of the characteristic Debye temperature, the lattice parameter and the atomic mass, and obtained a relation having the same form as the Lindemann model, but without any adjustable constants. It agrees with the melting parameters of many classical solids, almost as well as the empirical rule. It is remarkable that the same expression for melting temperature, with roughly comparable predictive success, emerges from two qualitatively different models: melting by atomic vibrations, and melting by the growth of dislocation pairs. But it is not accidental, because both models involve the disruption of molecular bonds, although with differing degrees of collectivity.

The theory of dislocation-mediated melting has been elaborated by Halpern, Nelson, and Young, and the mature theory is now known as the KTHNY model (Nelson 1980). It predicts that 2D melting can take place by two successive continuous transitions, each one so gradual that neither is marked by a discontinuity in any thermodynamic property or its derivatives. Thus, the new theory confirmed what Peierls' work had seemed to suggest long years before, that lowered dimensionality was the key to understanding melting.

8 Experimental tests

Many laboratories studied adsorbed monolayer films in the years following the initial studies in the 60's. As the quality of substrates improved, the melting anomalies that had been initially quite broad became much narrower. For several systems in which the films are incommensurate with the structure of the substrate, the transitions appear to be as abrupt as the experimental resolution allows (Fig. 3). The conclusion that they are first order phase changes is based on much evidence; regions of 2D solid-liquid coexistence, substantial differences between solid and liquid densities and mobilities, and sharp 2D triple points, where 2D vapor, liquid and solid coexist. In contrast, an extensive review (Strandburg 1988) found that no conclusive examples of continuous dislocation-mediated melting had been seen in any experimental systems. Since then, experiments on a few special systems; electrons trapped on the surface of liquid helium (Glattli 1988), charged colloidal particles (Murray 1990), liquid crystals (Chao 1996), and Langmuir-Blodgett films (Sikes 1997) have shown some features of the predicted behavior, yet no simple molecular systems have shown any indications of continuous melting.

10 Conclusions

Why have so few 2D systems shown the predicted behavior? Perhaps it is because continuous melting is preempted by a first order phase change. Before the solid can be heated to a temperature sufficiently high for breakdown by dislocations, its free energy becomes equal to that of the liquid state. Dimensionality seems to have little to do with it, for in both 2D and 3D, the liquid phase provides a lower free energy state than a disordered solid. And yet, there is a significant feature indicating that some sort of continuous melting does occur in most solids: there is no apparent hysteresis in melting. Hysteresis in phase transitions is associated with the birth of the new phase and the energy of the interface between it and the parent phase. This requires superheating or a temporary excursion of another thermodynamic variable. In contrast to condensation and freezing, the melting of nearly all solids begins without any detectable superheating.

This introduces the third level answer to the question posed in the Introduction. It was suspected long ago, according to reasoning described by Frenkel (1946):

> *It is well known that under ordinary circumstances an overheating (of a crystal), similar to the overheating of a liquid, is impossible. This peculiarity is connected with the fact that the melting of a crystal, which is kept at a homogeneous temperature, always begins on its free surface. The role of the latter must, accordingly, consist in lowering the activation energy, which is necessary for the formation of the liquid phase, i.e. of a thin liquid layer, down to zero.*

The idea was plausible, but it remained for modern experimental techniques to prove it, by the detection and measurement of the incipient liquid. The phenomenon is called *surface melting*, and the articles by David Oxtoby (1998) and John Wettlaufer (1998) describe its physics and importance.

Acknowledgements

It is a pleasure to thank Sam Fain, Michael Schick, and John Wettlaufer for insightful comments and corrections.

References

JH Bilgram 1987, Phys Repts **1**, 1–89.

PW Bridgman 1952, *The Physics of High Pressure*, G Bell and Sons, ch. 7.

CY Chao, et al. 1996, Phys Rev Lett **77**, 2750.

RMJ Cotterill, in *Ordering in Strongly Fluctuating Condensed Matter Systems*, NATO ASI Series B, Plenum 1979, p. 261.

JG Dash 1975 *Films on Solid Surfaces*, Academic Press.

RL Elgin and DL Goodstein 1973, in *Monolayer and Submonolayer Helium Films*, JG Daunt and E Lerner, eds, Plenum, p. 35.

J Frenkel 1946, *Kinetic Theory of Liquids*, Clarendon Press Oxford.

DC Glattli, EY Andrei and FIB Williams 1988 Phys Rev Lett **60**, 420.

DF Goble and LEH Trainor 1965, Phys Lett **18**, 122.

DL Goodstein, WD McCormick and JG Dash 1965, Phys Rev Lett **15**, 447.

DL Goodstein 1972, private communication.

PC Hohenberg 1967, Phys Rev **158, 383.**

GB Huff and JG Dash 1976, Low Temp Phys **24**, 155.

JM Kosterlitz and DJ Thouless 1972, J Phys C **5**, 124; 1973, J Phys C **6**, 1181.

D Kuhlmann-Wilsdorf 1965 *Theory of melting*, Phys Rev **140**, A1599.

JE Lennard-Jones and AF Devonshire 1939, Proc Roy Soc **A170**, 464.

ND Mermin 1968, Phys Rev **176**, 250.

FA Lindemann 1910, Z Phys **11**, 609.

JA Litzinger and GA Stewart 1980, in *Ordering in Two Dimensions*, SK Sinha, ed., North-Holland, p. 287.

CA Murray, WO Sprenger, and RA Wenk 1990, Phys Rev B**42**, 688.

DR Nelson 1983, in *Phase Transitions and Critical Phenomena, V. 7*, C.Domb and JL Lebowitz, eds., Academic Press, p.1.

D.W. Oxtoby 1998 in *Ice Physics and the Natural Environment*, Wettlaufer JS, Dash JG, and Untersteiner N, eds, NATO ASI **58** *Series I*, (Springer-Verlag, Heidelberg).

RE Peierls 1935, Ann Inst Henri Poincare **5**, 177–222.

RE Peierls 1966, UW solid state physics seminar, unpublished. A detailed calculation of the mean squared deviation in 1D and 2D lattices is given by JG Dash, *Films on Solid Surfaces*, Academic Press 1975.

RE Peierls 1968, private communication.

RE Peierls 1979, *Surprises in Theoretical Physics*, Princeton Univ. Press, 85–90.

M Planck 1882, Wied. Ann. **15**, 446.

HD Sikes and DK Schwartz 1997, Science **278**, 1604.

KJ Strandburg, Rev Mod Phys 1988, **60**, 161–207.

D Turnbull 1956, in *Solid State Physics; Research and Applications* v. 2, F Seitz and D Turnbull, eds., Academic Press, 226–306.

AR Ubbelhode 1965, *Melting and Crystal Structure*, Clarendon Press Oxford.

JD van der Waals 1873, *On the continuity of liquid and gaseous states* Thesis Univ. Leiden.

J.S. Wettlaufer 1998 in *Ice Physics and the Natural Environment*, Wettlaufer JS, Dash JG, and Untersteiner N, eds, NATO ASI **58** *Series I*, (Springer-Verlag, Heidelberg).

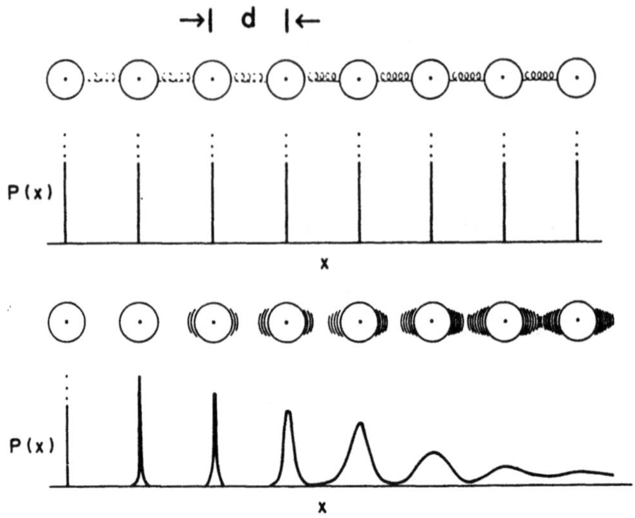

Fig. 1. Temperature above $T = 0$ destroys long range crystalline order in a one-dimensional lattice (Peierls 1935).

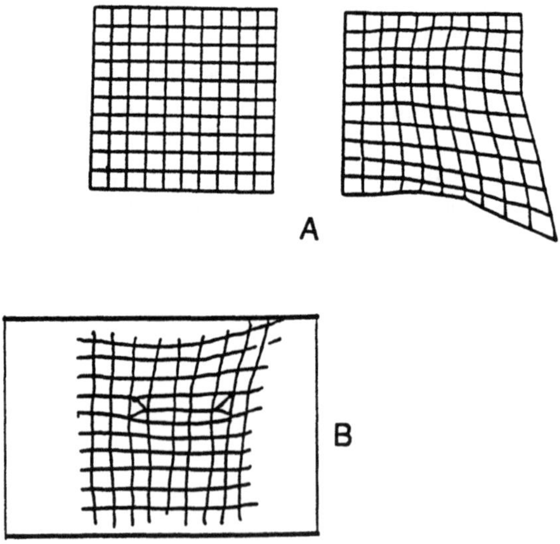

Fig. 2. Topological order is a second form of order that can survive finite temperature.
A. Elastic distortion does not destroy topological order.
B. The breakdown of topological order begins with the formation and growth of dislocations.

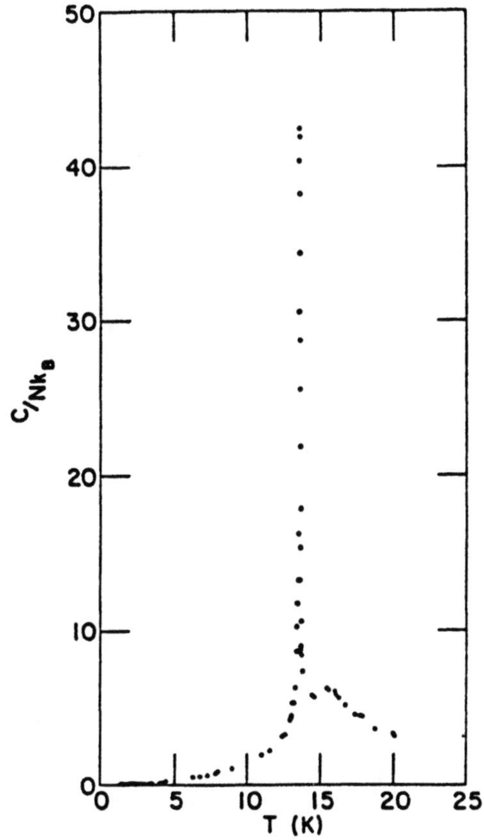

Fig. 3. The heat capacity of neon monolayers on graphite in the region of the 2D melting transition (Huff 1976).

Nucleation and Surface Melting of Ice

David W. Oxtoby

James Franck Institute and Department of Chemistry, University of Chicago
5640 South Ellis Avenue, Chicago IL 60637, USA

Abstract. The freezing of water and the melting of ice, phenomena that are extraordinarily familiar from the world around, are only poorly understood at a molecular level. Like all liquids, water can be undercooled without freezing; like most solids, ice cannot be superheated to a significant extent. The dynamics of these phase transitions, and what they tell about the molecular level structure of interfaces and critical nuclei, is revealed by experiment, computer simulation, and theory. In some regards, water resembles simpler liquids in the kinetics of its changes of state, but in others it is quite different. This chapter explores the microscopic dynamics of freezing and melting, with particular emphasis on the water-ice system.

1 Introduction

The freezing of liquid water to ice is a first-order phase transition that, according to thermodynamics, should take place at 0°C. Kinetics presents a different story, however, and metastable liquid water at temperatures as low as –30°C is well attested when the water is emulsified into small droplets; purified bulk samples are readily cooled below –10°C. In a first-order transition, two steps are necessary before a new phase becomes visible: nucleation and growth. In most cases, growth of crystals from the melt is rapid and the kinetic barrier to new phase formation is dominated by the first step, nucleation.

When ice is heated, quite a different situation obtains. Even though melting is also a first-order phase transition (like freezing, boiling, and condensation) it is unique among these four transitions in showing essentially no hysteresis or metastable phase formation. Like most solids, ice cannot be superheated to any significant extent; melting is initiated as soon as the temperature exceeds 0°C. The origin of this observation can be traced to the phenomenon of surface melting, the statement that an ice crystal is coated with a thin layer of liquid water (which can become thick as the melting point is approached from below). Melting thus can occur in a continuous fashion from the surface inward, rather than as a kinetically hindered nucleation process inside the bulk crystal.

In this chapter I describe theoretical and experimental approaches to the freezing and melting of crystals, with particular emphasis on the ways in which water and ice resemble and differ from simpler materials.

NATO ASI Series, Vol. I 56
Ice Physics and the Natural Environment
Edited by John S. Wettlaufer, J. Gregory Dash and Norbert Untersteiner

2 Homogeneous Nucleation of Crystals from the Melt

Small enough crystallites surrounded by the melt tend to shrink even in an undercooled liquid; large enough crystallites tend to grow because of the greater thermodynamic stability associated with the crystalline phase under these conditions. The crystallite that falls at the boundary between shrinking and growing is called the critical nucleus, because its rate of formation determines the overall rate of the phase transition.

2.1 Classical and Nonclassical Nucleation Theory

Consider the free energy to create a small spherical crystallite of radius R surrounded by the bulk metastable liquid. This excess free energy can be approximated as[1]

$$\Delta F(R) = -\frac{4\pi}{3} R^3 \, |\Delta F_{\mathrm{v}}| + 4\pi R^2 \gamma_{\mathrm{sl}}$$

The first term, proportional to the volume of the crystallite, is negative and reflects the driving force for crystallization (ΔF_{v} is the free energy difference per unit volume between crystal and melt). The second term, proportional to the surface area of the crystallite, is positive and reflects the fact that creating an interface between two phases costs free energy (γ_{sl} is the solid-liquid surface free energy per unit area). Such a free energy is positive for small values of R, reaching a maximum at

$$R^* = \frac{2g_{sl}}{|DF_v|}$$

at which point the free energy has the value

$$\Delta F^* = \frac{16\pi\gamma_{sl^3}}{3|DF_v|^2}$$

Although this has been presented as an equilibrium calculation of free energies, it is in fact a kinetic problem in which forward and backward rates for cluster sizes are related through an equilibrium constant that in turn can be expressed in terms of a free energy.[2] Clusters of radius smaller than R^* will have larger backward than forward rates and will tend to disappear; those larger will tend to grow. A steady-state rate expression can be written for the rate of formation of clusters large in size compared to R^*, and it has the form

$$J = J_{\mathrm{o}} \exp(-\Delta F^*/kT)$$

Here the kinetic prefactor J_{o} is the rate of attachment of particles to a cluster of the critical size, and depends on the growth kinetics across the solid-liquid interface.[3] This expression has the usual Arrhenius activated form for the kinetics of barrier crossing, with the important distinction that the barrier height ΔF^* is not a constant but rather a strong function of temperature. Its temperature dependence arises espe-

but rather a strong function of temperature. Its temperature dependence arises especially from the $|\Delta F_V|$ term, which varies as

$$\Delta F_V = \Delta H_V - T\Delta S_V = \Delta S_V (T_f - T)$$

where ΔH_V and ΔS_V are the bulk enthalpy and entropy change of crystallization per unit volume and T_f is the freezing temperature. The fact that this term occurs in the denominator and is squared means that the free energy barrier ΔF^* diverges as T approaches T_f (as expected, because the driving force for the transition disappears in that limit). The liquid-solid surface free energy also depends on temperature, but less strongly.

The nucleation theory just described is referred to as classical nucleation theory. It relies on the capillarity approximation, in which crystallites of microscopic size are treated as if they are macroscopic, and in which the kinetics is described as the stepwise attachment of single molecules across the crystal-melt interface. In fact, this approximation may not be valid under realistic conditions. A small crystallite may not achieve bulk properties at its center, and its interface may be so strongly curved that the planar value γ_{sl} no longer applies. The interface may be diffuse rather than sharp, so that the description of the kinetics as resulting from addition of solid particles one after another may not be valid; instead, a collective fluctuation may result in the simultaneous incorporation of a larger number of molecules in a loosely structured crystallite.

These observations form the basis of a new approach to nucleation theory built on density functional methods of statistical mechanics. The predictions and results of this nonclassical theory are described in greater detail in Ref. 1. The starting point is the observation that the average density of a crystal can be written as a sum over Fourier components:

$$\rho(r) = \rho_l[1+\mu_o + \Sigma_i\mu_i \exp(ik_i \cdot r)]$$

In this equation, ρ_l is the liquid density, μ_o is the average density change on freezing, and the k_i are vectors of the reciprocal lattice of the crystal. The parameters μ_i are amplitudes that describe the periodic structure in the crystal; they are zero in the liquid. In a critical nucleus, these order parameters μ_i depend on the radial distance from the center, approaching the value of zero at long distances in the bulk liquid phase. In density functional theory, the free energy is a functional of these order parameters:

$$F = F[\{\mu_i(r)\}]$$

This is a much more flexible theory than classical nucleation theory (which allows the free energy to depend only on a single quantity, the crystallite radius). Here, the interface can be diffuse instead of sharp, the crystallite can be strongly curved, and the order parameters can attain values different from the bulk at the center. At the same time, the nonclassical density functional approach goes over smoothly to the classical theory for large enough crystallites. Comparisons of the predictions of the classical and nonclassical nucleation theory of freezing will be given in Section 2.2 below. The nonclassical approach can also be extended to a theory of crystal growth.[4]

2.2 Experimental Studies of Ice Nucleation

The temperature at which a bulk sample of liquid water freezes is highly variable, depending on the extent of purification of the sample and the nature of the container containing the liquid; it can easily fall anywhere in the range from 0°C down to –15°C, and can be non-reproducible even when the same sample is repeatedly melted and frozen. The reason for this variability is that such a bulk sample freezes via heterogeneous nucleation either on impurity particles in the liquid or on the surface of the container. Controlled heterogeneous nucleation of ice on well-defined surfaces is an interesting subject in its own right (discussed in Section 4 below), but when it causes the nucleation of bulk water samples in an ill-defined fashion, it is a hindrance to making progress on the fundamental problem. Methods must be sought that avoid the heterogeneous nucleation of crystals; three approaches are described in this section.

The first method is to take a liquid sample and divide it into small droplets as an emulsion suspended in oil. If these droplets are small enough (micron-sized) then some will contain heterogeneous impurities and others will not. The entire suspension is then cooled and the freezing of individual droplets monitored by differential thermal analysis (monitoring the heat of fusion given off as each droplet crystallizes), volume measurements (taking advantage of the difference in density between crystal and liquid), or light scattering. As the temperature is lowered, droplets containing heterogeneous impurities of various kinds will crystallize over a range of temperatures, but those remaining will crystallize abruptly close to a single temperature, the homogeneous nucleation temperature. (To be more precise, the rate of crystallization changes from being very slow to very fast over a one to two-degree temperature range.) One caveat : it is possible that the nucleation is still not homogeneous but rather is catalyzed by the liquid-oil interface. This can be tested by trying a number of different types of oil.

Wood and Walton[5] applied the emulsion method to the nucleation of ice from water and found a homogeneous nucleation temperature close to –36°C. The rates were fifteen orders of magnitude larger than those predicted from classical nucleation theory, using the liquid-solid surface free energy of water measured at 0°C by an independent experiment. The authors suggested that a modest linear temperature dependence of the surface tension would account for this discrepancy, and in fact their measured temperature dependence between –35°C and –37°C extrapolated to a value close to the one measured at 0°C. However, this result was questioned by Butorin and Skripov[6] and more recently by Huang and Bartell[7] who showed that a more reasonable temperature extrapolation gives a surface free energy 30% lower than that measured directly at 0°C. The equilibrium surface free energy at 0°C is not accurately known (Ref. 8 gives 29 erg cm^{-2}, whereas earlier measurements gave values ranging from 25 to 45 erg cm^{-2}) so comparison with classical nucleation theory is difficult. In addition, it is not possible to separate the effect of temperature from the effect of curvature on the surface free energy: a critical nucleus at lower temperature will be smaller and thus more strongly curved.

A second way to ensure homogeneous nucleation is to work under conditions in which the growth rate is small compared to the nucleation rate. This leads to a very high density of nucleation centers, too high to be explained by heterogeneous impurity particles. MacFarlane et al.[9] used an emulsion suspension of aqueous lithium chloride in water to study crystallization near the glass transition, where the crystal growth is slowed down relative to nucleation. Under the conditions of their experiment, each droplet was estimated to contain many thousands of ice nuclei. However, this approach cannot be applied to the nucleation of pure water.

The third and most elegant method of studying homogeneous nucleation has been developed in recent years by Bartell and coworkers[7]. They used a supersonic jet expansion to cool gaseous water until it formed clusters of 4000 to 6000 molecules; further cooling in the jet caused the clusters to crystallize at about –73°C (200 K). The crystallization was monitored by electron diffraction, a technique that allows the structure of the crystals that first form to be identified. From the size of the clusters and the cooling rate, a nucleation rate of 10^{24} cm^{-3} s^{-1} was deduced, 20 orders of magnitude larger than that for the earlier emulsion experiments at –36°C. In spite of the extremely different temperature and nucleation rate, classical nucleation theory with a similar surface free energy to the earlier experiments was able to fit the data. These authors show that under these highly undercooled conditions water freezes not to the thermodynamically favored hexagonal ice Ih but to the kinetically favored cubic ice Ic. This difference may account for the lower values of surface free energies found in nucleation experiments compared to those measured at 0°C for the equilibrium interface. There are several advantages to the Bartell approach: purity is ensured in a small cluster formed from the gas phase, there is no oil-water surface that can cause heterogeneous nucleation, and the clusters are small enough and the crystallization rapid enough that it begins to approach conditions accessible to computer simulation of clusters.

2.3 Theory and Simulation of Crystal Nucleation

To date, experiments have not been able to test classical nucleation theory for freezing because there are not accurate measurements of the liquid-solid surface free energy. (This differs from the testing of nucleation theory for the gas-liquid transition, where the surface tension is routinely measured.) One way to obtain indirect information on this question is through theory: the density functional (nonclassical) approach described briefly above allows classical theory to be tested. We employed approximate density functionals[10,11] and used them to calculate both the free energy barrier in nonclassical theory and the parameters (surface free energy and bulk free energy difference) that appear in classical theory. The result of this self-consistent comparison is a nonclassical crystal nucleation rate for a Lennard-Jones (argon-like) fluid that is more than 10 orders of magnitude greater than the classical prediction under typical experimental conditions. The structure of the liquid-crystal interface that emerges is also interesting: periodic crystalline structure (as reflected in the order parameters μ_i) arises several angstroms before the density changes (order parameter μ_0) in moving from the liquid to the crystalline side of the interface. The interface calculated is relatively broad: several lattice spacings are typical, showing that sharp-interface models of crystal growth from the melt may have limited value.

Direct simulation of nucleation is difficult, because under realistic conditions nucleation is a rare event. Two approaches have been developed that give very useful information about the critical nucleus in crystallization. Baez and Clancy[12] start with crystalline clusters of varying sizes surrounded by liquid, and follows them on the computer to see whether they shrink or grow; this gives a direct estimate of the size and other properties of the critical nucleus. Ten Wolde and Frenkel[13] use an umbrella sampling Monte Carlo technique to select preferentially (via a weighting factor) clusters intermediate between pure liquid and pure crystal, and thus to map out the free energy pathway over the nucleation barrier. They then obtain the kinetic prefactor in the nucleation rate by starting with the critical nucleus and measuring the rate at which it grows.

The theoretical and simulation approaches just described have been applied to Lennard-Jones fluid crystallization, but not to models for water. There is nothing in principle to prevent this, however, except for the fact that water calculations are slower than calculations on simple spherical potentials. Ding et al.[14] applied a density functional approach to the bulk crystallization of water, and obtained reasonable results, although they did not look at nucleation. Karim et al.[15] used two different potentials to simulate the ice-water interface and found that the results were robust to changes in the potential. Svishchev and Kusalik[16] simulated liquid water in an electric field and showed that it crystallized preferentially into the cubic ice Ic form. Although none of these calculations looked at nucleation, this is possible now and will be done in the future for water.

One result that is found for the crystallization of Lennard-Jones fluids may have particular relevance to the nucleation of ice from liquid water. Ten Wolde and Frenkel[13] found significant body-centered cubic (bcc) ordering at the interface between the face-centered cubic (fcc) crystal and the liquid. Shen and Oxtoby[17] studied this using density functional theory and showed that a nearby metastable bcc phase had a large effect on the barrier to nucleation and the nature of the critical nucleus, which for large undercoolings can be described as a bcc crystallite that only later is converted into an fcc crystallite. It is likely that metastable phases also play a large role in the crystallization of water. The complexity of the phase diagram means that there are a number of such phases surrounding the liquid phase. In particular, the experiments of Bartell[7] suggest that deeply undercooled water crystallizes into cubic ice Ic for kinetic reasons, rather than the thermodynamically favored hexagonal ice Ih, and these authors suggest that the same phase may be initially formed at –36°C as well, converting into hexagonal ice only later. (Recall that the freezing of water in an electric field is also found in computer simulation to take place in the cubic phase.[16]) Further simulations and density functional calculations on water will shed light on the role of metastable phases in the kinetics of crystallization.

3 The Singular Behavior of Undercooled Water

Simple liquids show no evidence for a low temperature spinodal in the undercooled state. In this, freezing differs from the other three types of phase transition (melting, boiling, and condensation). The spinodal is the point at which the original phase becomes unstable with respect to small fluctuations of the new phase, so that the barrier to nucleation vanishes entirely and a new mechanism of phase change (spinodal decomposition) arises, with the new phase appearing and growing spontaneously throughout the sample. Such an instability in the liquid would be signaled by a divergence in the equilibrium structure factor $S(k)$, and there is no evidence for such behavior in the case of simple liquids. Anomalies in undercooled liquids instead appear in the dynamics: diffusion can slow down and viscosities can increase by many orders of magnitude as a fluid approaches the glass transition. Such a state of thermal arrest can be attained even for a simple liquid that is cooled rapidly enough, and when it occurs nucleation of crystallites is suppressed.

Water differs from simple liquids in that it shows certain tantalizing indications of a low temperature singularity in equilibrium as well as dynamic properties. Speedy and Angell[18] measured the isothermal compressibility of undercooled liquid water and showed that it could be fit to a power law divergence at –45°C. They demonstrated that other properties, both static and dynamic, could also be fit to power laws about the same limiting temperature, giving for example a divergence of the heat capacity and the viscosity and a vanishing of the diffusivity and even of the density at this same temperature. Such behavior was suggested to be evidence for a mechanical instability of the spinodal type in water, associated with cooperative behavior in its network of hydrogen bonds. The fact that the density is decreasing with decreasing temperature (the opposite of the behavior for simple liquids) may be relevant to the existence of such a singularity.

In a recent paper, Pruppacher[19] carried this analysis still further in order to analyze the results of field experiments on ice nucleation in clouds. He took four additional properties for which there is no evidence of singular behavior at –45°C (heat of melting, heat of evaporation, surface tension, and liquid-solid surface free energy) and fit them to power laws that led to a vanishing of the first and the last and a divergence of the second and third of these properties. With these functional forms, he then took nucleation rate data and used them to fit the only other parameter, the activation free energy for mass transfer across the liquid-solid interface. The result was that this quantity increased as the water was cooled down to –30°C and then began to decrease sharply.

The results of Bartell on undercooled small droplets (discussed in Section 2.1 above) raise real questions about the existence of a spinodal singularity, and especially about the type of analysis used by Pruppacher. They show that nucleation can be avoided in liquid water all the way down to –73°C, with no sign of an anomaly at –45°C (however, a small droplet has a high enough internal pressure that any anomaly may be suppressed by that effect alone). Recent simulations have also raised questions about the spinodal hypothesis. Poole et al.[20] simulated water at low temperatures and over a range of pressures to explore the phase diagram. They suggested that the low temperature singularity is not caused by a spinodal but rather by a critical point between two amorphous phases. In this picture, there is a continuity of states between liquid water and the high-density amorphous solid water seen experimentally at cryogenic temperatures, while a separate metastable phase transition to low-density amorphous solid water (also found experimentally) terminates in a critical point that gives rise to the apparent singularities in the behavior of undercooled water. Subsequently Tanaka[21] suggested that this critical point may lie at negative pressures. The experimental observation[22] that ice can be converted to an amorphous phase by the rapid imposition of pressure (pressure-induced amorphization) must also be understood in this context. It is clear that the anomalous behavior of undercooled water is only poorly understood and that there is room for much progress in the future in both experiment and theory.

4 Heterogeneous Nucleation

The homogeneous nucleation of ice from water can be achieved in the laboratory only by breaking up the sample into tiny droplets; otherwise the process is always heterogeneous, taking place on or around foreign particles. To describe freezing in the physical environment, it is therefore necessary to understand the heterogeneous nucleation of liquids, including water, on well-defined solid substrates.

4.1 Classical and Nonclassical Theory

Consider a small droplet of a new crystalline phase (β) in contact with both the original liquid phase (α) and with a flat solid substrate (Fig. 1). The free energy change ΔF to create this new phase from the original metastable phase involves volume and surface terms, just as in the case of homogeneous nucleation. The negative bulk term (proportional to the volume of the new phase present) once again depends in classical nucleation theory on the free energy difference per unit volume driving the phase transition. There are now three interfacial free energies, however, arising from the

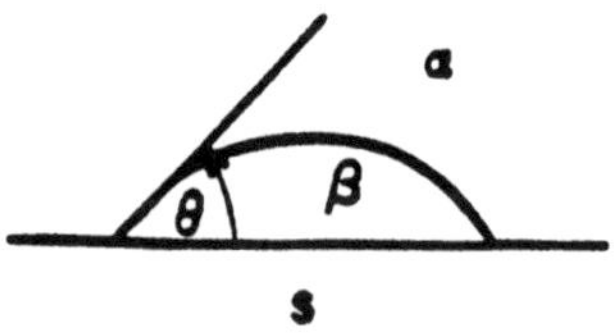

Fig. 1 Contact angle of two phases α and β with substrate, s. In heterogeneous nucleation of crystallization, α is the melt and β is the crystal. transition. There are now three interfacial free energies, however, arising from the new interfaces between substrate and liquid and between substrate and crystal, as well as that considered already between liquid and crystal.

new interfaces between substrate and liquid and between substrate and crystal, as well as that considered already between liquid and crystal.

The mathematical (variational) problem to be solved is to obtain the crystallite shape that minimizes the free energy for a given nucleus volume. As before, we ignore any variation of surface free energy with crystalline facet. In this case, it is straightforward to show that the shape adopted by the minimum free energy crystallite is a spherical sector, which makes contact with the solid substrate at an angle θ. This angle is related to the three interfacial free energies via the equation[23]

$$\cos\theta = \frac{\gamma_{s\alpha} - \gamma_{s\beta}}{\gamma_{\alpha\beta}}$$

The free energy of such a heterogeneous nucleus is closely related to the corresponding homogeneous nucleus free energy:

$$\Delta F_{het}^{*} = \Delta F_{hom}^{*} f(\theta)$$

where the function $f(\theta)$ is defined as

$$f(\theta) = \frac{(2+\cos\theta)(1-\cos\theta)^2}{4}$$

and the homogeneous free energy refers to a full sphere with the same radius R as that of the spherical sector in the heterogeneous case.

Two limiting cases are of interest here. As the angle θ approaches 180°, the interfacial free energies approach the condition

$$\gamma_{s\alpha} - \gamma_{s\beta} \leq -\gamma_{\alpha\beta}$$

and the spherical sector turns into an entire sphere that loses contact with the solid substrate. In this limit, $f(\theta)$ approaches 1 and the nucleation becomes homogeneous; the solid substrate has no effect on nucleation because it raises, rather than lowers, the free energy of formation for a nucleus. The other limiting case occurs as θ approaches 0. Here the inequality becomes

$$g_{ab} \text{ £ } g_{sa} - g_{sb}$$

The system can lower its free energy in this case by inserting a thin layer of the β phase between the α phase and the substrate, and we say that β "wets" the substrate. (The language comes from the spreading of a liquid over a solid substrate when both are in contact with the vapor phase.) In this limit, $f(\theta)$ approaches 0 and the barrier to nucleation vanishes. One way to describe the appearance of bcc ordering in the middle of the fcc-liquid interface (see Ref. 17 and Section 2.2 above) is to say that the bcc phase "wets" the fcc-liquid interface.

The discussion in this section has focussed entirely on perfectly flat solid substrates. Surface roughness, curvature, crevices, or impurities can significantly promote or inhibit heterogeneous nucleation in real systems.

4.2 Heterogeneous Nucleation of Ice

Different substrates differ in their power to nucleate crystals from the melt, with the largest reduction in nucleation barrier (and thus the largest increase in rate) coming when the substrate-crystal surface free energy is as low as possible. One way to make this interaction more favorable is to choose a substrate whose periodicity is in registry with that of the crystal that is nucleating, in order to minimize any strain energy between crystal and substrate and maximize the attractive interactions. A well known example is the nucleation of ice on silver iodide crystals, a result that has led to the use of such crystals for seeding rain clouds.

A heterogeneous substrate behaves reproducibly if upon repeated slow cooling experiments the liquid always crystallizes at the same temperature. This does not mean that the lag-time until a crystal becomes visible at fixed temperature is constant. Because nucleation is a random event, there should be an exponential distribution of times to nucleation under fixed external conditions. This type of distribution has been demonstrated in a recent study of ice nucleation on silver iodide.[24] If a reproducible lag-time is seen, it is not correct to define the nucleation rate as the inverse of the lag-time; it is more likely in this case that what is being measured is the time delay for the crystal to grow to observable size.

Ice can also be nucleated heterogeneously at liquid-liquid interfaces[25] and by alcohol monolayers adsorbed on the surface of water.[26] In the former study, undercooled water at an interface with terpenes or chlorobenzenes in carbon tetrachloride showed a large range of ice nucleation temperatures, ranging from –7.4°C down to –25°C, suggesting that the interface can either favor or inhibit nucleation. The alcohol monolayer study revealed an interesting odd-even asymmetry: those aliphatic alcohols with an odd number of carbon atoms give nucleation temperatures that approach 0°C with increasing length, while those with an even number give consistently lower nucleation temperatures and level off at –8°C. The relative role of substrate periodicity and of electric fields induced by dipoles at the surface is not known; recall the simulations[16] that show a large effect of electric field on crystallization into cubic ice.

Living species adapt to freeze-thaw cycles in their environment either by avoiding freezing of the water in their tissues (lowering the nucleation temperature) or by freezing and thawing reproducibly at a high temperature (avoiding the flash freezing that can be lethal if a deeply undercooled state is reached).[27] There are therefore uses both for proteins that accelerate nucleation and for those that retard it. It seems likely that the surface of a nucleating protein can adopt a periodic structure that favors heterogeneous crystallization, whereas one that hinders nucleation may act by surrounding small crystallites as soon as they appear, preventing them from growing.

5 Melting

In many liquids, crystallization is difficult to achieve and in some cases is avoided entirely by rapid cooling to below the glass transition. By contrast, melting is difficult to avoid, and only in special circumstances can any significant degree of superheating be attained. To understand this difference, we first consider the difference in bulk nucleation behavior of the two transitions, and then turn to a still more important factor: the role of a free surface in initiating melting.

5.1 Nucleation of the Melting Transition

There has been much less attention paid to homogeneous nucleation of melting than to freezing. In classical nucleation theory, the two processes are completely parallel, and the same expression for the critical radius and critical nucleus free energy result, suggesting that it should be possible to superheat a crystal by the same amount as one can undercool a liquid. This symmetry is broken in the nonclassical theory, however, because of the presence of a high-temperature spinodal: the point at which the crystal becomes mechanically unstable to small fluctuations. As discussed above, there is no evidence for a corresponding spinodal in the freezing of liquids. The result is that nucleation barriers fall to zero with increasing temperature, and the possible superheating is thereby reduced.[11]

Experiment shows, however, that not even a small amount of superheating seems to be possible for most crystals. Thus an alternative explanation to homogeneous nucleation must be sought. One possibility is the unavoidable presence of defects in crystals; for example, there is a significant degree of disorder at a grain boundary or a dislocation, and that may provide a site for nucleation of the liquid phase. Another way in which nucleation of melting differs from nucleation of freezing is that the phase surrounding an incipient droplet is rigid for melting, making density relaxation more difficult. This would suggest that the nucleation may occur at constant volume rather than constant pressure; in a case where the liquid density is lower than that of the crystal, the liquid droplet would be compressed, while in the reverse case (such as the ice-water transition) it would be under negative pressure. This effect might change the height of the nucleation barrier, but would certainly not lower it to zero.

The currently accepted explanation for the difficulty associated with superheating is that a crystal with a free surface will melt from the surface inward, with no barrier to be overcome in the transition. This mechanism, referred to as surface melting, will be described further in the sections that follow. It immediately suggests that the only way to superheat a crystal is to heat the interior above the melting point while keeping the surface below that point. This has been achieved for ice[28] and gives rise to negative pressures and compression fractures as internal melting begins.

5.2 Theory of Surface Melting

Surface melting is a special case of surface wetting, in which a liquid wets the surface of its own crystal. In the language of surface free energies, this occurs when

$$\gamma_{sl} + \gamma_{lv} < \gamma_{sv}$$

so the system can lower its free energy by having a solid-liquid and a liquid-vapor interface instead of a single solid-vapor interface. (In this case, the quantity γ_{sv} becomes a purely hypothetical quantity, because an equilibrium solid-vapor interface cannot exist.) Of course, the liquid layer at a finite temperature below the freezing point cannot have a macroscopic thickness. This can be seen both from the Gibbs phase rule (solid, liquid, and vapor can only coexist in bulk at the triple point), and is also understandable from a free energy point of view (below the melting point, it costs free energy to convert solid to liquid). However, a liquid layer of molecular thickness is possible, and as the solid approaches its equilibrium melting point the free energy cost to melt solid to liquid decreases. The result is a divergence in the thickness of the liquid layer at the melting point.

The nature of this divergence in surface melting can be analyzed theoretically. One way to think of it is as an effective interaction between the two interfaces (solid-liquid and liquid-vapor) which balances the cost in free energy to increase the amount of liquid present. For a short-ranged potential the layer thickness diverges logarithmically:

$$l(T) = k\,|\ln(T_m - T)|$$

For a long-ranged (power law) potential, there is a crossover close to T_m to a power-law divergence; for example, the variation asymptotically close to T_m for an R^{-6} dispersion interaction is

$$l(T) = k\,(T_m - T)^{-1/3}$$

These predictions from mean field lattice models[29] are consistent with density functional calculations of the crystal surface density profile.[30]

Experimental measurements of surface melting have been carried out for noble gases and for metals. Sequential melting of surface layers of thin supported films of argon has been studied by neutron diffraction[31] and through thin film heat capacities.[32] The latter method applied to neon gave evidence for a transition from a logarithmic to a power-law divergence of the melt layer thickness, in agreement with theoretical predictions.[29] The surface melting of lead has also been studied extensively by Rutherford backscattering of high-energy protons[33] and a logarithmic divergence of the melted layer thickness was seen, as expected for the short-ranged interactions characteristic of a metal. Subsequent work showed that surface melting depended on the crystal face studied;[34] liquid lead, for example, wets the {001} but not the {111} surface of solid lead. Metois and Heyraud[35] have shown that lead crystallites containing only {111} facets can be superheated by several degrees, a clear demonstration of the role of surface melting in preventing superheating of solids.

5.3 Surface Melting of Ice

Dash et al.[36] recently reviewed the surface melting of ice and its environmental effects, and that reference provides a large number of references to earlier work. They present evidence for surface melting based on a wide variety of experiments, including proton backscattering, ellipsometry, and x-ray diffraction. However, they also call attention to other studies in which direct visual observation just below the triple point suggests the presence of standing drops of water on the surface of ice, with a very small but finite contact angle. This latter observation would indicate the absence of true surface melting.

Elbaum and co-workers have resolved these discrepancies and provided both theoretical and experimental evidence for an unusual form of incomplete surface melting in the case of ice. Elbaum and Schick[37] carried out a theoretical calculation of surface melting in ice, using frequency dependent polarizabilities and retarded polarization forces. Thin water film stability depends on the high-frequency polarizability, which favors wetting, but as the film becomes thicker, lower frequencies become important, and these disfavor wetting. Whereas a calculation omitting retardation gives the usual –1/3 power divergence of the melting layer, the inclusion of retardation causes the film thickness to level off between 150 and 200 angstroms close to the melting point. This would give rise to incomplete surface melting in this limit and allow droplets with finite contact angles to appear just below the triple point. Elbaum et al.[38] then used optical reflectometry and interference microscopy to verify this prediction of incomplete wetting experimentally. They found that the surface layer of water on the basal plane of ice increased rapidly in thickness from –2°C up to a few hundredths of a degree below bulk melting, when stable droplets suddenly appeared. Thus both theory and experiment are consistent with a picture of incomplete surface melting just below the equilibrium bulk melting temperature.

The review by Dash et al.[36] also discusses interfacial melting of ice in contact with other substances and the many environmental consequences of ice premelting, including frost action and thunderstorms.

5.4 Melting of Small Particles

Up to this point, the melting transitions considered have involved bulk samples with free surfaces. Small particles can melt as well, however, and the nature of that transition depends strongly on particle size. Buffat and Borel[39] present some experimental and theoretical results for small gold particles that show that the melting temperature can be lowered from 1340 K in bulk to below 600 K for 25-angstrom size particles. Their data can be fit by either of two theoretical models, both of which show that the lowering of the melting point is proportional to the curvature of the particles (and thus inversely proportional to their radius r_s). The first approach is simply based on thermodynamics and calculates the variation of chemical potential with pressure and temperature. They note that the pressure P_l inside a liquid drop is related to the external pressure P_{ext} via the Kelvin equation

$$P_l = P_{ext} + 2\gamma_{lv}/r_l$$

with a corresponding equation for the pressure in the solid crystal. Setting the two chemical potentials equal and solving for the melting temperature $T(r_s)$ gives

$$\frac{T_m - T(r_s)}{T_m} = \frac{2}{r_s L r_s}\left[g_{sv} - g_{lv}\left(\frac{r_s}{\eta}\right)^{2/3}\right]$$

where ρ_s and ρ_l are the densities of solid and liquid and L is the latent heat of fusion. A second approach is to assume that the solid is wet by a thin layer of the liquid below the melting point, so there is a solid-liquid and a liquid-vapor interface. In this case, the result for the lowering of the melting temperature becomes

$$\frac{T_m - T(r_s)}{T_m} = \frac{2}{r_s L}\left[\frac{g_{sl}}{r-d} + \frac{g_{lv}}{r}\left(1 - \frac{r_s}{\eta}\right)\right]$$

where δ is the thickness of the liquid surface layer.

Small particles are also observed to melt in computer simulations at temperatures well below bulk melting. Rare gas clusters as small as 13 particles show apparent phase transitions,[40] although of course at small sizes these are rounded rather than sharp transitions and resemble conformational equilibria more than collective many-body processes. A simulation of gold particles[41] revealed a three-step process for the melting of the larger particles: onset of surface diffusion, formation of liquid patches at the surface, and abrupt melting of the whole cluster. The 477-atom cluster melted at 800 K and the 219-atom cluster below 600 K.

Small ice particles also demonstrate surface melting, and this fact has important geophysical consequences. As Baker and Dash[42] have shown, if two particles collide that have different liquid layer thicknesses, there will be a differential pressure across the two that will lead to mass transfer from the particle with the thicker film to the one with the thinner. Because of the surface orientation of water molecules, there is preferential adsorption of negative ions near the free surface, so that along with the mass transfer there will also be a net charge transfer of negative ions. The estimated magnitude of the charge transfer is in agreement with laboratory experiments, and gives a plausible mechanism for the electrification of thunderclouds. (There are several alternative theoretical explanations for charge separation at growing ice-water interfaces, however, and at present experiments are not able to resolve this important question.) Beaglehole[43] has extended the work of Baker and Dash to show the effect of small concentrations of dissolved impurities in favoring the surface melting of small particles, another important effect in the atmosphere where many ice particles form by nucleation on small soluble aerosol particles.

References

1. D. W. Oxtoby, *J. Phys. Cond. Matter* 4, 7627 (1992).
2. J. L. Katz and F. Spaepen, *Phil. Mag. B* 37, 137 (1978).
3. D. Turnbull and J. C. Fisher, *J. Chem. Phys.* 17, 71 (1949).
4. D. W. Oxtoby and P. R. Harrowell, *J. Chem. Phys.* 96, 3834 (1992).
5. G. R. Wood and A. G. Walton, *J. Appl. Phys.* 41, 3027 (1970).
6. G. T. Butorin and V. P. Skripov, *Sov. Phys.-Crystallogr.* 17, 322 (1972).
7. J. Huang and L. S. Bartell, *J. Phys. Chem.* 99, 3924 (1995).
8. S. C. Hardy, *Phil. Mag.* 35, 471 (1977).
9. D. R. MacFarlane, R. K. Kadiyala, and C. A. Angell, *J. Chem. Phys.* 79, 3921 (1983).
10. P. R. Harrowell and D. W. Oxtoby, *J. Chem. Phys.* 80, 1639 (1984).
11. Y. C. Shen and D. W. Oxtoby, *J. Chem. Phys.* 105, 6517 (1996).
12. L. A. Baez and P. Clancy, *J. Chem. Phys.* 102, 8138 (1995).
13. P. R. ten Wolde, M. J. Ruiz-Montero, and D. Frenkel, *J. Chem. Phys.* 104, 9932 (1996).
14. K. Ding, D. Chandler, S. J. Smithline, and A. D. J. Haymet, *Phys. Rev. Lett.* 59, 1698 (1987).
15. O. A. Karim, P. A. Kay, and A. D. J. Haymet, *J. Chem. Phys.* 92, 4634 (1990).
16. I. M. Svishchev and P. G. Kusalik, *Phys. Rev. Lett.* 73, 975 (1994).
17. Y. C. Shen and D. W. Oxtoby, *Phys. Rev. Lett.* 77, 3585 (1996).
18. R. J. Speedy and C. A. Angell, *J. Chem. Phys.* 65, 851 (1976).
19. H. R. Pruppacher, *J. Atmos. Sci.* 52, 1924 (1995).
20. P. H. Poole, F. Sciortino, U. Essmann, and H. E. Stanley, *Nature* 360, 324 (1992).
21. H Tanaka, *Nature* 380, 328 (1996).
22. O. Mishima, *Nature* 384, 546 (1996).
23. D. W. Oxtoby, in *Inhomogeneous Liquids*, ed. by Douglas Henderson (Marcel Dekker, New York, 1992), Chapter 10.
24. W. Barlow and A. D. J. Haymet, *Rev. Sci. Inst.* 66, 2996 (1995).
25. J. Rosinski and A. Lecinski, *J. Phys. Chem.* 85, 2993 (1981).
26. M. Gavish, R. Popovitz-Biro, M. Lahav, and L. Leiserowitz, *Science* 250, 973 (1990).
27. K. B. Storey and J. M. Storey, *Sci. Am.* (December, 1990), p. 92.
28. C. A. Knight and N. C. Knight, *Science* (Nov. 10, 1972).
29. R. Lipowsky, *Phys. Rev. Lett.* 49, 1575 (1982).
30. R. Ohnesorge, H. Lowen, and H. Wagner, *Phys. Rev. E* 50, 4801 (1994).
31. J. Z. Larese and Q. M. Zhang, *Phys. Rev. Lett.* 64, 922 (1990).
32. D. Zhu and J. G. Dash, *Phys. Rev. Lett.* 60, 432 (1988).
33. J. W. M. Frenken and J. F. van der Veen, *Phys. Rev. Lett.* 54, 134 (1985).

34. B. Pluis, A. W. Denier van der Gon, J. W. M. Frenken, and J. F. van der Veen, *Phys. Rev. Lett.* 59, 2678 (1987).
35. J. J. Metois and J. C. Heyraud, *J. Phys. France* 50, 3175 (1989).
36. J. G. Dash, H. Fu and J. S. Wettlaufer, *Rep. Prog. Phys.* 58, 115 (1995).
37. M. Elbaum and M. Schick, *Phys. Rev. Lett.* 66, 1713 (1991).
38. M Elbaum, S. G. Lipson, and J. G. Dash, *J. Cryst. Growth* 129, 491 (1993).
39. P. Buffat and J.-P. Borel, *Phys. Rev. A* 13, 2287 (1976).
40. T. L. Beck, J. Jellinek, and R. S. Berry, *J. Chem. Phys.* 87, 545 (1987).
41. F. Ercolessi, W. Andreoni, and E. Tosatti, *Phys. Rev. Lett.* 66, 911 (1991).
42. M. B. Baker and J. G. Dash, *J. Cryst. Growth* 97, 770 (1989); *J. Geophys. Res.* 99, 10621 (1994).
43. D. Beaglehole, *J. Cryst. Growth 112*, 663 (1991).

Crystal Growth, Surface Phase Transitions and Thermomolecular Pressure

J.S. Wettlaufer

Applied Physics Laboratory and Department of Physics,
University of Washington, Seattle, Washington, 98105-5640, USA.

Abstract. The most abundant crystal growth problems in the natural environment involve ice, yet in nearly every setting a crucial roadblock to our progress concerns how microscopic processes influence macroscopic behavior. Recent advances in understanding the surface phase transitions that alter the equilibrium and near equilibrium interfacial structure are central to many pattern formation problems during ice crystal growth, the geometric evolution of polycrystals, and the dynamics of frost heave. This chapter focuses on surface, size and impurity effects in the melting, growth and interfacial evolution of ice crystals. The primary focus in melting and growth concerns how the nature of the microscopic interfacial structure at ice/vapor, ice/gaseous-atmosphere and ice/water interfaces influences the growth shapes. The treatment of impurities in solidification dynamics begins at the molecular level and is traced to the long range transport effects that drive interfacial instabilities. Finally, we examine the dynamical implications of taking subfreezing interfaces into a weakly nonequilibrium regime wherein the underlying causes of frost heave are revealed. Experimental, theoretical and computer simulation techniques all play important roles in our evolving understanding of both the basic phenomena under scrutiny and their environmental implications. The fundamental issues are observed in nearly all materials, but their environmental manifestations are most striking in the case of ice, lying as they do at the heart of the evolving shape of a snowflake, the origin of the forces driving cryoturbation in soils and the microscopic basis of charge transfer inside icy clouds.

Keywords. Ice, Surface Melting, Crystal Growth, Intermolecular forces

1 Introduction

Several surface-specific structural phase transitions are observed in ice. We will focus here on complementary aspects of the basic phenomena of interfacial melting, surface roughening, and molecular kinetics and their application to the equilibrium and growth structures of ice crystals. A central issue is how the interfacial structure at ice/vapor, ice/gaseous-atmosphere and ice/water interfaces influences the adsorption potential, the growth shapes and surface transport properties. The environmental relevance concerns the fact that ice dominates the crystal growth phenomena we observe

NATO ASI Series, Vol. I 56
Ice Physics and the Natural Environment
Edited by John S. Wettlaufer, J. Gregory Dash and Norbert Untersteiner

on the surface of the earth and in the atmosphere. The broad scientific relevance concerns the fact that surface phase transitions occur in a host of materials, so that what we learn in ice may be important in many other settings. Underlying these problems is the interplay between phase behavior and dimensionality, a topic that developed in the context of quantum materials (Dash et al., 1994), and a parallel development that grew out of Nakaya's studies of ice crystal growth (Nakaya 1954; Furukawa 1997). The long search for a microscopic theory of melting displays how our focus has moved from the bulk to the surface of a crystal. The formation of a crystal from a vapor or a melt phase has its own history in the field of nucleation. Both topics, discussed in complementary chapters by Greg Dash and David Oxtoby, come together in crystal growth.

2 Surface Structure and Growth

2.1 Equilibrium Crystal Shapes

It is not widely appreciated that the plethora of naturally occurring solids we encounter, ranging from mineral formations to snowflakes, have experienced many of the same basic growth processes during their evolution. Their most striking feature is their overall shape. Therefore, it is important to examine the extent to which this shape can be understood in terms of equilibrium thermodynamics. The concept of a global geometry that arises from an extremum principle is broadly relevant in many areas of condensed matter science and mathematics, and a great deal of our contemporary approach to such questions originated in Wulff's goal to understand the shapes of naturally occurring crystals (Wulff, 1901). Experiments, exact solutions of microscopic models, and mean-field theory have demonstrated that the equilibrium shape of a crystal depends on temperature (excellent reviews include Weeks and Gilmer 1979, Lipson and Polturak 1987, and Balibar et al., 1993). Classical theories have been sufficient to demonstrate the main qualitative features: An ideal dislocation free crystal is fully faceted at absolute zero, and becomes more rounded, or locally rough, as its temperature increases.

The surface free energy of a crystal, $\gamma(\boldsymbol{n})$, depends on the orientation of the unit normal $\boldsymbol{n}$ of the surface relative to the underlying crystalline lattice. The crystal itself is built by a replication of unit cells. When the dominant intermolecular interactions in the material are short ranged, one can estimate the bond energies by considering only nearest neighbors in the lattice. We can then think of $\gamma(\boldsymbol{n})$ as the sum of the energies of all the bonds broken per unit area in the creation of the surface. The energy itself depends on where, relative to the underlying crystalline lattice, the surface is created[1]. If one knows a-priori, the surface free energy of the crystal as a function of

---

[1] Note that such a treatment of the surface free energy ignores the role of curvature in local distortion of the surface, an effect with increasing importance as a crystal becomes smaller (e.g., Oxtoby 1998).

all orientations present, then the geometric Wulff construction determines the equilibrium crystal shape (e.g., Wulff, 1901; Herring, 1951; Cahn and Carter, 1996). Conversely, were one to measure the shape of a crystal in the absence of a growth drive[2], the Wulff construction will provide the surface free energy for all orientations present on that shape (e.g., Heyraud and Mètois, 1986).

We suppose that $\gamma(\boldsymbol{n})$ is known for all crystal faces, and the construction determines uniquely the equilibrium crystal shape by minimizing the orientation-dependent total surface free energy per unit area for the volume enclosed by the crystal surface:

$$I_S = \int_S \gamma(\boldsymbol{n})dA. \tag{1}$$

We can consider Wulff's construction as a transform on the surface free energy itself, acting to minimize the above integral at constant volume, temperature, and chemical potential;

$$\boldsymbol{n} \bullet \boldsymbol{r} = \lambda\gamma(\boldsymbol{n}). \tag{2}$$

Here $\boldsymbol{r}$ defines a radial vector from the origin to the equilibrium crystal surface, which is given by the inner envelope of a set of planes perpendicular to radial rays, intersecting a polar plot of $\gamma(\boldsymbol{n})$ as shown in Fig. 1. The boundary of the equilibrium *Wulff* shape is given by equation (2) and is geometrically similar at any size determined by λ above (e.g., Taylor et al., 1992). Dinghas (1944), Herring (1951) and Landau (1965), among others, have provided proofs that Wulff's construction does indeed minimize I_S. An example with which we have some familiarity is that of a soap bubble, wherein $\gamma(\boldsymbol{n})$ is independent of $\boldsymbol{n}$ and in the absence of gravity the minimization provides the spherical equilibrium shape. The effect of gravity is only important when the body under consideration becomes large (e.g., Avron et al., 1983).

Cusps in the polar plot of $\gamma(\boldsymbol{n})$ locate high symmetry crystallographic planes that are molecularly smooth. We call these *facets* or *singular surfaces*, the latter referring to the nonanalyticity in the $\gamma(\boldsymbol{n})$ plot. Note that because some cusps have a lower free energy than others, not all cusps appearing in the $\gamma(\boldsymbol{n})$ plot will be represented by a facet on the Wulff shape. Surfaces of a crystal normal to which there is a gradual variation over the scale of a number of molecular planes are called *molecularly rough* or *diffuse* surfaces. An equilibrium crystal shape may be fully faceted or completely rough, or may have facets and rough regions coexisting.

---

[2]The growth drive for crystallization is defined in Equation (3).

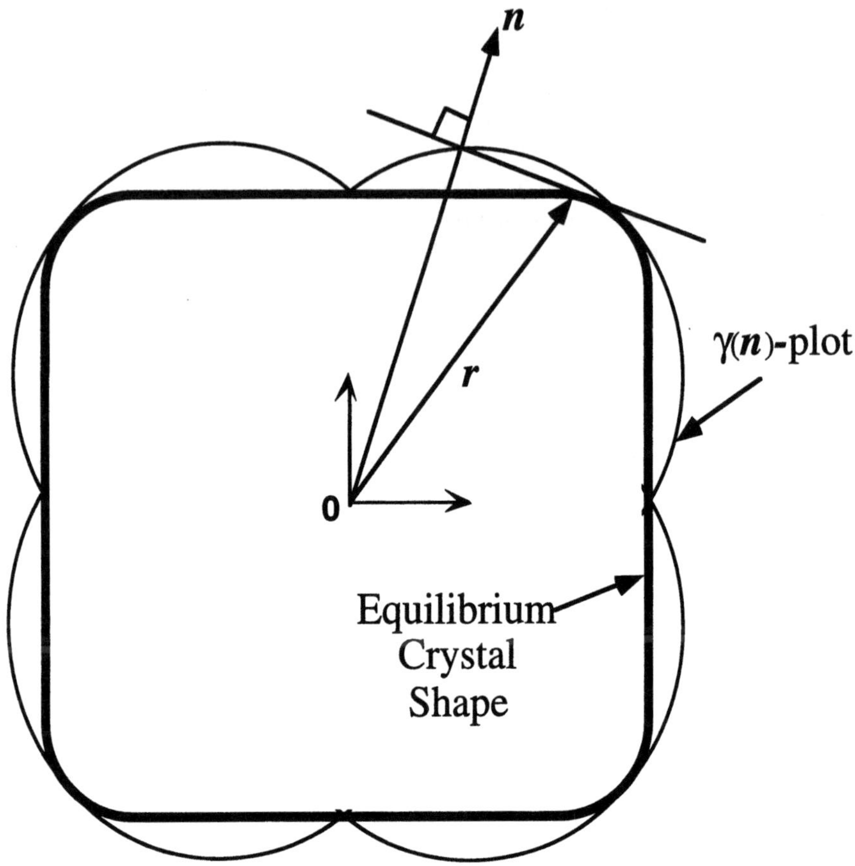

Fig. 1. The boundary (bold lines) of the equilibrium crystal shape with cubic symmetry formed from the Wulff construction, which is the interior envelope of the set of perpendiculars to radial rays intersecting the polar plot of surface free energy (lighter lines).

2.2 Surface Roughening

Two thermodynamic interfacial transitions are relevant here. The first involves a structural rearrangement of solid material. A crystal surface is characterized by the free energy of the various interfacial configurations: faces, ledges, corners, edges, and point-defects such as admolecules and vacancies. The energetic cost of any state is reckoned with the sum of the formation enthalpies of the sites, and the benefit toward lowering the total free energy comes from increasing the configurational entropy of the surface. The latter is the driving force for the *thermodynamic roughening transition*. As the temperature increases, thermal fluctuations give rise to the continual formation of steps. Creating a step is tantamount to slightly tilting the surface, which requires doing some work against the periodic potential of the underlying crystalline array. As the temperature T increases, thermal agitation facilitates step formation and the free energy cost of a step, $\sigma_S(T)$, decreases. The coherence length of the surface, $\eta=\gamma(\boldsymbol{n})/\sigma_S(T)$, is a measure of the distance (relative to the mean orientation of the surface) over which fluctuations at two points are correlated (e.g., Lipson and Polturak 1987). For an infinite two-dimensional surface $\sigma_S(T) \longrightarrow 0$ at the *roughening temperature*, T_r, thermal agitation liberates the surface from the ordering influence of the underlying crystalline lattice and fluctuations are correlated on all length scales so that the surface roughens. Strictly speaking, one-dimensional interfaces are always rough, and three-dimensional (and higher) interfaces are never rough.

Any *real* crystal has facets of finite size, and as the temperature increases η approaches the facet size and they roughen at $T = T_f < T_r$. If $T > T_f$ for all facets present, an equilibrium shape will be completely rough and rounded.[3] There is a substantial literature in the field, and while theoretical treatments are in qualitative agreement, predictions of T_r and T_f are strongly model dependent. Because the roughening transition is responsible for marked interfacial restructuring, the process has a profound influence on the dynamics of molecules on the surface, and ultimately on near equilibrium crystal growth shapes. This point reminds us of an intimately related phenomenon to be described in a subsequent section; the *kinetic* or *dynamic* roughening transition, which occurs when the growth rate exceeds a critical value. There are other distinct, but closely related transitions associated with surface *reconstruction* that do not concern us here but which are fundamentally important and often observed in other materials (e.g., den Nijs, 1997).

There have been few definitive studies of these transitions as they occur in ice. The roughening transition of the prism facet of ice I_h has been observed under vapor conditions by Elbaum (1991). A beautiful series of recent experiments has been able to address these transitions over a tremendous thermodynamic range for ice in contact with water (Maruyama et al., 1997). Maruyama, has demonstrated both the roughening transition of the prism facet of ice I_h and the equilibrium crystal shape by carefully controlling the pressure along the solid-liquid coexistence line. Using a high pressure optical cell, he observed shape changes from near the triple point to -21°C at 200 MPa. He observed the prism facets to undergo a thermodynamic roughening transition at approximately -16°C at 165 MPa. The technique has the ability to accurately control the growth drive very near equilibrium and hence can test theoretical predictions of crystal shapes growing and melting near equilibrium.

2.3 Interfacial and Surface Melting

The second transition to be discussed is a kind of lower dimensional melting. The description of the melting of a solid as a first-order phase transition is a consequence of the discontinuous change in *bulk* quantities at the transition point. However, every crystal is finite, and bounded by its own surface area where the process of melting may actually be initiated: If there were a layer of liquid at the surface, at temperatures below the bulk melting transition, then there is little need to activate the melting proc-

---

[3]Note that η diverges logarithmically with crystal size.

ess. Such a deduction was made by Stranski (1942) on the basis of the inability to superheat crystals. However, since it is well known that liquids can be supercooled, the melting process exhibits an asymmetry about the transition point. The chapters by Dash and Oxtoby alleviate the need to develop any more detail than is necessary to insure that this presentation is reasonably self contained.

Recent studies have borne out the intuition of Stranski. The main issues are embodied in the effect of *surface melting*, which takes place at the surface of a crystal in contact with its vapor and is therefore extremely important in environmental ice science (Dash et al., 1995). Surface melting occurs when a liquid wets its own solid in the form of a stable thin liquid layer disjoining the solid/vapor interface at temperatures below the bulk transition temperature T_m. (Dash 1998; Oxtoby 1998). It is an equilibrium phenomenon driven by the tendency to reduce the interfacial free energy of the system. The process occurs in many classes of solids, including metals, semiconductors, solid rare gases and molecular solids. Typically surface melting begins at $T < 0.9\ T_m$, with a film thickness of one or two monolayers, which thickens gradually with increasing T, and diverges at T_m. In a modern framework, we think of the problem as a special case of wetting (e.g., Schick, 1988). In surface melting the underlying solid consists of the same material as the liquid above. Whether or not a layer of melt beads up, or spreads to form a thin film on its own solid depends on a competition between the attraction of the liquid to the solid, *adhesion*, and the attraction of the liquid to itself, *cohesion*. The origin of the competition may be long range intermolecular dispersion forces, which have unique consequences in ice systems (Elbaum and Schick, 1991). If we replace the vapor phase by a wall of a different material, the process is referred to as *interfacial melting* and the mean field thermodynamic description is essentially the same. In this case unscreened long ranged surface ionization forces are important (Wilen et al., 1995). Depending on the nature of the substrate and the temperature, system specific short ranged electrical interactions may also be present (Petrenko, 1993).

Most relevant to pattern formation during the growth of ice crystals is the surface melting of ice at interfaces with vapor and gaseous atmospheres (Furukawa, 1997). There have been several laboratory experiments, with extremely variable and complex results. We believe that the variability lies in both the great sensitivity to the atmosphere and conditions of preparation, and in the inherent differences in measurement technique. The evidence of surface melting comes from studies by proton backscattering, ellipsometry, optical reflectometry and interference microscopy, X-ray diffraction and glancing angle X-ray scattering. Because the techniques themselves are designed to measure different structural aspects of the surface there is still active debate in the community regarding the interpretation of the various results throughout the entire temperature range (Dash et al., 1995). A particularly interesting effect observed by Elbaum et al., (1993) is worth noting here. It is called *incomplete surface melting*, and is distinguished by the coexistence of water droplets with a thin water film at the ice/vapor interface. The effect is due to the attenuation with film thickness of the interactions that favor adhesion. Regardless of the complexity of the results, what is certain is that sufficiently close to the bulk melting transition, the disordered surface, whether liquid-like or taking some other structure, the growth kinetics will be substantially altered (Kuroda and Lacmann, 1982; Sei and Gonda 1989; Furukawa and Kohata, 1993; Nada and Furukawa, 1996).

2.4 Near Equilibrium Crystal Growth

In addition to the aesthetic appeal of crystal growth shapes, their understanding informs many areas of technology, from computer chip manufacturing to powder metallurgy. Furthermore, because growth shapes are often cast in the form of a free boundary problem, there are a myriad of theoretical analogues in hydrodynamic, chemical and biological systems (e.g., Kirkaldy 1992, Cross and Hohenberg 1993). The problems of equilibrium crystal shapes and growth shapes merge when one begins to address nucleation and subsequent solidification (Frank 1958; Chernov 1984; Elbaum and Wettlaufer 1993; Bagdassarian and Oxtoby 1994; Wettlaufer et al., 1994). In the laboratory, crystal growth very close to equilibrium can be achieved (e.g., Heyraud and Mètois 1983, 1986; Carmi et al., 1987; Elbaum et al., 1993; Maruyama et al., 1997). When the growth drive is removed, a crystal growth shape must eventually relax to the equilibrium form. The relaxation time depends on the mechanism facilitating the redistribution of material. For example, when surface diffusion dominates the relaxation times are typically extremely long (Cahn and Taylor 1994). Crystals observed in Nature are often growth shapes that are still undergoing slow relaxation toward equilibrium. Indeed, nearly all the shapes we observe depend in some manner on the history of the object.

David Oxtoby (1998) describes how a seed of the solid phase is nucleated. Subsequent to nucleation, the manner in which growth proceeds depends on the state of the crystal surface and the thermodynamic conditions of the nutrient phase. Our main goal in crystal growth is to determine the evolution of the *unknown shape* of the solid. We must take care in understanding the dominant processes through which a molecule in the nutrient phase becomes part of the solid. For example, the shape evolution depends on whether the interfacial motion is controlled by long-range diffusion or by local interfacial processes. We first consider single crystals growing from a pure parent phase and define the two common growth regimes. Second, we add impurities and discuss a number of the phenomena that control the structure of sea ice and other natural ice forms.

The growth drive for crystallization is the difference in the chemical potential between a molecule in the parent phase and one in the crystal. Molecules move in such directions as will lower chemical potential. When bulk solid and liquid phases are in coexistence at temperature T_m and pressure P_m the chemical potentials of both phases are equal: $\mu_s(T_m, P_m) = \mu_\ell(T_m, P_m)$, where the subscripts denote the solid (S) and liquid (ℓ) phases. Hence, there is zero growth drive for crystallization: $\mu_\ell(T_m, P_m) - \mu_s(T_m, P_m) \equiv \Delta\mu = 0$. However, were we to adjust the pressure by ΔP and/or the temperature by ΔT to extend into the solid region of the bulk phase diagram, then the solid becomes the stable phase and its chemical potential is reduced relative to the liquid phase, and the growth drive for crystallization becomes positive. Expanding the growth drive about the point (T_m, P_m) we find

$$\Delta\mu(T_m+\Delta T, P_m+\Delta P) = -\frac{q_m}{T_m}\Delta T-(v_s-v_\ell)\Delta P\,. \tag{3}$$

Experimentally it is ideal to create either isothermal or isobaric control. For ice and water very well controlled growth can be achieved in the range $0 \leq T_m \leq -20°C$. With either $-\Delta P = 0.1$ MPa or $-\Delta T = 0.01°C$ one can obtain $\Delta\mu/k_bT = 10^{-3}$ (e.g., Maruyama et al., 1997).

We do not know in detail the relationship between a *nucleated shape* and an *equilibrium shape*. However, it is reasonable to consider the latter as one that experiences the least activation barrier, is the most likely to be nucleated and hence is the initial shape for growth studies. The manner in which molecules attach to the existing solid at a specific orientation is loosely referred to as *growth kinetics*. These kinetics depend on the local interfacial free energy in a particular crystallographic orientation. We can distinguish two broad regimes of growth: *interface-controlled* and *diffusion-limited*. The former regime is best treated in the context of *geometric* theories wherein the interfacial growth velocity is controlled only by local surface parameters that are decoupled from diffusional or other long-range influences.[4] More technical descriptions are given by Taylor et al., (1992) Yokoyama and Sekerka, (1992), and Wettlaufer et al., (1994).

For interface controlled growth, the goal is to formulate a theory for evolution of the overall crystal shape by constructing the local velocity $V(n)$ normal to an interfacial point. The theory should be based on our knowledge concerning molecular attachment kinetics in single interfacial states, faceted or rough, or the transition between them (Burton et al., 1951; Cahn 1960; Cahn et al., 1964; Weeks and Gilmer 1979; Chernov 1984). In 1958 Charles Frank initiated this type of study in a paper that is highly recommended to any student of the topic. The essential point of the simplest class of geometric models is that the local interfacial velocity will depend on local interfacial processes rather than on long-range processes such as diffusion in the bulk. A number of familiar situations wherein this is the case include the early stages of atmospheric ice growth, when the mean free path in the vapor is larger than the characteristic size of the crystal, the growth of electronic materials using the techniques of molecular beam epitaxy, the growth of ferromagnetic or ferroelectric domain walls, grain growth, and stress-driven-zone migration. It is important to stress that a number of instabilities associated with diffusional growth cannot manifest themselves in this regime. Within a geometric framework, we have been able to explain some general features of transient anisotropic growth observed in metals, ^{4}He, and ice. We first review basic growth kinetics and then ask how a well defined crystalline surface, either faceted or rough, moves normal to itself under the imposition of a *weak* growth drive $\Delta\mu$ that is spatially homogeneous and steady.

---

[4]The diffusion-limited regime can only be treated by a geometric theory in special limiting regimes.

2.4.1 Growth Kinetics

2.4.1.1 Growth from the Melt

Consider a facet in equilibrium ($\Delta\mu = 0$). The addition of any integral number of identical lattice planes provides the same, minimum free energy, equilibrium configuration The surface itself (except for its size) is identical after the completion of each layer. A small growth drive will create some fractional coverage of molecules on the surface, thereby increasing the free energy, until a complete plane is added and the low energy surface is regained. Hence, in analogy with nucleation, a facet is metastable *during* growth. Continuous growth normal to a facet relies primarily on the lateral motion of steps across the surface (Burton et al., 1951, Cahn 1960). The term *step* is most often reserved for a single molecular layer, but less restrictive definitions are appropriate because it is observed often in semiconductors (Liu et al., 1997), and can be inferred from optical measurements on ice (Hallett 1961), that steps consisting of many molecular layers propagate across a growing facet. I view Cahn's (1960) definition to be the clearest: A step describes the transition between two adjacent, parallel, and structurally identical regions of a surface that are separated by an integral number of lattice planes. Defect- and dislocation-free facets have available only one step generation mechanism: the two-dimensional nucleation of solid clusters. Other distinct step generation mechanisms such as screw dislocations, or Frank-Read sources (a ledge separating a pair of oppositely signed screw dislocations) are often observed. Each mechanism has a specific form relating the growth drive to the growth rate, and which mechanism dominates depends on the material and the principal facets being studied. Geometric theories can treat a variety of step generation mechanisms.

When the primary facets of ice are defect free, the relevant step generation mechanism is the formation of two-dimensional nuclei. The process is driven by thermally activated fluctuations of the liquid phase that create two-dimensional heterophase clusters with solid-like structure. The most probable cluster geometry is that of a pill-box, for which the edge-to-surface free energy ratio will favor spreading at a given drive. In analogy with three-dimensions, the nucleation frequency $\boldsymbol{I}$ per unit facet area A is of a Maxwell-Boltzmann form, $\boldsymbol{I} \propto exp(-\pi\ \sigma_n^2/\Delta\mu\ k_b T)$, where σ_n^2 is the free energy of a critical nucleus on the facet (e.g., Weeks and Gilmer 1979). If lattice plane spacing is given by â then we can write the normal growth rate at faceted orientations as $V_f(\boldsymbol{n}) = \hat{a}\ \boldsymbol{I}\ A$. This growth mechanism has been observed on the basal planes of ice growing from water over a large range of undercoolings from 0.03 to 0.2 K (Hillig 1958; Wilen and Dash 1995; Hodgkin et al., 1998). The observation of this step generation mechanism displays the structural perfection of the basal plane facets for pure samples. It should be noted that this growth mode can be frustrated by surface active impurities. By contrast, rounded surfaces are molecularly rough in that steps are already present on the surface so that growth occurs by random incorporation of nutrient molecules onto the surface. Therefore, the entire interface continuously advances normal to itself in the presence of any finite driving force, and the velocity is linearly related to the drive, $V_r(\boldsymbol{n}) \propto \Delta\mu$ (Burton et al. 1951, Cahn 1960), with some orientation-dependent factor accounting for local anisotropy.

2.4.1.2 Growth from the Vapor

The linear relationship is most simply shown by analogy with the *Wilson-Frenkel* theory of growth from the vapor (e.g., Chernov, 1984). The total growth rate V_T is the difference between the deposition and evaporation rates. Thus, the actual deposition rate v_{eq}^{+} is proportional to the vapor pressure: $v_{eq}^{+} = K_{eq} exp(\Delta\mu / k_b T)$ where K_{eq} is the equilibrium deposition rate. In a detailed balance $V_T = 0$, but during growth the simplification is made that the evaporation rate takes the equilibrium value; $K_{eq}^{-} = K_{eq}$. The *Wilson-Frenkel* rate is

$$V_{WF} = v_{eq}^{+} - K_{eq}^{-} = K_{eq}\left(exp(\Delta\mu / k_b T) - 1\right). \tag{4}$$

For $T \gg T_r$. to leading order in $\Delta\mu$, the rate becomes $V_{WF} = K_{eq}\left(\Delta\mu / k_b T\right)$, which crudely shows that the interfacial velocity is linearly related to the drive. It is physically plausible that due to the homogeneity of surface evaporation and the large surface diffusivity at high temperatures, the linearized *Wilson-Frenkel* expression may be quantitatively accurate. It should be stressed that this is a purely illustrative example, for it is well known that it serves solely as an upper bound.

2.4.2 Kinetic Roughening

During *kinetic* or *dynamic* roughening a large growth drive may eliminate the ordering effect of the underlying lattice (e.g., Balibar et al., 1993) and thereby change the nature of the interfacial attachment kinetics. Here, the competition is between a "crystallization fluctuation" which may arise from the formation and subsequent evaporation/melting of subcritical surface nuclei, or a growth/melt wave induced by thermal fluctuations, and a periodic lattice potential of the form $\Phi cos(2\pi z/\hat{a})$, where z is the height of the surface and the lattice plane has thickness $\hat{a}$. Such a fluctuation carries with it a latent heat which must diffuse. The dissipation occurs over a characteristic time scale τ_F which depends on the size and the diffusivity of the fluctuation. If the interface is moving at $\partial z/\partial t = V_f$ then the addition of a lattice plane takes a time $\hat{a}/V_f$. If the fluctuation dissipates before a layer is added, i.e., $\tau_F < \hat{a}/V_f$, then the lattice potential influences the entire history of the disturbance. However, if growth is fast enough so that $\tau_F > \hat{a}/V_f$, then the influence of the lattice potential on the fluctuation is time periodic and therefore averages to zero. Hence, the growth process itself renders the anchoring effect of the lattice ineffective, and the surface *roughens dynamically*. Let us be clear in our distinctions: For a zero growth drive, an equilibrium form is faceted up to the *thermodynamic roughening* transition T_r. For growth at a fixed temperature and fixed coupling strength to the underlying lattice, a facet will dynamically roughen within a range of growth drives wherein $\tau_F \approx \hat{a}/V_f$,. Roughening is suppressed as the strength of the lattice coupling increases, and thermodynamic roughening is "smeared out" under nonequilibrium conditions, effectively shifting the transition to lower temperatures.

A different kinetic realization of dynamic roughening is as follows: If the rate of step-spreading is much greater than the nucleation rate, then a single layer can cover the facet before the subsequent nucleation event occurs. As the driving force is increased, this condition is more difficult to satisfy, and the facet becomes decorated with multiple supercritical nuclei simultaneously. Over a time period τ_S the nuclei spread and join to complete the layer. At still larger growth drives, the increase in the surface density of molecules increases the nucleation rate, and during a time less than τ_S, secondary nuclei may form on the existing islands. The picture is that of islands growing upon islands and effectively roughening the surface. The process also results in a crossover from activated, or layer by layer growth, to continuous growth (e.g., Cahn 1960 and Nozières and Gallet 1987) and is related to some growth modes observed in molecular beam epitaxy of metals (Kunkel et al., 1990). Detailed theories of kinetic roughening have been given by Josè et al., (1977), Chui and Weeks (1978), and Nozières and Gallet (1987).

2.4.3 An Example: Evolution of an Equilibrium Seed

We are now in a position to explore the evolution of an equilibrium seed. Different surface structures coexist on a partially faceted equilibrium crystal shape (Fig. 1). By considering such an initial shape we are able to perceive a kind of intrinsic growth anisotropy. We impose a small growth drive, so that the shape is uniformly bathed by nutrient molecules. The relaxation rate at rough orientations is much more rapid than that on facets, and if the drive itself is less than that necessary to induce kinetic roughening on the facets, then at faceted orientations there will be slow normal growth by nucleation and spreading of monolayers. Relatively fast normal growth will occur at molecularly rough orientations. Finally, diffusion of admolecules on the facets causes them to serve as catchment areas, and the diffusing material influences the growth rate in vicinal orientations.[5].

There are several experimentally realizable limits that are particularly instructive. We can loosely distinguish these in terms of the magnitude of the growth drive $\Delta\mu$. The first regime occurs for $\Delta\mu \leq 10^{-3}$, wherein the accretion of mass is much slower than all of the available relaxation processes, so that the crystal surface is indistinguishable from the equilibrium crystal shape. This regime is called "thermodynamically slow" or "shape preserving" growth and can describe the growth shapes via a continuous expansion of the scale of the Wulff diagram (Elbaum and Wettlaufer, 1993). Growth shapes in this regime have been observed at temperatures above, below and very near the roughening transition of the prism facet growing from water (Fig. .3 of Maruyama et al., 1997). The result is also trivially correct for iso-

---

[5]A vicinal orientation is one with a small angle deviation from a singular orientation, characterized by broad terraces separated by monomolecular steps.

tropic surface free energy; a crystal everywhere above its roughening transition, or for example in a liquid drop. One then simply expands a sphere.

In the second regime $\Delta\mu$ is larger, but nonetheless smaller than the activation barrier for two-dimensional nucleation on the facets. Therefore, the facets are pinned and the rough orientations accrete mass to take the Wulff shape of an "equilibrium" crystal of increasing size (Elbaum and Wettlaufer, 1993). Ultimately the rough orientations grow themselves out of existence leaving a fully faceted growth form behind[6]. Theoretical growth shapes are generated by two methods. The first method expands the polar surface free energy plot uniformly by increasing λ of equation (2), and redrawing the perpendiculars construction subject to the additional constraint that the facets *do not* move normal to themselves. Hence, because of the lack of an activation barrier in the rough orientations, the crystal surface only remains in "contact" with the polar surface free energy plot at the rough orientations. The latter contribute a decreasing fraction of the total surface of the crystal shape with time (see Fig. 1 of Elbaum and Wettlaufer, 1993), but the curvature of these regions *decreases* with time. The method is not thermodynamically rigorous, but it is intuitive and qualitatively reproduces experimental behavior. It lacks thermodynamic rigor because we cannot of course apply a global extremum principal locally. It is intuitive because the rough orientations can "keep up" with the growth drive and are therefore near equilibrium (i.e., the free energy polar plot), whereas, the facets fall behind.

The second method is a geometric model that predicts the evolution of every point of the crystal surface. As described above, this class of models requires us to construct the local velocity $V(\boldsymbol{n})$ normal to an interfacial point. In trying to understand the evolution of partially faceted equilibrium shapes, such a function must explicitly account for the variation in growth kinetics with location on the surface. Our approach has been to include activated growth on facets, nonactivated growth in rough regions and their modification in the vicinal orientations (Wettlaufer et al., 1994). An observer moving on the surface of the crystal will see activated growth at singular orientations; having to "jump up" as a monomolecular step runs across the surface. Walking away from singular faces, the observer will see a transition to the rough growth kinetics, from, in the vicinal regions, surface migration of admolecules away from facets, to the eventual rough terrain where growth occurs via random incorporation of molecules into the surface. Such effects can be modelled as: $V(\boldsymbol{n},\Delta\mu) = V_f(\boldsymbol{n},\Delta\mu)\,\xi(\boldsymbol{n}) + V_r(\boldsymbol{n},\Delta\mu)\,(1-\xi(\boldsymbol{n}))$, where $\xi(\boldsymbol{n})$ describes the observer's transition between facet-like and rough-like growth. The necessary properties of $\xi(\boldsymbol{n})$ are that it is periodic in $2\pi/n$, that $0 \leq \xi(\boldsymbol{n}) \leq 1$, and $\xi(\boldsymbol{n}_f) = 1$, $\xi(\boldsymbol{n}_f + \pi/n) = 0$, where $\boldsymbol{n}_f$ defines the normal at a facet. We choose the simplest form of the function satisfying these properties and predict the evolution of the same initial shape as previously (Fig. 2). We find the same overall evolution as in the previous model, but here the facets are allowed to move normal to themselves when the critical supercooling for two-dimensional nucleation is exceeded, albeit the rate is rather slow relative to nonsin-

---

[6]The growth and relaxation of a fully facetted shape is of interest in its own right (Carter et al., 1995), but here we focus on the details of how the shape arises after which a host of other effects become important.

gular orientations. We observe the spreading of the facets, and the change in character, from smooth to abrupt, of the joint between facets and rounded parts. These qualitative features have been observed on crystals of solid He growing into superfluid, of ordinary water ice growing into pure water vapor (Elbaum et al., 1993) and into liquid water (Maruyama, pers. comm. 1997), and on small metal crystals growing into metal vapor (Heyraud and Mètois, 1987).

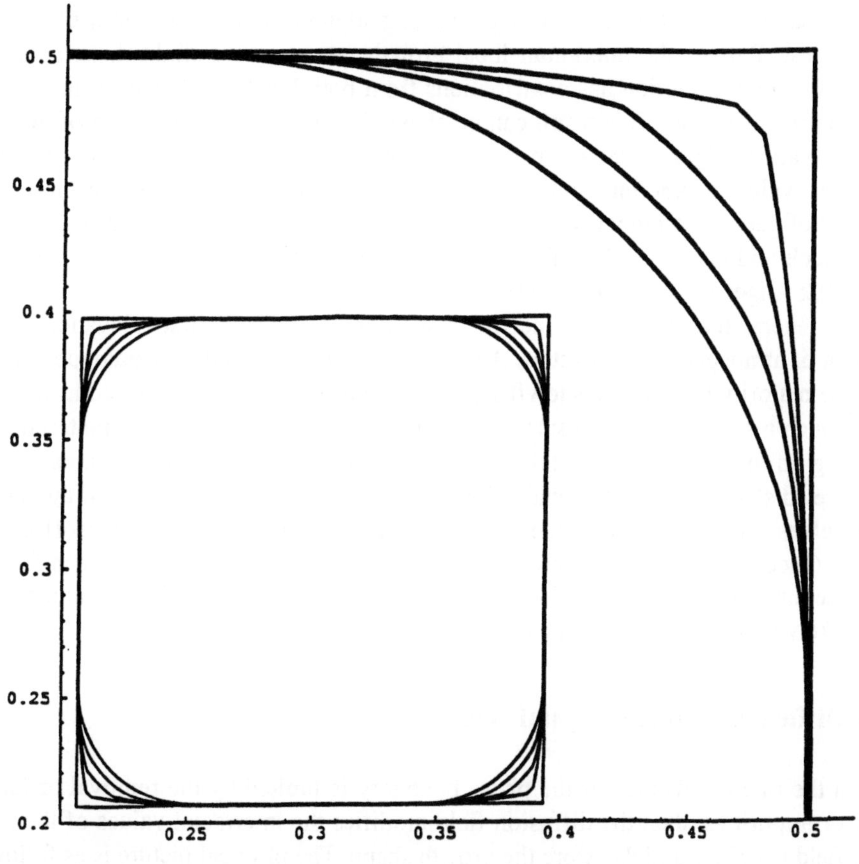

Fig. 2. A sequence of growth shapes at dimensionless times t = 0, 0.018, 0.038, 0.068, 0.35 of the upper right quadrant of a crystal with cubic ($n = 4$) symmetry. The units are arbitrary. The initial (inner) equilibrium shape is that is formed by the Wulff construction shown in Fig. 1. The inset shows the full crystal shape at the same times. Note that the rough orientations grow out of existence with a *decreasing* curvature, and that there are sharp joints where the vicinal regions join the rough regions. The curvatures at the rough "corners" where the angle θ defining $\boldsymbol{n}$ relative to the positive x-axis is $\theta = \pi/4$ for t = 0, 0.018, 0.038, 0.068 are 4, 3.7, 3.4, 3.1 (the numerical values agreeing with the exact analytic solutions to one significant figure Wettlaufer et al., 1994). As growth progresses, facet orientations move substantially slower than the rough regions, and the crystal surface loses orientations until it is fully faceted and possesses only four orientations.

The rough regions of the surface take a *decreasing* curvature as growth progresses between the equilibrium and fully faceted shapes. Recall that equilibrium forms become more faceted as the temperature decreases. This growth process is intrinsically transient and we refer to it as "global kinetic faceting", in order to emphasize that (a) it is the global (i.e., the entire closed surface) effect of local dynamics, and (b) it should be distinguished from equilibrium faceting or local kinetic faceting. The decrease in curvature at rough orientations during this transition is consistent with notions of the heuristic model above, but has a kinetic basis in that the local normal growth rate is not, as is commonly supposed, a maximum where the density of steps is the largest. Rather, the maximum local growth rate occurs a slight distance away where the normal motion has contributions from both local attachment of molecules and surface diffusion. This result can be shown by an explicit calculation of the surface curvature. Figure 2 shows an example from our own studies, and there are other forms of velocity functions both within the context of our model and in the larger context of geometric models of interface motion (Langer, 1987; Kessler et al., 1988; Ben-Jacob and Garik, 1990; Cahn et al., 1991; Goldstein & Petrich, 1991; Taylor et al., 1992; Angenent & Gurtin, 1994).

The general features of this faceting transition are of particular relevance to the growth of atmospheric ice crystals. A study of Nakaya's snowflake catalogue reveals that the majority of platelike snowflakes have evolved from a fully-faceted hexagonal seed. Such a seed contains no surface which can easily accept accreting material. Depending on the size of the crystal relative to the diffusion length in the vapor phase, the faceted state defines a reasonable lower bound for kinetic roughening of the facets themselves, or the onset of either shape instabilities or oscillations (Tersoff et al., 1993). Once the crystal size approaches the diffusion length, the corners reach farther into the nutrient phase and tend to trigger changes in growth morphology and mode. One likely morphological change is discussed presently.

2.5 Diffusion-Limited Crystal Growth

When the rate of advance of the phase boundary is limited by the removal of latent heat or impurities, then the diffusion field modifies the interfacial values of the relevant field variables and therefore the growth shape. The physical picture is as follows: As a phase boundary advances, latent heat and/or impurities are emitted at the front. Their build up occurs over a characteristic dimension, or boundary layer, that depends on their respective diffusivities and the rate of advance. The motion itself is limited by the speed with which this excess is depleted by diffusion. These *diffusion-limited* regimes are generally studied using non-geometric formulations.[7] The reason is that the

---

[7]Rarely can diffusion-limited growth be thought of as local and so be described geometrically. One exception occurs when the characteristic scale of the boundary layer is much less than the radius of curvature of the interface, and one can view the rate limiting part of the diffusion field as being "localized". This, so-called boundary layer hypothesis, was invoked to make diffusion-limited growth problems tractable theoretically (e.g., Langer 1987; Kessler et al., 1988; Ben-Jacob and P. Garik 1990). However, physical realizations consistent with the hypothesis are

motion at a point depends on the diffusion field near that location, which itself depends on the history of the field both there, and at other points. More commonly we say that the growth is controlled by a non-local retarded field. Non-geometric theories generally treat growth on surfaces that are everywhere molecularly rough, and the anisotropy is introduced into the interfacial conditions of the particular free boundary problem being studied. For example, anisotropy is ascribed to an orientation dependence in the surface free energy or some other interfacial coefficient. Because of the general importance and environmental applicability, I describe a morphological instability characteristic of *impurity* diffusion limited growth.

In nearly every natural setting, phase coexistence will be altered by the presence of impurities, and in the case of water substance we most often encounter electrolyte solutions. It is common lore that ice rejects salt, and it is observed that in equilibrium the solubility of most impurity species in ice is much less than that in solution. This observation is commonly represented by the *equilibrium distribution coefficient* for that species;

$$k_{eq} = \frac{C_S}{C_L}, \tag{5}$$

where C_S (C_L) is the equilibrium concentration of solute (impurity) in the solid (liquid). In a subeutectic solution, $k_{eq} < 1$ and decreases in magnitude with the degree of lattice mismatch between the solute and the crystal of solvent. In most aqueous solutions $10^{-4} < k_{eq} < 10^{-1}$ (e.g., Gross 1987; Gross and Svec 1997), so that ice grown *very slowly* from solution may remain nearly pure. In practice a myriad of effects complicate this distillation effect, one of which we shall describe. In the pure case, especially in ice, growth is highly anisotropic (e.g., Elbaum et al., 1993; Furukawa and Kohata, 1993), and we expect anisotropy in solute segregation which translates into a facet dependence of k_{eq}. However, in other materials such an effect has only been observed in semiconductors (e.g., Tiller 1991). Additional complications, some of which have plagued the interpretation of experimental data, include, (i) Dependence of segregation on valency (ii) Samples are often modified by convection. (iii) During growth there is a possibility of phase boundary instabilities (discussed below) which can result in the trapping of impurities in bulk. (iv) Some species show a clear concentration dependence of k_{eq} and others do not. (v) Measurements near the solubility limit or the eutectic point are difficult to interpret. (vi) In almost all cases, above some threshold growth rate, the impurity distribution will depend on growth rate (e.g., Aziz 1982). Therefore, rather than observing an equilibrium quantity, it is more common to actually measure an *effective* distribution coefficient k_{eff}, which represents the net effect of some or all of these processes in terms of a measurement of the bulk concentrations before and after an experiment.

---

difficult to achieve. In the common case of a free dendrite growing into an undercooled melt, the radius of curvature at the tip is much smaller than the boundary layer unless very large, and experimentally awkward, undercoolings are imposed.

We consider solutions as being *ideal* when the concentration of solute molecules is low enough that they do not interact. At constant pressure, the impurity-induced shift in the equilibrium temperature of a solution is given by Raoult's law, which is written in terms of the number of moles of solute n_j dissolved in n moles of solvent;

$$T_C = T_m\left(1 - \frac{RT_m}{q_m}\frac{n_j}{n}\right), \tag{6}$$

where R is the gas constant, and q_m is the heat of fusion of the solvent. More commonly however, T_c is related to the solute concentration in the liquid C_L by an empirical liquidus slope M determined from the appropriate binary phase diagram: $T_c = M\,C_L$. Many salts go into solution as ions, so we expect a deviation from the conditions of an ideal solution. Depending on the impurity, the ideal slope may often differ from the empirical value, so we often rely on the latter. Ion–ion interactions and more complicated solution chemistry effects complicate theoretical treatments so the detailed experimental values have been tabulated by electrolyte chemists (Partanen 1991) are of great utility. The earth's most pervasive and important solution is the ocean, and the physical chemistry of seawater a vast field in itself (e.g., Millero 1974; Andersson 1998), and a notable feature is that the relative concentration of the inorganic components that dominate by weight are constant throughout the world oceans: Na/Cl = 60, K/Cl = 228, Mg/Cl = 240. The basic physics of the ocean freezing that is occurring daily in one of the polar regions can be understood by vastly simplifying the detailed chemistry of seawater and treating it as a simple two component system described by an empirical phase diagram. More complicated electrolyte effects and their applications are discussed in the summary.

The freezing point of small solid particles convex into their melt phase is reduced from that of the bulk. The effect is associated with Gibbs and Thomson and for a spherical particle is written as

$$T_{GT} = T_m\left(1 - \frac{\gamma_{s\ell}T_m}{\rho_s q_m}\frac{2}{r}\right), \tag{7}$$

where ρ_s is the density of the solid, q_m is the latent heat of fusion, and r is the radius of the particle.

The three important ingredients for understanding morphological instabilities driven by diffusion fields are now in hand: Solute segregation, and solute and curvature depression of the freezing point. Consider a planar ice/solution interface in bulk two-phase, two-component coexistence. The stationary isothermal system is at the compositionally determined freezing point given by equation (6), and there is a jump in the solute concentration at the solid/liquid interface determined by equation (5). A unidirectional constant temperature gradient is instantaneously imposed perpendicular to the ice/solution interface. It is cold in the ice and warm in the solution and the interface moves toward the latter at constant speed. Solute, let us call it salt, is rejected into the solution and builds up at the interface to take a value determined by the rate of advance (which depends on the imposed temperature gradient), the magnitude of the segregation coefficient and the value of the (negative) liquidus slope M. The region of solution adjacent to the interface is enriched with salt, and its freezing point is depressed relative to the value far into the liquid. The enriched region has a spatial extent that depends on the diffusivity of salt in water which is several hundred times slower than that of heat. Depending on the magnitude of the actual temperature gradient and the rate of motion of the interface, the possibility exists that, because of the slow diffusion of salt, a region of fluid will be *constitutionally supercooled.* This occurs when the gradient in the liquidus temperature is greater than the actual temperature gradient. Therefore, a fluctuation of the planar phase boundary that is convex into the solution will encounter a region of metastable fluid and tend to grow, creating a cellular interfacial geometry that traps regions of highly concentrated solution between regions of nearly pure ice. The fluctuation can be stabilized by the curvature depression of the freezing point. This *morphological instability* is prevalent in metallurgy (Rutter and Chalmers, 1953; Mullins and Sekerka 1963), and is responsible for the bulk two-phase properties of sea ice which generally grows under unstable conditions (Wettlaufer 1992). Morphological instabilities are of the most active present interest in materials science, wherein there is wide practical applicability and a long history (e.g., Tiller, 1991; Sekerka 1993). This instability lies in the molecular diffusion field, but it does not act alone. Rather, it is often coupled to fluid mechanical effects, and the interaction of convection and solidification is an important influence in the phase behavior and structure of sea ice (Wettlaufer et al., 1997; Worster and Wettlaufer, 1997)

3 Thermomolecular Pressure and Frost Heave

As discussed in section 2.3, both *surface melting* and *interfacial melting* are treated as wetting phenomena at the interface between a solid and its vapor phase, or a wall of a different material, at temperatures below the bulk freezing point T_m. The melting process may be *complete* or *incomplete.* In the former case the film thickness diverges as T_m is approached from below, and in the latter case the film growth is truncated at finite undercooling, which can be best understood as being due to the influence of retarded potential effects. Convincing interfacial melting measurements have been made in ice (Furukawa and Ishikawa, 1993; Beaglehole and Wilson, 1994)), and a primary motivation for understanding the process concerns the important role it can play in a host of environmental problems ranging from polar stratospheric cloud chemistry to frost heave in porous media (Dash et al., 1995).

The dramatic deformation of water saturated soils in cold climates is known generically as "frost heave" and the chapters by Greg Dash and Peter Williams stress the important implications of the phenomenon from various perspectives. Yet it is rather more appropriate to refer to frost heave as the collective action of a number of phenomena each of interest and intense study in its own right. Although frost heave is clearly not caused by the volume expansion of water during solidification, for many years it has been known to be associated with the existence of stable liquid water at subfreezing interfaces (Williams and Smith 1989). However, the varied causes of this water have obscured attempts to extract the fundamental mechanisms driving frost heave. In a porous medium, curvature, confinement, interfacial roughness and disorder, impurities and interfacial premelting *all* contribute to the finite liquid fraction at subfreezing temperatures. Individually, these effects have distinct temperature dependencies, but their combined effects have thus far limited frost heave research to semi-empirical treatments. It is for this reason that we have focused on isolating the role of interfacial melting in the existence and mobility of unfrozen interfacial water. The intermolecular origin of the films, the effect of external forcing and the geometry of the confining wall are the main features of our studies. We test our theories using experiments on isolated crystalline surfaces.

The approaches to the problem range from microscopic to macroscopic (e.g., Gilpin 1980; Dash 1988; Löwen 1994; Wettlaufer and Worster 1995; Netz and Andelman 1997; Oxtoby 1998), and here I need only to repeat the aspects necessary to study dynamical behavior and refer the reader to Oxtoby's chapter for the remainder of the topic. We begin by writing down the mean field grand potential Ω for the system which is the natural free energy for describing interphase pressure differences. The total free energy for a layered wall (w), film (ℓ), solid (s) system of area A_i is written as a combination of bulk and surface terms:

$$\Omega = -P_\ell V_\ell - P_s V_s + \Im(d)\,. \tag{8}$$

Here P and V denote pressure and volume and the liquid film has a thickness d. The interfacial term $\Im(d) = \left(\Delta\gamma f(d) + \tilde{\gamma}_{sw}\right)A_i$ is a phenomenological representation of the potential between the three layers. By definition, $\Delta\gamma = \gamma_{s\ell} + \gamma_{\ell w} - \tilde{\gamma}_{sw}$, where the γ's are the solid–liquid ($s\ell$) liquid–wall (ℓw) and solid–wall (sw) interfacial free energies. We assume that the volume-volume interactions are of a power law form represented by $f(d) = 1 - (\sigma / d)^{\nu-1}$ where σ is on the order of a molecular diameter,[8] and ν depends on the nature of the interactions. The relevant types include $\nu = 3$ ($\nu = 4$) for nonretarded (retarded) van der Waals interactions[9], and $\nu = 2$ ($\nu = 3/2$) for

---

[8]Note that the short range cut off defined by the value of σ is intimately wed to the effective nature of the solid–wall coefficient $\tilde{\gamma}_{sw}$ which is a non-equilibrium interfacial energy (see the Appendix in Wettlaufer and Worster, 1995).

[9]Various forces arise between charged, uncharged and dipolar molecules. Briefly, intermolecular forces can be *Coulombic* arising between charges and permanent multipoles. Dipole moments "induced" in molecules by local electric fields or charges are responsible for *polarization forces* between molecules. Chemical bonding itself is controlled by *quantum mechanical forces*. *Van der Waals forces* are quantum mechanical in origin and act between all molecules. Even a neutral molecule has a finite dipole moment due to the fluctuations in the electron positions, and the dipole creates an electric field that polarizes a molecule in its vicinity. The dipoles of the two molecules interact to give rise to a force that depends on the frequency of the

long range (short range) electrostatic interactions. By electrostatic interfacial interactions we refer to a simple treatment in which the substrate possesses a surface charge density and the confinement of counterions present in the liquid creates a repulsive force across the layer (Wilen et al., 1995). The electrical conductivity of the ice surface may also be calculated (Petrenko and Ryzhkin 1997) and influence the short range behavior. This form of $\mathfrak{I}(d)$ captures the essential features of more detailed treatments of complete interfacial melting under the influence of dispersion or electrostatic forces, and allows us to develop a dynamical theory for arbitrary power law interactions and to make experimentally relevant predictions. At fixed temperature and chemical potential, minimizing Ω with respect to d provides two very useful, interaction dependent, results. To first order in the reduced temperature $t_r = (T_m - T)/T_m$ the film thickness is

$$d = \left(-\frac{(\nu-1)\sigma^{\nu-1}\Delta\gamma}{\rho_s q_m} \right)^{1/\nu} t_r^{-1/\nu} \equiv \lambda_\nu t_r^{-1/\nu} , \tag{9}$$

where ρ_s and q_m are as previously defined. The coefficient λ_ν depends on the exponent ν, for example when $\nu = 3$, $\sigma^2 \Delta\gamma = A / 12\pi$, where A is the Hamaker constant. The pressure in each phase is uniform but the interfacial interactions create a pressure difference between the film and the bulk solid

$$P_\ell - P_s = \Delta\gamma(\nu-1)\sigma^{(\nu-1)} d^{-\nu} ; \tag{10}$$

the solid repels the wall through the liquid, in analogy with the *disjoining pressure* of wetting. When taken together these results yield a thermodynamic statement that is *independent* of the nature of the interactions and helps to describe the main ingredient of premelting dynamics:

$$P_\ell = P_s - \rho_s q_m t_r . \tag{11}$$

We imagine that the pressure in the solid is fixed by contact with a reservoir. Therefore, the pressure in the liquid film increases with temperature and, by imposing a temperature gradient *parallel* to the interface, the resulting *thermomolecular pressure gradient* will drive premelted liquid from high to low temperature. Thermomolecular pressure-driven flow as a mechanism for frost heave captures our attention because the maximum pressure in the case of water is approximately 11 atm per degree The film thins as the temperature decreases, so continuity requires that liquid convert to solid during its transit toward lower temperatures.

In order to anticipate the behavior of an ongoing experiment, we made our first calculations in an idealized geometry (Wettlaufer and Worster 1995). Soon thereafter Larry Wilen and Greg Dash (1995) *observed* premelting dynamics at ice single crystal interfaces formed against a flexible polymer membrane with a more complicated ge-

---

fluctuations and is attenuated with distance. When two molecules are an appreciable distance apart the strength of the interaction can be weakened by the time it takes for the field to travel between the two molecules. The effect is known as "retardation".

ometry than that in our model. We immediately set to work constructing a theory that accounted for the membrane geometry, and allowed us to back out the relevant types of interactions in the system (Wettlaufer et al., 1996). For brevity, the discussion will be limited to the latter study, but for a more detailed and complete treatment see Wettlaufer et al. (1997).

In analogy with the frost heave terminology we call the pressure exerted by the confining wall the "overburden pressure". If it is greater than the maximum pressure in the film then flow will cease, but if it is less than the maximum, flow will persist. The thermomolecular pressure gradient is written as

$$\nabla P_\ell = -\rho_\ell q_m \nabla t_r + \frac{\rho_\ell}{\rho_s} \nabla P_s \, , \tag{12}$$

where P_s is the external pressure exerted on the solid by the confining wall/membrane. In the observed regime, the radius of curvature of the deformed membrane is large compared with *d*, so the membrane exerts a pressure deep within the solid. We assume that any solid deformation is slow relative to the film dynamics. The configuration is shown in Fig. 3. The film thickness depends solely on the temperature, relaxing to the value determined by equation (9) on a time scale negligible relative to that for heat conduction. The confining wall evolves in time according to the spatial variation in the thin film volume flux $Q \propto -d^3 \nabla P_\ell$. Wilen and Dash (1995) imposed a constant temperature gradient G parallel to the ice/membrane interface.

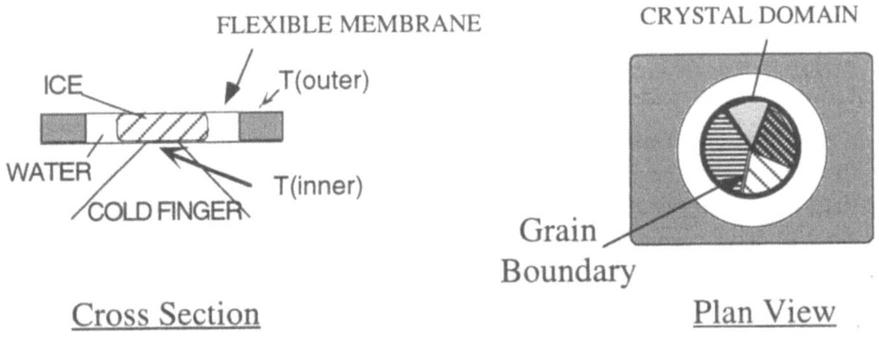

Fig. 3(a). Cross section and plan view of the Wilen and Dash (1995) experimental cell. The temperatures at the cold finger T(inner) and at the outer rim T(outer) are independently controlled.

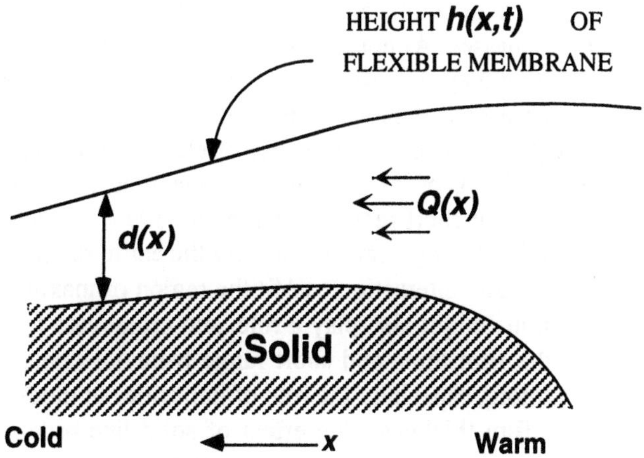

Fig. 3(b). The local configuration of a radial frost heave cell treated in a one-dimensional slab geometry (Wettlaufer et al., 1996). The membrane is locally under tension, and exerts a restoring force proportional to its *curvature*. The temperature gradient G is constant so $t_r = G\,x/T_m$, and the film thickness d depends only on x. The thermomolecular pressure gradient drives the evolution of the membrane height $h(x,t)$ relative to an initial reference height $h(x,t_0)$. The three power laws $t^{1/2}$, $t^{6/11}$, and $t^{3/5}$ arise from the short and long range electrostatic ($\nu = 3/2$ and $\nu = 2$) and nonretarded Van der Waals ($\nu = 3$) interactions respectively. For each interaction we fit $\left((\nu-1)\sigma^{\nu-1}\Delta\gamma\right)^{-\left(\frac{9}{4\nu+3}\right)}$, related to λ_ν as defined in Eq. (9). a, b, and c are material constants, and $\lambda_{\nu=3/2} = 0.0337$ Å, $\lambda_{\nu=2} = 0.2101$ Å, and $\lambda_{\nu=3} = 1.3759$ Å.

The main qualitative features of the steady state temperature gradient experiments are as follows: The height of the membrane, $h = h(x,t)$, was measured relative to an initial reference height $h(x,t_0)$. The bulk solid–liquid interface is located at the fixed position x. A constant temperature gradient drives a volume flux gradient, and thermomolecular pressure continually draws more fluid into the film where it freezes and deforms the membrane. This response to the flow of premelted water occurs in a narrow temperature range near T_m, with a *relative* lack of membrane deformation at lower temperatures. The position of maximum deformation x_{max} moves towards colder temperatures and the maximum displacement h_{max} increases with time. In principle, large deformations can be achieved by increasing G and decreasing the strength of the membrane. However, the migration of the peak toward lower temperatures will be most dramatic when G is small and the membrane strength is large.

The experiment clearly demonstrates the phenomenon of interfacial melting, the viscous nature of the premelted liquid and its dynamical consequences. Initially the observed *relative* lack of membrane deformation at lower temperatures was thought to be due to an abrupt jump to $d = 0$. Such an interpretation is consistent with an interfacial free energy that decreases monotonically with d at long range, but which possesses a local minimum at shorter range. The theory has allowed us to show that the data can be described with a monotonically decreasing interfacial free energy over the entire range of film thicknesses. Although the low temperature dynamics are slow, non-local effects insure that they still play a role, and that they will be especially important in nature where seasonal time scales are very long. The agreement between theory and experiment (Fig. 4) is best in the case of electrostatic interactions which always dominate van der Waals at long range and, depending on the surface charge density, may also dominate at short range (Wilen et al., 1995). Nonetheless, the observations cannot be fitted by a single interaction over the entire range of deformation, although a simple electrostatic interaction can fit the region of maximum deflection. Of course other interactions might also play a role, and crossovers from one type to another may occur as the film thins and short range behavior comes into its own. There is much left to explore, especially the role of impurities in modifying the interactions that control the film thickness, the effect of solid-like ordering due to the proximity of the interfaces and other dynamically significant issues.

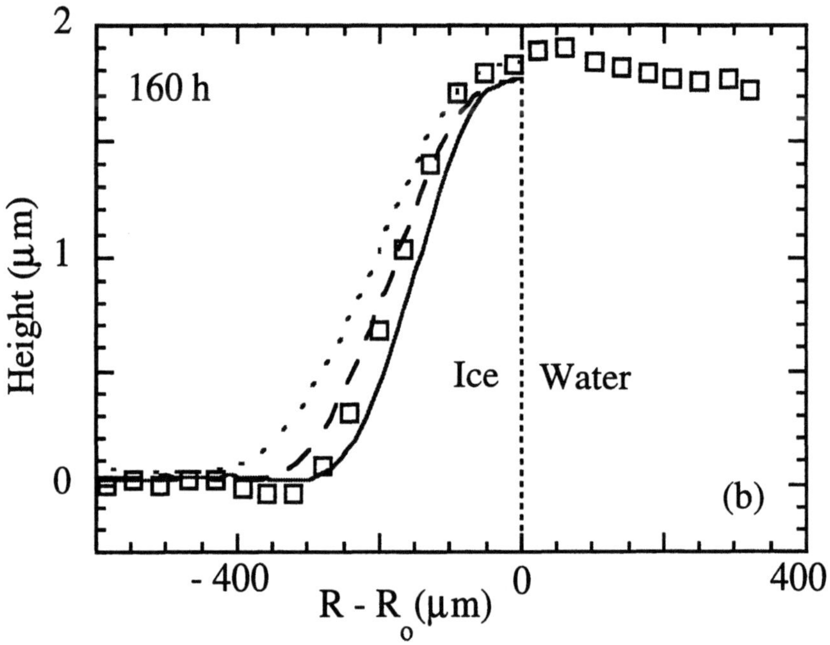

Fig. 4. Comparison of the theoretical predictions and the experimental values described by Wilen and Dash (1995) for the membrane height $h(x,t)$ at 160 hours. $R - R_O$ is the experimental x-coordinate. At the *bulk* ice/water interface, $R = R_O$ and G = 0.92 K/cm. The predictions for the short and long range electrostatic ($\nu = 3/2$ and $\nu = 2$), and the nonretarded Van der Waals ($\nu = 3$) interactions are shown by the solid, dashed, and dotted lines respectively. For orientation "Ice" ($R - R_O < 0$) and "Water" ($R - R_O > 0$) refer to regions where the *bulk* phases are stable. For $R - R_O < 0$ an *interfacial* water film coexists between the membrane and *bulk* ice.

4 Summary and Future Directions

The phenomena described in this chapter centered on interfacial disorder, sometimes liquid like disorder, of both a thermodynamic and dynamic origin, and the redistribution of both the disordered material and impurities at solid-liquid interfaces. Because of the change in the surface structure created by both surface roughening and surface melting, they have a profound influence on the local attachment kinetics, and ultimately on pattern formation in ice crystals. We have a clear picture of several issues, namely (i) that roughening is a crucial aspect of growth anisotropy, and (ii) that interfacially melted water at subfreezing interfaces is mobile and responds in a thermodynamically consistent and predictable manner. Less clear however are a number of other issues as discussed below.

4.1 Surface Melting and Crystal Growth

The role of surface melting in the growth of crystals is a fertile area still in its infancy. On isolated surfaces thermodynamic roughening, kinetic roughening and surface melting have been extensively studied, but their coexistence on a single crystal surface results in highly anisotropic global growth shapes (Furukawa and Kohata, 1993; Wettlaufer et al., 1994; Nada and Furukawa, 1996). In the case of surface melting, the communication with the parent phase is mediated by a mobile thin film (Sei and Gonda, 1989; Kuroda and Lacmann, 1982). The difference in the vapor-solid and the vapor-liquid/liquid-solid accretion mechanisms, and the change in these with temperature as the film thickness changes will influence the growth rate. But what are these temperature dependencies? Some important and telling experiments have already been done.

Sei and Gonda (1989) grew polyhedral ice crystals at several percent supersaturation. From their observations of growth rates versus supersaturation, they concluded that, from the triple point to –2°C, the basal and prism faces grow by a mechanism of vapor/quasi-liquid/solid attachment kinetics, and from –2 to –30°C, they grow by the classical kinetic mechanisms described above which operate in the absence of a quasi-liquid.

Furukawa and Kohata (1993) studied *negative* crystals formed by creating cavities in ice single crystals. They observed receding crystallographic (basal and prism) facets in a fixed evaporation rate. Below –2°C, they found that a negative crystal took the form of a hexagonal prism, but changed habit to a plate and back to a prism exactly as does a *positive* ice crystal growing from the vapor or a gaseous atmosphere. However, at temperatures above –2°C the hexagonal prism was observed to change into a sphere truncated only by basal faces.

Furukawa and Nada (1997) have combined a number of interesting observations In his ellipsometric studies of surface melting on the basal and prism facets a crossover was observed in the film thicknesses at about –1°C. At temperatures greater than this, the basal plane film was thicker, but the prism plane film was thicker at lower temperatures. From studies of the anisotropic "growth" of negative crystals Furukawa and

Kohata (1993) inferred from the growth kinetics that the prism facet undergoes a roughening transition at approximately –2°C. However, since surface melting is initiated at –4°C, this transition occurs between the prism facet and a quasi-liquid layer tens of nanometers thick. Recall that Maruyama et al., (1997) observed the prism facets to undergo a thermodynamic roughening transition into bulk water at approximately –16°C at 165 MPa. Finally, using molecular dynamics simulations they found qualitative agreement with the facet dependence of surface melting. However, such simulations exhibit phase behavior that is very sensitive to the details of the model potential, and it is not surprising that the simulated films were substantially thinner than the experimental films, and the crossover temperature substantially lower. The qualitative behavior clearly demonstrates the sensitivity of the equilibrium and growth forms to anisotropy in the surface structure. Much caution, as noted by Furukawa and others, must be administered in deducing quantitative behavior. Despite the great care taken by individual groups, there is still a substantial spread in the film thickness versus undercooling curves for the vapor interface[10].

The question of the roughening transition of the surface melted prism face is an interesting one in its own right. Maruyama's experiment in bulk water shows a very low roughening temperature; 16°C at 165 MPa. Yet at the surface melted prism facet Furukawa infers a roughening temperature of ice against a quasi-liquid layer of approximately –2°C. Hence, the proximity of the quasi-liquid/vapor interface must influence the fluctuations at the ice/quasi-liquid interface, suggesting a kind of "proximity induced roughening transition" but a quantitative explanation is still lacking. Simulations are notoriously difficult with water/ice systems, and they are an excellent intuitive guide, but until they adequately reproduce the bulk phase behavior, they will remain a tool to drive experiment and theory, much of which remains to be done on this topic.

Were success forthcoming in the area just discussed, an outstanding question of general interest concerns the role of surface melted films in the morphological stability of crystals growing in the diffusion limit. Although there is no experimental verification, Löwen and Lipowsky (1991) predict a dynamical thickening of the surface melt during evaporation. During growth, if the solid orientation underlying the surface melted film is a proper facet, then due to the lack of surface attachment impedance at the film/vapor interface, and activated growth at the solid/film interface, we would expect a *dynamical thinning* of the surface melted film. The actual film thickness as a function of growth drive would be exceedingly difficulty to measure directly. However, it might be best investigated by observing growth anisotropy in a combined experimental-theoretical study. Once an understanding of the film thickness versus growth drive is in hand, its effect on the morphological stability of crystals growing

---

[10]See, for example, Fig. 1 of Elbaum et al. (1993) although it doesn't include more recent glancing angle X-ray scattering data taken by Dosch et al. (1995). More work using a combination of techniques, including Second-Harmonic-Generation, light scattering, ellipsometry, X-ray scattering and force microscopy should be embraced to resolve existing discrepencies.

under varying drives and temperatures will shed some light on the pattern formation problems in the growth of snow crystals.

4.2 Electrolyte Effects

There are a myriad of interfacial effects associated with electrolytes that are pervasive in the natural environment but with which we have only dealt crudely herein. *All* electrostatic fields in any medium containing free charges will become *screened* by the polarization of the charges. A screened electric field decays approximately exponentially with distance x; $exp(-\kappa_D x)$, where $1/\kappa_D$ is the *Debye screening length.* When charges preferentially adsorb onto an interface adjacent to an aqueous solution they are balanced by counterions creating an *electric double layer*. For an aqueous NaCl solution $1/\kappa_D = 30.4$ nm at 10^{-4} M, 0.96 nm at 0.1M, and in pure water of pH 7, $1/\kappa_D$ is about one micron (e.g., Israelachvili 1992). With this in mind, the interfacial dynamics at an ice/solution interface can become quite complicated, and our studies of premelting dynamics might require consideration of a continuous variation in the interaction potential depending on the redistribution of ions. Preferential ion incorporation is best demonstrated in this context of solid/liquid solute distribution introducing ionic coefficients; $k_{eq}^{+,-}$. A simple example for a monovalent electrolyte solution is the electrical imbalance associated with the preferential Na^+ rejection during the growth of ice from a dilute aqueous NaCl solution. Within the framework of a theory of Bronshteyn and Chernov (1991), we imagine the trapping of a line of negative charge density parallel to a newly formed *planar* freezing front. The electric field set up between the solid and the rejected liquid cations, creates a Debye double layer on the liquid side of the interface. (Of course, $1/\kappa_D$ will be a dynamical quantity due to the moving crystallization front.) Charge redistribution is driven by a number of well known processes: in the solid it is accommodated by the mobility of Bjerrum and ionization defects, and in the liquid it is determined by the dissociation rates and diffusion coefficients. The higher concentration of intrinsic ions in the liquid, and their much larger mobility in the solid, result in strongly phase dependent neutralization rates. Bronshteyn and Chernov (1991) predict that a freezing potential arises only when the product of the impurity concentration at the interface and the mean growth rate exceed a pH dependent critical value. The comparison of the theory with experiment is limited by the indetermination of the morphological stability of the solidification front. Chernov (1993) has recently reviewed growth from aqueous solutions. Electric fields at the ice/solution interface observed during growth have been associated with Workman and Reynolds (1950) (see also Hobbs 1974, Gross and Svec, 1997 and refs. therein), and combined with the thermoelectric effect (currents driven by temperature gradients) in ice, and ion dependent phase segregation, these phenomena might be central to the issue of ice particle collision-driven charge transfer in thunderstorm electrification (Mason and Dash 1998).

4.3 Additional Effects

In the field or in a laboratory setting we must think of frost heave as the collective action of a number of phenomena. In a porous medium, the distinct effects of curvature, confinement, interfacial roughness and disorder, impurities and interfacial premelting *all* contribute to the finite liquid fraction at subfreezing temperatures. One approach is to measure their combined effect on the liquid fraction and to construct a system specific constitutive relation. Many of these phenomena have been studied in other materials in the context of phase transitions and critical phenomena, and their temperature dependencies can, in principle, be deconvoluted. This suggests that it will fruitful to study these effects in turn, thereby building up a fundamental understanding of their importance. In this manner, one can construct a more general constitutive relation between temperature and liquid fraction that would be broadly applicable.

The study of dislocations is central to understanding the mechanical behavior of ice, although their role in ice crystal growth is still evolving. How and when impurities create or pin dislocations is important in growth from the vapor or the melt phase.

I hope that it is clear that the microscopic phenomena discussed here are both interesting in their own right, and have macroscopic implications with real environmental consequences. Progress requires serious efforts using well developed methods from within the disciplines, and continuous communication of these efforts across disciplinary boundaries. These and other key questions that are impeding progress can be resolved by working jointly through direct interaction, the sharing of different perspectives, data and by visiting the respective laboratories. The challenges are substantial, but the rewards are great.

It is a pleasure to acknowledge extremely useful conversations with J.W. Cahn, J.G. Dash, M. Elbaum, S.C. Fain, Y. Furukawa, J.F. Nye, M. Maruyama, M. Schick, L.A. Wilen and M.G. Worster, on many aspects of the work described herein. I am fortunate to have as colleagues Greg Dash, Larry Wilen and Grae Worster who make the efforts of research so enjoyable. Our work has been generously supported by the National Science Foundation, the Office of Naval Research, and the Natural Environment Research Council of the U.K.

References

Andersson, L., in *IAPSO Advanced Study Institute-Summer School: Physics of Ice-Covered Seas*, ed. by M. Leppäranta (University of Helsinki Press, Helsinki, 1998)
Angenent, S., & Gurtin, M.E., *J. Reine Angew. Math.* **446**, 1–114 (1994).
Avron, J.E., Taylor, J.E. & Zia, R.K.P., *J. Stat. Phys.* **33** 493–522 (1983).
Aziz, M.J., *J. Appl. Phys.* **53**, 1158–68 (1982).
Balibar, S., Gallet, F., & Rolley, E., *Physica Scripta* **T29**, 265–9 (1989).
Bagdassarian, C.K. and Oxtoby, D.W., *J. Chem. Phys.* **100**, 2139–48 (1994).
Beaglehole, D., & Nason, D., *Surf. Sci.* **96**, 357–365 (1980).
Beaglehole, D., & Wilson, P., *J. Phys. Chem.* **98**, 8096–100 (1994).
Ben-Jacob, E. & Garik, P., *Nature* **343**, 523–30 (1990).

Burton, W.K., Cabrera, N., & Frank, F.C., *Phil. Trans. R. Soc.* **A 243**, 299–358 (1951).
Cahn, J.W., Taylor, J.E., & Handwerker, C.A., in *Sir Charles Frank, OBE, FRS, An eightieth birthday tribute* eds. R.G. Chambers et al., p. 88–118 (Hilger, New York, 1991)
Cahn, J.W. & Carter, W.C., *Metall. and Mat. Trans.* **27A** 1431–40 (1996).
Carter, W.C., Roosen, A.R., Cahn, J.W., & Taylor, J.E., *Acta Metall.* **43**, 4309–23 (1995).
Chernov, A.A., *Modern Crystallography III, Crystal Growth* (Springer, Berlin, 1984), p. 122.
Chernov, A.A., *Prog. Cryst. Growth & Char. Mat.* **26**, 121–51 (1993).
Chui, S.T., & Weeks, J.D., *Phys. Rev. Lett.* **40**, 733–6 (1978).
Cross, M.C. & Hohenberg, P.C., *Rev. Mod. Phys.* **65** 851–1112 (1993).
Dash, J.G., in *Proceedings of the Nineteenth Solvay Conference,* eds F.W. de Wette, (Springer-Verlag, Berlin, 1988).
Dash, J.G., in *Ice Physics in the Natural and Endangered Environment,* NATO/ASI eds J.S. Wettlaufer, J.G. Dash and N. Untersteiner, (Springer-Verlag, New York, 1998).
Dash, J.G., Fu, H-Y, & Wettlaufer, J.S., *Rep. Prog. Phys.* **58**, 115–67 (1995).
Dash, J.G., Schick, M., & Vilches, O.E., *Surf. Sci.*, **299–300**, 405–14 (1994).
den Nijs, M., *J. Phys. A*, **30**, 397–404 (1997).
Dinghas, A., *Z. Krist.* **105**, 304–14 (1944)
Dosch, H., Lied, A., & Bilgram, J.H., *Surf. Sci.* **327**, 145–64 (1995).
Elbaum, M., *Phys. Rev. Lett.* **67**, 2982–5 (1991).
Elbaum, M., & Schick, M., *Phys. Rev. Lett.* **66**, 1713–6 (1991).
Elbaum, M., Lipson, S.G., & Dash, J.G., *J. Cryst. Growth* **129**, 491–505 (1993).
Elbaum, M., & Wettlaufer, J.S., *Phys. Rev. E* **48**, 3180–3 (1993).
Frank, F.C., in *Growth and Perfection of Crystals* eds., R.H. Doremus, B.W. Roberts, and D. Turnbull (Wiley, New York, 1958).
Furukawa, Y., Yamamoto, M., & Kuroda, T, *J. Cryst. Growth* **82**, 665–77 (1987).
Furukawa, Y., & Ishikawa, I. *J.Cryst. Growth,* **128**, 1137–42 (1993).
Furukawa, Y., & Kohata, S. *J.Cryst. Growth,* **129**, 571–581 (1993).
Furukawa, Y., & Nada, H., *J. Phys. Chem. B* **101**, 6167–70 (1997).
Gay, J.M., J. Suzanne, J.G. Dash & H-Y. Fu, *J. Cryst. Growth,* **125**, 33 (1992).
Gross, G.W., Gutjahr, A., & Caylor, K. *J de Physique Coll.* **48** 527–533 (1987).
Gross, G.W. & Svec, R.K., *J. Phys. Chem. B* **101**, 6282–4 (1997).
Gilpin, R.R., *Water Resources Res.* **16**, 918–30 (1980).
Goldstein, R.E., & Petrich, D.M., *Phys. Rev. Lett.* **67**, 3203–6 (1991).
Golecki, I., & Jaccard, C., *Phys. Lett. A* **63**, 374 (1977).
Hallett, J., Phil. Mag. **6**, 1073–87 (1961).
Herring, C., Phys. Rev. **82**, 87—93 (1951).
Heyraud, J.C. & Mètois, J.J. *Surf. Sci.* **177**, 213–20 (1986).
Hobbs, P.V. *Ice Physics*, (Clarendon, Oxford, 1974, p. 187).

Hodgkin, V.A., Wettlaufer, J.S., & Dash, J.G., in *Ice Physics in the Natural and Endangered Environment,* NATO/ASI eds J.S. Wettlaufer, J.G. Dash and N. Untersteiner, (Springer-Verlag, New York, 1998).

Ishizaki, T., Y. Furukawa, M. Maruyama & J.G. Dash, *J. Cryst. Growth* **163**, 455–60 (1996).

Israelachvili, J.N., *Intermolecular and Surface Forces* (Academic, London, 1992).

Josè, J.V., Kadanoff, L.P., Kirkpatrick, S., & Nelson, D.R., *Phys. Rev. B* **16** 1217–30 (1977).

Kessler, D.A., Koplik, J., & Levine, H., *Adv. Phys.* **37** 255–339 (1988).

Kirkaldy, J.S., *Rep. Prog. Phys.* **55** 723–95 (1992).

Kunkel, R., Poelsema, B., Verheij, L.K, & Comsa, G., *Phys. Rev. Lett.* **65**, 733–6 (1990).

Kuroda, T. & Lacmann, R., *J. Cryst. Growth* **56**, 189–205 (1982).

Landau, L.D., in *Collected Papers of L.D. Landau,* (Pergammon, Oxford, 1965), p. 540.

Langer, J.S., in *Chance and Matter* eds. J. Souletie, J. Vannimenus, R. Stora p. 629–711 (North Holland, New York, N.Y. 1987).

Lipson, S.G. & Polturak, E., *Prog. Low. Temp. Phys.* **11**, 128–188 (1987).

Liu, D.-J., Weeks, J. D. & Kandell, D., *Surface Rev. and Lett.*, **4** 107–13 (1997).

Löwen, H., *Phys. Rep.* **237**, 249–324 (1994).

Löwen, H, and Lipowsky, R., *Phys. Rev. B* **43**, 3507–13 (1991).

Maruyama, M., Bienfait, M., Dash, J.G., & Coddens, G., *J. Cryst. Growth,* **118**, 33 (1992).

Maruyama, M., Kishimoto, Y., & Sawada, T., *J. Cryst. Growth* **172**, 521–27 (1997).

Mason, B., & Dash, J.G., in *Ice Physics in the Natural and Endangered Environment,* NATO/ASI eds J.S. Wettlaufer, J.G. Dash and N. Untersteiner, (Springer-Verlag, New York, 1998).

Millero, F.J., *Ann. Rev. Earth & Planet. Sci.* **2**, 101–50 (1974).

Mullins, W.W., & Sekerka, R.F., *J. Appl. Phys.* **35**, 444–451 (1964).

Nada, H. & Furukawa, Y., *J.Cryst. Growth,* **169**, 587–597 (1996).

Nakaya, U., *Snow Crystals* (Harvard Univ. Press, 1954).

Netz, R.R., & Andelman, D., *Phys. Rev. E*, **55** 687–700 (1997).

Nozières, P. & Gallet, F., *J. Physique* **48**, 353–67 (1987).

Oxtoby, D.W., in *Ice Physics in the Natural and Endangered Environment,* NATO/ASI eds J.S. Wettlaufer, J.G. Dash and N. Untersteiner, (Springer-Verlag, New York, 1998).

Partanen, J.I. and Lindström, M.J., *Acta Chem. Scand.* **45**, 172–6 (1991).

Petrenko, V.F., *Electrical Properties of Ice*, Cold Regions Research and Engineering Laboratory Special Report 93-20, Hanover, NH (1993).

Petrenko, V.F., & Ryzhkin, I., *J. Phys. Chem. B* **101**, 6285–9 (1997).

Rutter, J.W., & Chalmers, B., *Can. J. Phys.* **31**, 15–39 (1953).

Schick, M., in *Liquids at Interfaces*, Les Houches Session XLVIII eds., J. Charvolin, J.F. Joanny, J. Zinn-Justin p. 415, Elsevier, 1990.

Sei, T., & Gonda, T., *J. Cryst Growth* **94**, 697–707 (1989).

Sekerka, R.F., *J. Cryst Growth* **128** 1–12 (1993).

Stranski, I.N., *Die Naturwissen.* **28**, 425–30 (1942).
Taylor, J.E., Cahn, J.W., & Handwerker, C.A., *Acta Metall.* **40**, 1443 (1992).
Tersoff, J., Denier van der Gon, A.W., & Tromp, R.M., *Phys. Rev. Lett.* **70,** 1143–6 (1993)
Weeks, J.D. & Gilmer, G.H., *Adv. Chem. Phys.* **40**, 157–228 (1979)
Wettlaufer, J. S., *Europhys. Lett.* **19**, 337–42 (1992).
Wettlaufer, J. S., Jackson, M., & Elbaum, M., *J. Phys. A* **27**, 5957–67 (1994).
Wettlaufer, J. S. & Worster, M. G., *Phys. Rev. E* **51**, 4679–89 (1995).
Wettlaufer, J. S., Worster, M. G., Wilen, L.A., & Dash, J.G.,*Phys. Rev. Lett.* **76**, 3602–5 (1996).
Wettlaufer, J. S., Worster, M. G., & Huppert, H. E., *J. Fluid Mech.* **344**, 291–316 (1997)
Wettlaufer, J. S., Worster, M. G., & Wilen, L.A., *J. Phys. Chem. B* **101**, 6137–41 (1997).
Wilen, L.A. & Dash, J.G., *Phys. Rev. Lett.* **74** 5076–9 (1995).
Wilen, L.A., J. S. Wettlaufer, Elbaum, M., & Schick, M., *Phys. Rev. B* **52**, 12,426–33 (1995).
Williams, P.J. & Smith, M.W., *The Frozen Earth: Fundamentals of Geocryology* (Cambridge Univ. Press, Cambridge, 1989).
Workman, E., & Reynolds, S., *Phys. Rev.*, **78** 254–9 (1950).
Worster, M.G. & Wettlaufer, J. S., *J. Phys. Chem. B* **101**, 6132–36 (1997).
Wulff, G., *Z. Krist.* **34**, 449–530 (1901).

Some Aspects of the Physics of Glaciers

W. S. B. Paterson
Box 303, Heriot Bay, British Columbia V0P 1H0, Canada

Abstract. The three modes of glacier flow (ice deformation, sliding, bed deformation) are described. The pattern of flow-lines is deduced from the principle of mass conservation; the temperature distribution by considering heat sources and heat transfer by conduction and advection. The mechanisms of surging and of the tidewater-glacier instability are outlined. The interpretation of Heinrich Events as evidence of surges of the Laurentide Ice Sheet is critically reviewed.

Keywords. Flow law (ice), flow model (glacier), glacier flow, glacier instability, glacier physics, glacier sliding, glacier surge, glacier temperature, Heinrich Event, ice sheet, ice stream, iceberg calving, sediment deformation, till deformation

1 Introduction

Much of the present interest in glacier studies stems from two aspects of the relation between glaciers and climate. (1) Cores from the polar ice sheets provide records of atmospheric temperature and composition for periods of the order of 10^5 years. Calculation based on flow models is one frequently-used method of dating these records. It is the only one available for initial planning. (2) Predicting the effect of global warming on glaciers and ice sheets. How will this change world sea level? May the Antarctic Ice Sheet respond unstably?

A glacier may be defined as a large mass of perennial ice that originates on land by the compaction and recrystallisation of snow and shows evidence of present or past flow. Glaciers may be divided into four main classes: (1) glaciers that are continuous sheets flowing outwards in all directions, called *ice sheets* if they are the size of Antarctica, Greenland, or the ice-age ice masses, and *ice caps* if they are smaller; (2) glaciers whose paths are confined by the surrounding bedrock, called *mountain* or *valley glaciers* according to their location; (3) regions in the ice sheets, with no visible rock boundaries, in which the ice flows much faster than in the areas on either side, called *ice streams*; and (4) glaciers whose lower parts spread out on flat ground or float on the ocean, called *piedmont glaciers* and *ice shelves* respectively. Combinations of these types occur frequently; an ice cap may be drained by several valley glaciers, for example.

The amount of snow that falls on the upper part of a glacier during a year exceeds that lost by melting and evaporation. In the lower part, all the previous

NATO ASI Series, Vol. I 56
Ice Physics and the Natural Environment
Edited by John S. Wettlaufer, J. Gregory Dash and Norbert Untersteiner

winter's snow and some glacier ice are removed by melting and run-off. Calving of icebergs into the sea or a lake may also cause significant loss of mass. The profile of the glacier does not change much from year to year, however, because ice flows under gravity from the *accumulation area* to the *ablation area* (Figure 1). The boundary between the two at the end of summer is called the *equilibrium line.* Its position varies from year to year according to the weather. In Antarctica, the equilibrium line is close to sea level and icebergs account for 85% of the annual mass loss.

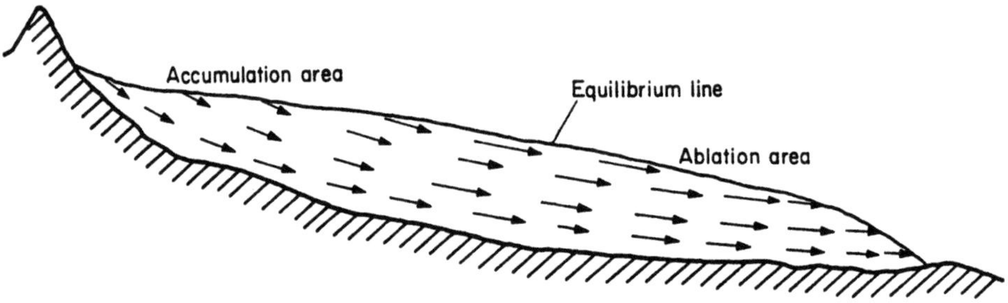

Fig. 1. Velocity vectors in an idealised glacier. Flow lines and particle paths are the same if the glacier is in a steady state.

To maintain constant thickness in the accumulation area , ice must flow downwards relative to the surface. This velocity component decreases steadily with increase of depth and reaches zero at the bed if there is no melting there. Similarly, ice flows upwards relative to the surface in the ablation area.

Consider a cross-section through the glacier perpendicular to the surface and to the direction of flow. If the profile is to remain unchanged, the annual flux of ice through any such cross-section in the accumulation area must be equal to the total mass of snow that has collected during the year on the area above the cross-section. Similarly, the annual flux through any cross-section in the ablation area must balance the mass loss between that cross-section and the terminus. Thus this flux increases steadily from zero at the head of the glacier to a maximum at the equilibrium line and from there decreases steadily to zero at the terminus. The velocity component parallel to the surface should vary in a similar way, provided that the glacier's width and thickness do not change greatly throughout its length. It follows that the ice is stretched longitudinally in the accumulation area (*'extending flow'*) and compressed longitudinally in the ablation area (*'compressing flow'*). Because ice is incompressible, except in so far as the volume of any air bubbles in it can be reduced, longitudinal extension must be accompanied by an equal compression in the direction perpendicular to the surface. This can maintain constant thickness in spite of accumulation at the surface. Similarly, longitudinal compression in the ablation area is accompanied by extension in the perpendicular direction. This argument applies only to a valley glacier where the valley walls prevent lateral extension. The general condition is that the algebraic sum of the deformation rates in any three mutually perpendicular directions is zero.

In real glaciers, variations in thickness, width, and other factors produce many deviations from this simple flow pattern. But these are important ideas to keep in mind. Extending and compressing flow also enable a glacier to flow past large bumps in its bed. A quantitative derivation of flow lines would provide a time scale for an ice core. This cannot be based solely on mass conservation as were the previous arguments, which dealt only with depth-averaged velocities. Stresses and deformation rates have to be calculated.

Deformation rates and stresses are related by the so-called *'flow law'*. Although its form can be derived from dislocation theory, the 'law' is essentially an empirical fit to laboratory and field data for the loading conditions and range of stresses encountered in glaciers.

2 Deformation of Glacier Ice

As a simple model of a glacier, consider a parallel-sided slab of thickness h frozen to a plane of slope α (Figure 2). The length and width of the slab are large compared with h. It deforms in simple shear under its own weight. Consider a column of ice perpendicular to the plane and of unit cross-section. The component of its weight parallel to the plane, the *driving stress*, is $\rho g h \sin\alpha$. (Here ρ is density and g gravitational acceleration). This is balanced by the *basal drag*, the shear stress across the base of the column:

$$\tau_b = \rho g h \sin\alpha. \tag{1}$$

If the slab is of uniform thickness, this relation holds at every point. In a real glacier h varies with x and so there will be places where one term exceeds the other. In other words there are longitudinal stresses as well as shear:

$$\tau_b = \rho g h \sin\alpha + \int_b^s \frac{\partial \sigma_x}{\partial x}\, dz. \tag{2}$$

Here s and b denote surface and bed. At most places except near an ice divide and near the terminus, the second term on the right-hand side is small compared with the first, especially if α is interpreted as the surface slope averaged over a distance of about $10h$. At this scale, flow is in the direction of maximum surface slope.

At most places in glaciers, τ_b, calculated from Eq. 1, lies between 50 and 150 kPa. So ice can be regarded, to a first approximation, as a perfectly plastic material with a *yield stress*, τ_0, of about 100 kPa. (If a gradually increasing stress is applied to a perfectly plastic material, there is no deformation until the yield stress is reached, at which point the deformation rate becomes large.) This assumption leads to a simple picture of glacier flow: the thickness adjusts itself so that at every point the shear stress at the base is equal to the yield stress.

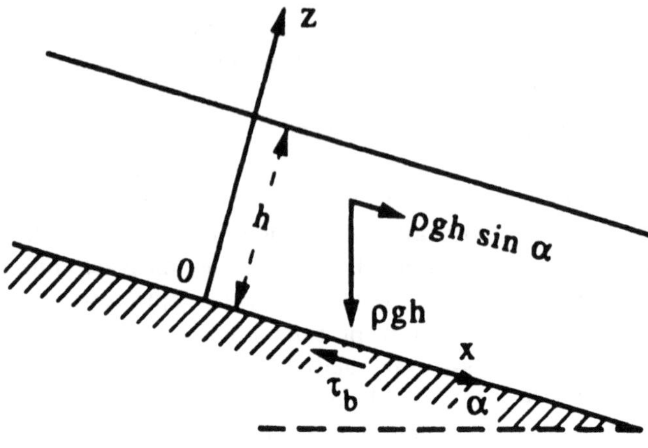

Fig. 2. Coordinate system for parallel-sided slab.

For an ice sheet on a horizontal bed, the surface slope $\alpha = dh/dx$ and Eq. 1 can be integrated to give the surface profile

$$h^2 = (2\tau_0/\rho g)\,x, \tag{3}$$

where x is distance from the edge, measured along the direction of the maximum surface slope. This is a parabola. This formula is often used to calculate the thickness and volume of ice-age ice sheets from evidence of their extent provided by glacial deposits.

Laboratory tests and field measurements have shown that, for stresses in the range 50 to 200 kPa, the relation between shear stress τ_{xz} and deformation rate is

$$\dot{\epsilon}_{xz} = A\,\tau_{xz}^n. \tag{4}$$

Here n is a constant, usually taken as 3, and A varies with absolute temperature T according to the Arrhenius relation

$$A = A_0 \exp(-Q/RT). \tag{5}$$

Here R is the gas constant (8.314 J $\mathrm{mol}^{-1}\,\mathrm{K}^{-1}$) and Q the activation energy for creep. For $T < -10$°C, $Q = 60$ kJ/mol, which implies that the deformation rate for a given stress at −10°C is about five times that at −25°C. The Arrhenius relation breaks down at temperatures above −10°C because water at grain boundaries introduces additional deformation mechanisms. Current practice is to retain Eq. 5 but increase the value of Q. Experiments show that the value of A_0 depends on crystal fabric and impurity content, but not on hydrostatic pressure provided that T is measured relative to the appropriate melting temperature.

Perfect plasticity corresponds to $n \to \infty$, as may be seen by writing $A = A_0\tau_0^{-n}$ in Eq. 4.

Because several stresses act simultaneously at most places in glaciers, the flow relation for uniaxial stress has to be generalised. The ice is assumed to be isotropic. Normal stress components are denoted by σ_x, σ_y, σ_z, shear stress components by τ_{xy}, τ_{xz}, τ_{yz}, and strain-rate components by $\dot\epsilon_x$, $\dot\epsilon_y$, $\dot\epsilon_z$, $\dot\epsilon_{xy}$, $\dot\epsilon_{xz}$, $\dot\epsilon_{yz}$. The ***stress-deviator*** components are

$$s_x = \sigma_x - \tfrac{1}{3}\left(\sigma_x + \sigma_y + \sigma_z\right) \tag{6}$$

and two similar relations. The shear-stress components are unchanged. A relation between strain rates and stress deviators is independent of hydrostatic pressure. A flow relation for complex stress systems must connect quantities that describe the overall state of stress and strain rate; it cannot be affected by the way the coordinate axes are drawn. Two assumptions are made: (1) At any point, each strain-rate component is proportional to the corresponding stress-deviator component. This is plausible for an isotropic material. (2) The flow relation is assumed to be a relation between only the second invariants of the strain-rate and stress-deviator tensors, namely

$$\dot\epsilon = A\tau^n, \tag{7}$$

with

$$2\dot\epsilon^2 = \dot\epsilon_x^2 + \dot\epsilon_y^2 + \dot\epsilon_z^2 + 2\left(\dot\epsilon_{xy}^2 + \dot\epsilon_{xz}^2 + \dot\epsilon_{yz}^2\right), \tag{8}$$

$$2\tau^2 = s_x^2 + s_y^2 + s_z^2 + 2\left(\tau_{xy}^2 + \tau_{xz}^2 + \tau_{yz}^2\right). \tag{9}$$

This is a drastic assumption although some experimental evidence supports it. The most general relation would include third invariants as well. The third invariants are the determinants of each tensor. (The first invariants are zero by Eq. 6 and assumption 1).

In two dimensional analyses (plane strain), which are frequently used in glacier studies, the third invariants are zero. Failure of the second assumption would imply that a sample in simple shear would expand or contract in the direction perpendicular to the shear plane and, because ice is incompressible, this would be accompanied by contraction or expansion in the shear plane. A plane-strain model could not then be used.

In terms of components, the generalised flow relation is

$$\dot\epsilon_x = A\tau^{n-1}s_x\,; \qquad \dot\epsilon_{xz} = A\tau^{n-1}\tau_{xz}, \tag{10}$$

and similar equations. Velocity components (u, v, w) are related to strain rates by

$$\dot\epsilon_x = \frac{\partial u}{\partial x}\,; \qquad \dot\epsilon_{xz} = \frac{1}{2}\left(\frac{\partial u}{\partial z} + \frac{\partial w}{\partial x}\right), \tag{11}$$

and similar equations.

Equations 9 and 10 show that a given stress acting in combination with others produces a greater deformation rate than it would when acting alone. For example, a tunnel dug in a glacier closes up under the pressure of the ice above it. The ice at the foot of an icefall is subject to strong longitudinal compression. Thus a tunnel there is expected to close more rapidly than one at the same depth elsewhere. This has been observed. This interaction between stresses, which would not occur if ice were Newtonian viscous ($n = 1$), greatly complicates analyses of glacier flow. The interaction between ice flow and heat flow is an additional complication; deformation rate increases with temperature and ice flow transfers heat within the glacier.

Glacier flow by ice deformation is a combination of shear, which predominates near the bed, and longitudinal extension or compression (according to accumulation and ablation) near the surface. For this case, $s_x = -2\,s_y = -2\,s_z$, $\tau_{xy} = \tau_{yz} = 0$, and Eq. 10 with $n = 3$ reduces to:

$$\dot{\epsilon}_{xz} = A\left(\tfrac{1}{3}\sigma_x^2 + \tau_{xz}^2\right)\tau_{xz}. \tag{12}$$

This applies to isotropic ice. However, much of the ice at depth in the polar ice sheets has developed a crystal fabric (c-axes clustered in vertical direction) which favours deformation in the direction of the applied shear. Such ice is "hard" with respect to vertical compression. In this case, the value of A should be increased but the weighting assigned to σ_x^2 should be much less than 1/3. Failure to take account of anisotropy makes questionable the results of many analyses of ice-sheet flow.

By an argument similar to that used to derive Eq. 1, the shear stress τ_{xz} at depth z is

$$\tau_{xz} = \rho g z \sin\alpha. \tag{13}$$

From Eqs. 11 and 12 with σ_x neglected, the horizontal velocity component u is expected to increase as the fourth power of the depth. Figure 3 shows velocity profiles obtained by measuring the tilting rates of initially-vertical boreholes. Differences in shape between the profiles result from differences in temperature, longitudinal stress, and anisotropy of the ice.

The stress components are related by the stress-equilibrium equations:

$$\begin{aligned}
\frac{\partial\sigma_x}{\partial x} + \frac{\partial\tau_{xy}}{\partial y} + \frac{\partial\tau_{xz}}{\partial z} &= -\rho g_x \\
\frac{\partial\tau_{xy}}{\partial x} + \frac{\partial\sigma_y}{\partial y} + \frac{\partial\tau_{yz}}{\partial z} &= -\rho g_y \\
\frac{\partial\tau_{xz}}{\partial x} + \frac{\partial\tau_{yz}}{\partial y} + \frac{\partial\sigma_z}{\partial z} &= -\rho g_z
\end{aligned} \tag{14}$$

These three equations plus six equations of the form 10 with strain rates expressed as velocity gradients by Eq. 11, plus boundary conditions, are sufficient to determine the six stress components and three velocity components. This is for an isothermal glacier, an unrealistic case. In general the heat transfer equation (Eq. 17 below) has to be added to determine the temperature. The interaction between ice flow and heat flow greatly complicates the analysis.

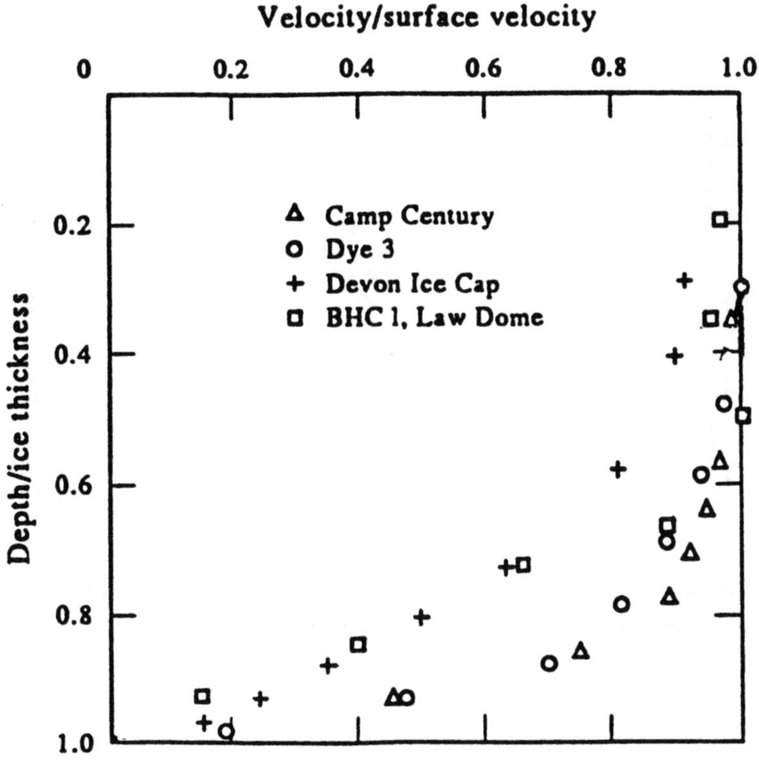

Fig. 3. Measured variation of horizontal velocity with depth in polar ice caps. Adapted from Paterson (1994) by permission of Butterworth–Heinemann.

If the bedrock configuration, ice thickness at the head of the glacier, and the distribution of accumulation, ablation and surface temperature are given, the distribution of ice thickness, velocity, and temperature that will keep the glacier in a steady state can be calculated. The parameters in the flow relation and the geothermal heat flux must also be given. Flow-lines, the travel time of the ice along them, and a time scale for an ice core can be calculated from the velocity distribution. The non-steady problem of how the system reacts to changes in accumulation, ablation, and temperature can also be solved.

3 Basal Motion

Figure 4 shows driving stress, calculated from ice thickness and surface slope, and surface velocity along a flow-line in the southern part of the Greenland Ice Sheet. The curves have the expected shape. The driving stress increases from zero at the ice divide to about 100 kPa at a distance of 50 km, a value that is maintained until the ice thins near the terminus. The maximum velocity is at the equilibrium line.

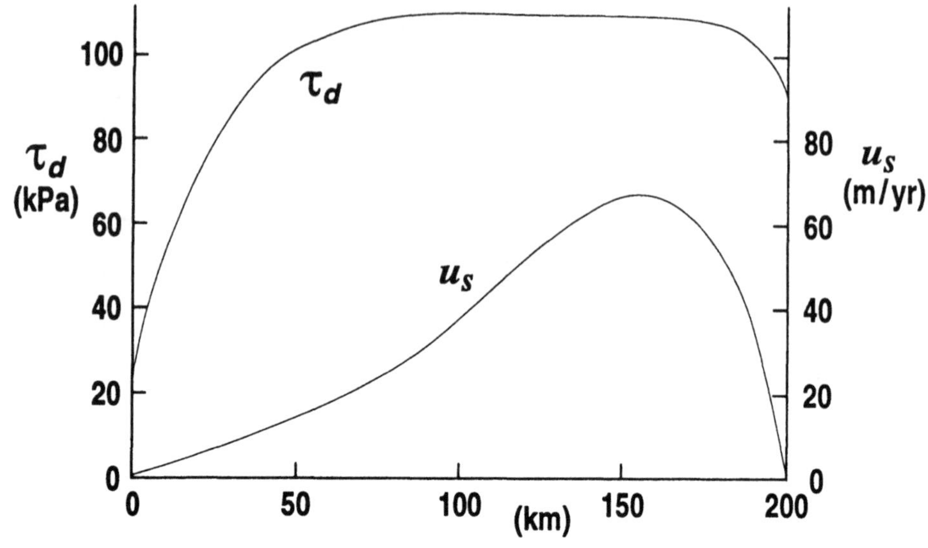

Fig. 4. Computed driving stress and surface velocity along a typical flow-line in south Greenland. The equilibrium line is at 165 km from the ice divide. From Dahl-Jensen (1989).

Figure 5 shows the same quantities along a flow-line in West Antarctica. This flow-line includes Ice Stream B, one of five ice streams that drain ice from the interior into the Ross Ice Shelf. These ice streams are a few hundred kilometres long, 30–50 km wide, and 1–1.5 km thick. Typical velocities are several hundred metres per year apart from in Ice Stream C which appears, from the depth of buried crevasses, to have stopped moving rapidly about 150 years ago. The ice streams do not occupy bedrock depressions but are separated from slow-moving ice on either side by 5-km wide shear zones. They have very small surface slopes so that driving stresses are low, less than 20 kPa on Ice Stream B. Nevertheless, velocities range from 50 to 825 m/yr. Because the driving stress is so low, ice deformation in these ice streams must be insignificant.

These ice streams are believed to move by deformation of water-saturated sediments beneath the ice. Seismic sounding has shown that the ice is underlain by about 6 m of water-saturated sediment (Blankenship *et al.*, 1986). Drilling of boreholes confirmed that the basal material was glacial till and water levels in the boreholes indicated subglacial water pressures of about 95% of those needed to float the ice (Engelhardt *et al.*, 1990). Although glacial geologists have long known that glacial sediments can, under certain conditions, deform, this is a new idea for glaciologists. The turning point came when Boulton and

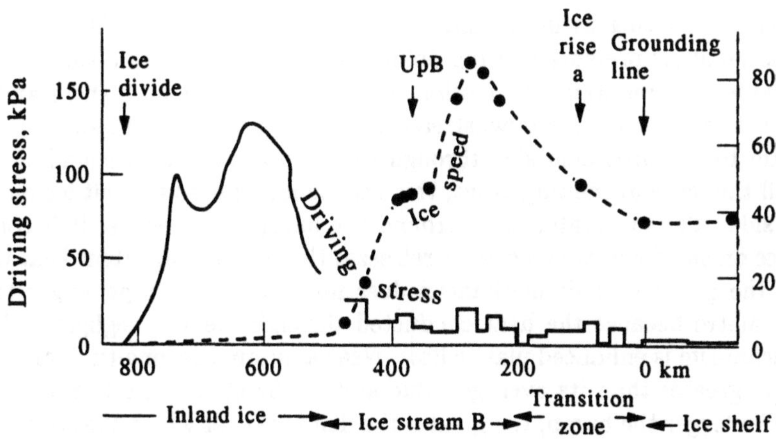

Fig. 5. Driving stress and surface velocity along a flow-line in the West Antarctic Ice Sheet. The flow-line starts in the inland ice, runs along Ice Stream B, and then through a "transition zone" into the floating Ross Ice Shelf. From Alley and Whillans (1991) by permission of the American Association for the Advancement of Science.

Hindmarsh (1987) made measurements in a tunnel under the terminus of a glacier in Iceland. The form of the flow relation is in dispute. Most workers have assumed that it is Newtonian viscous. However, values of viscosity, obtained either by measurements or by using viscosity as an adjustable parameter in flow models, range over three orders of magnitude (Paterson, 1994, p. 170). This recalls the situation in glacier mechanics, fifty years ago; little progress was made until the assumption that ice behaved as a viscous fluid was replaced by a realistic non-linear flow relation. At the other extreme, Kamb (1991) believes that the flow relation for till is non-linear with a value of n of the order 100. In this case, deformation would be confined to a thin layer at the base of the sediments.

Most of the outflow in both Greenland and Antarctica is channelled into fast-flowing ice streams or outlet glaciers, the behaviour of which largely controls the state of the ice sheet. In the Ross Sea sector of West Antarctica, the channelling is caused by the nature of the subglacial material and the presence of water at high pressures. The more usual situation, as in Greenland and the East Antarctic side of the Ross Sea, is large outlet glaciers flowing rapidly through gaps in coastal mountains. Most of these glaciers have driving stresses in the normal range and ice deformation appears to be their mechanism of flow.

Sliding of ice over its bed is the third possible mechanism of glacier motion. Sliding velocities span a much greater range than is possible by ice deformation and can change over periods of hours to days. However, sliding is possible only when the basal ice is at melting point. What enables a glacier to move past bumps in its bed? Observations confirm that there are two mechanisms. The first is pressure melting as in the classic regelation experiment. All the ice is assumed to be at its pressure melting point. Because the upstream side of each bump provides the resistance to motion, there is excess pressure there and so the ice is colder than on the downstream side. Heat therefore flows from the downstream to the upstream side through the bump and the surrounding ice. Because all the ice is at melting point, this heat must be used to melt ice on the upstream side. The meltwater flows around the bump and, because it is colder than the ice on the downstream side, it refreezes there, releasing latent heat that maintains the process. This mechanism does not work for bumps longer than about one metre because the heat conducted through them is negligible. The second mechanism is enhanced plastic flow. Near a bump, the longitudinal stress in the ice is greater than its average value and so the strain rate is also above average. The larger the bump, the greater the distance over which the stress and deformation are enhanced. This mechanism is therefore more effective for large bumps than for small ones. The glacier moves over its bed by a combination of these two processes; neither is sufficient by itself. It follows that bumps of some intermediate size, probably between 0.05 and 0.5 m, provide most of the resistance to sliding. Mathematical analysis leads to an equation for the sliding velocity

$$u_b = B\,\tau_d^{(n+1)/2} R^{-n-1}. \tag{15}$$

Here τ_d is the driving stress, R the bed roughness (amplitude of bumps divided by their spacing), n (=3) the index in the flow relation and B a constant that depends on the thermal and mechanical properties of the ice. This equation cannot be used for prediction because bed roughnesses are seldom, if ever, known.

The principle objection to Eq. 15 is that the effect of water at the bed is ignored. Observations of increased velocity during the melt season and after heavy rain and of diurnal fluctuations show that sliding velocity is affected by water at the bed. This has been taken into account empirically by writing

$$u_b = C\,\tau_d^m / (\rho g h - P)^p. \tag{16}$$

Here P is subglacial water pressure, h is ice thickness, C is a "constant" which now includes the bed roughness, and m and p are positive integers. This equation is also of limited value because P can vary during the course of a single day and is seldom measured. A further objection is the dubious assumption that sliding at any point is unaffected by water conditions up- and down-stream.

An equation for basal velocity (sliding or bed deformation) is an essential lower boundary condition in numerical models of glacier flow. There is some sketchy evidence that the velocity of bed deformation also obeys an equation of the form 16 (Boulton and Hindmarsh, 1987). The difficulty of measuring bed roughness and water pressure and uncertainties in the value of m and p give wide scope for tuning these models.

4 Temperatures in Glaciers and Ice Sheets

The temperature distribution in a glacier results from the interaction of many processes. Temperatures near the surface are controlled by the surface energy balance. Seasonal variations are damped out in the uppermost 15 m. Depth of penetration increases with the period of the change; the 10^5 year ice-age cycle penetrates through the deepest ice sheets into bedrock. In Antarctica and the higher parts of Greenland, where the air temperature never rises to 0°C, the temperature at a depth of 15 m is equal to the mean annual air temperature. In other regions, latent heat released by the refreezing of percolating meltwater in the snow pack can raise the near-surface temperature in the accumulation area several degrees above the mean annual air temperature. Thermal conditions at the base are determined mainly by the geothermal heat flux and by frictional heat if the ice is sliding. If the basal temperature is below the melting point there is no sliding and all the geothermal heat is conducted into the ice. An average geothermal flux produces a temperature gradient of 0.025 deg/m in ice. If the basal temperature is at melting point, some of the heat is used to melt ice and the remainder conducted into the ice. In a *temperate glacier* (ice at melting point throughout) or if there is a basal layer of finite thickness at melting point, the pressure melting gradient prevents the upward conduction of any heat. Ice deformation, most of which takes place near the bed, produces heat within the glacier.

Heat is transferred within the glacier by conduction and ice flow. The vertical velocity component in the accumulation area, the amplitude of which decreases with increasing depth, carries cold ice downwards from the surface. In the upper layers, it is a much more important means of heat transfer than conduction. In a stagnant ice sheet, with no accumulation or ablation, the temperature would increase steadily with depth at a rate equal to the geothermal gradient. In the accumulation area, the temperature distribution is still like this near the bed, where the vertical velocity is very low, but downward advection of cold surface ice may make the temperature of the upper layers nearly uniform. The greater the accumulation rate the stronger the effect, as shown in Fig. 6. In the ablation area, because ice flows upwards relative to the surface and carries warm basal ice with it, the temperature gradient in the upper layers exceeds the geothermal. However, neglect of the horizontal velocity component, as in the calculations underlying Fig. 6, is difficult to justify.

The pattern of particle paths in a glacier (Fig. 1) shows that ice at depth has originated at progressively higher elevations and thus colder temperatures. Although heat conduction changes the temperature, flow from upstream often results in the coldest ice in a borehole being encountered some distance below the surface. This effect becomes more important as distance from the head of the glacier increases. Such a temperature minimum could not occur as a steady state in a static medium. However, recent climatic warming could also explain such observations.

It follows that measurements of the present temperature distribution in a borehole can be inverted to give a record of past surface temperature, provided

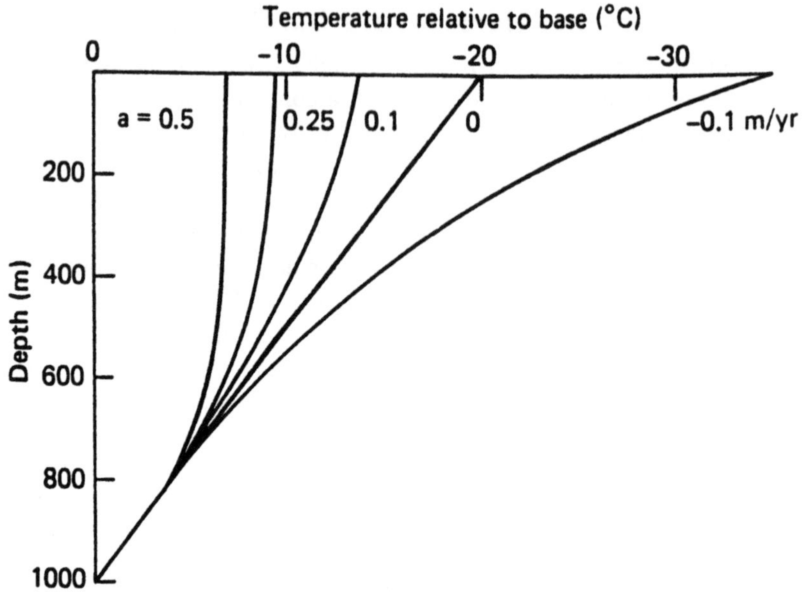

Fig. 6. Computed temperature distributions in a steady-state ice sheet with a frozen bed for different values of ice accumulation rate (positive) and ablation rate (negative). From Paterson (1992) by permission of Academic Press.

that flow from upstream is allowed for or eliminated by drilling the hole at an ice divide. The resulting record is heavily smoothed however, and increasingly so the further back in time it extends. Oxygen-isotope profiles from ice cores (e.g. Jouzel *et al.*, 1993) provide a much more detailed record of past surface temperatures. Nevertheless, borehole temperature profiles have proved valuable for calibrating such records, that is, converting a given change in oxygen-isotope ratio to a temperature change. Such analyses have shown that temperatures in central Greenland during the last ice age were 15 to 20 deg colder than at present (Cuffey *et al.*, 1995; Johnsen *et al.*, 1995).

The equation of heat transfer, in two dimensions, can be written

$$\frac{\partial T}{\partial t} = \frac{K}{\rho c}\left(\frac{\partial^2 T}{\partial x^2} + \frac{\partial^2 T}{\partial z^2}\right) - u\frac{\partial T}{\partial x} - w\frac{\partial T}{\partial z} + \frac{f}{\rho c}. \tag{17}$$

Here $T(x, z, t)$ is temperature, t time, K thermal conductivity, ρ density, c specific heat capacity, and $f = 2\dot{\epsilon}\tau$ the heat of deformation. Equations 8 and 9 define $\dot{\epsilon}$ and τ. The initial temperature distribution and the surface temperature as a function of time are prescribed. The basal boundary condition is either $T_b = T_m$, the melting temperature, or $(\partial T/\partial z)_b = -G/K$, with G the geothermal heat flux ($\approx$50 mW/m^2) if $T_b < T_m$. This form of the equation neglects the dependence of K on T and ρ. Inclusion of a term $\partial K/\partial T$ makes the equation non-linear.

5 Glacier Surges

A *surge* is the best known type of unstable behaviour of glaciers. Surging glaciers alternate between an active period, usually 1 to 3 years, during which they move at speeds of 10 to 1000 times normal, and a quiescent period roughly 10 times as long. Surging velocities are so high (tens of metres per day in some cases) that they must result from basal motion. A surge transfers a large volume of ice from a reservoir area, where the surface drops 50 to 100 m, to a receiving area where the ice thickens by the same amount. These are not the same as the accumulation and ablation areas; some surges do not extend to the head of the glacier and a few are confined to the ablation area. The ice in the lower part of the glacier moves forward several kilometres during a surge. This, however, is reactivation and thickening of stagnant ice rather than an advance of the terminus. During the quiescent period, ice flow cannot keep pace with accumulation and ablation. The upper part thickens while the lower part melts down and stagnates, until some critical profile, which appears to trigger the next surge, is reached. The medial moraines on many surging glaciers are deformed into large bulb-like loops by surges of tributaries. The regularity of these patterns, and the regular interval between surges of the same glacier, suggest that they result from some internal instability.

Surging glaciers are found only in certain areas - the Alaska range, central Asia, Chile, Iceland, Svalbard, arctic Canada and Greenland. There are none in the Alps or Rockies. Even in these areas with surging glaciers, only a few surge and these appear to be non-randomly distributed. Parts of ice caps in Iceland and Svalbard have surged; glaciers draining the same ice cap do not necessarily surge at the same time. The majority of surging glaciers are in relatively young mountain ranges undergoing rapid erosion.

A surge apparently results from disruption of the plumbing system that drains water from beneath the glacier. Build-up of water pressure permits rapid sliding or bed deformation or both. During the surge of Variegated Glacier in Alaska subglacial water pressures were always within 0.5 MPa of the ice-overburden pressure (3.6 MPa) and occasionally exceeded it. Dye tracing experiments showed that the transit time of water beneath the glacier decreased from an average of 90 hr during the surge to 4 hr afterwards. The end of the surge coincided with a large flood in the outflow stream at the terminus and the ice surface dropped by about 0.2 m (Kamb *et al.*, 1985). Because Variegated Glacier is temperate (at melting point throughout) a thermal mechanism of surging (alternate melting and freezing of the glacier bed) cannot apply to it.

There is a thermal barrier beneath Trapridge Glacier, a small surge-type glacier in the Yukon, at present in its quiescent phase. The bed beneath a central area is at melting point, but it is frozen beneath the upper and terminal areas and along the sides. This type of temperature distribution is found in many non-surging glaciers as well. The boundary between the frozen and melting bed upstream of the terminus is marked by a 40 m bulge in the ice surface. Subglacial water is not being trapped here, however; the bed consists of sediments the lower layers of which are unfrozen and the water is draining away through them (Clarke *et al.*, 1984).

Whether all surges have the same cause is unclear. At least the details of the water-trapping mechanism must depend on whether the glacier rests on bedrock or sediments. Most of the water under Variegated Glacier during its surge is believed to be in cavities on the downstream side of bedrock bumps. Such cavities cannot exist in a sediment bed. The most important point for my subsequent discussion is that no known surge has a thermal mechanism.

6 Instability of Tidewater Glaciers

Meier and Post (1987) believe that glaciers that end in fjords go through a cycle, not directly related to climatic change, namely: stable position with terminus at the head of the fjord, slow advance, stable extended position, rapid retreat. Their arguments rest on the plausible hypothesis that the rate of iceberg calving is proportional to the water depth, irrespective of whether the terminus is floating or grounded.

If a glacier terminus is grounded on a moraine shoal at the mouth of a fjord, a small retreat moves it back into deeper water, calving increases, and so retreat continues, possibly to the head of the fjord where the retreat is likely to stop. A moraine shoal gradually builds up there. This reduces calving, the mass balance becomes positive, and the glacier starts to advance. The advance can continue into increasingly deep water, however, only if the moraine moves forward to restrict the calving rate. This is achieved by erosion of material from the upstream face and deposition on the downstream side. Although such an advance is slow, it can continue to a place where the fjord widens, or possibly right to its mouth. The substantial increase in calving rate at the mouth of a fjord can trigger catastrophic retreat even in the absence of climatic forcing (Alley, 1991). Meier and Post estimate that such an advance may last for about 1000 yr, whereas the retreat phase may take only 100 yr. Nearby glaciers may behave asynchronously as a result of differences in fjord topography and erosion and deposition rates. This process can explain the catastrophic retreats, in one case amounting to 100 km, of many tidewater glaciers in Alaska since the first recorded observations in the 18th century, while some other tidewater glaciers in the area have advanced.

7 Unstable Flow in Ice Sheets

The major outlet glaciers and ice streams in Antarctica and Greenland flow at surging speeds. Whether these can be maintained indefinitely, or whether they undergo a regular cycle of alternating fast and slow flow is unclear. The flow of Jakobshavns Isbræ in West Greenland, the fastest glacier in the world, has been maintained for at least the past 30 years. However, such a glacier may be able to draw ice from an increasing area of the ice sheet, whereas the reservoir area of a valley glacier is limited. Thus the active period of ice-sheet outlets is expected to be much longer than the one to three years typical of surging glaciers. Moreover, because of low precipitation, the length of the quiescent period (the time taken to refill the "reservoir") of an ice-sheet outlet may be hundreds or even thousands of years rather than the tens of years typical of valley glaciers. The near-stagnation of Ice Stream C in Antarctica some 150 years ago might suggest cyclic behaviour. It has, however, been attributed to "piracy" of either ice or subglacial water by the adjacent Ice Stream B as a result of changes in the topography of the ice surface (Anandakrishnan and Alley, 1997). This suggests an unusual event rather than one with a regular period.

There is a major difference between hydrological conditions in Antarctica and those in surging glaciers in temperate and sub-polar regions. A large seasonal influx of surface meltwater to the bed appears to play an important role in the mechanism and timing of known surges. This is lacking in Antarctica; all the water at the bed is generated there at what is presumably a constant rate. In Greenland, although there is extensive surface melting in summer on the lower parts of the large outlet glaciers, the lack of seasonal variations in velocity of Jakobshavns Isbræ, the only one to have been observed in detail, suggests that surface water does not penetrate to the bed.

There is some evidence of unstable behavour of the Laurentide Ice Sheet in North America during the last ice age. Whether this can be described as surging is perhaps a matter of definition. A characteristic feature of surges in present-day valley glaciers is their regular period, a consequence of fixed reservoir areas and stable precipitation rates. However, the ice-age climate in the North Atlantic region was characterized by the so-called Dansgaard-Oeschger Events: cycles of slow cooling followed by rapid warming with periods of 1000 to 3000 years and amplitudes of about half the glacial-interglacial temperature difference (Dansgaard *et al.*, 1993; Bond *et al.*, 1993). There were corresponding changes in precipitation rate. A constant interval between surges is not expected in these circumstances. Distinguishing between an internal instability in the ice sheet and a non-linear response to the climate forcing is also difficult.

The sediment record from the North Atlantic between latitudes 40° and 55°N contains six layers that are rich in ice-rafted debris and poor in foraminifera. These layers, which contain relatively high concentrations of detrital carbonate, appear to have been deposited rapidly at times when the surface water was exceptionally cold and had low salinity. They span the period from about 66 kyr B.P. (Layer 6) to 14 kyr B.P. (Layer 1). In addition to the presence of carbonate, the distribution of the layers (thick in the Labrador Sea and western North Atlantic, thinning steadily towards the east) suggests that the material in them originated in Hudson Strait. The deposition of most of the layers has been attributed to exceptionally large discharges of icebergs (*"Heinrich Events"*) from the Hudson Bay sector of the Laurentide Ice Sheet. Layer 3 is an exception; its distribution and composition suggest multiple origins, possibly including East

Greenland and Scandinavia. Each event probably lasted no more than 1000 years. Comparison with the climate record in Greenland ice cores shows that Heinrich Events occurred towards the end of exceptionally cold periods and were followed by rapid warmings. The above information is contained in papers by Heinrich (1988), Andrews and Tedesco (1992), Bond *et al.* (1992, 1993), Broecker *et al.* (1992), Grousset *et al.* (1993), and Broecker (1994).

The relatively high concentration of ice-rafted debris in each Heinrich Layer could result either from an increase in the deposition rate of this debris or a decrease in the deposition rate of other sediments. Both are expected after strong cooling. Glaciers would advance, possibly to the edge of the continental shelf, and iceberg calving would increase. The extent of sea ice would also increase, decreasing the productivity of the ocean. Evidence for increases in concentrations in ice-rafted debris from Fennoscandia and Iceland, coincident with Heinrich Events (Fronval *et al.*, 1995; Revel *et al.*, 1996), is consistent with either possibility and so cannot distinguish between them. However, the distribution of the Heinrich Layers, the carbonate in them, and the evidence that they were deposited rapidly, have led many authors to conclude that they resulted from unstable flow in the Hudson Bay sector of the Laurentide Ice Sheet. Others (e.g. Bond and Lotti, 1995) have favoured a climatic trigger.

The postulated mechanism of the instability is the periodic warming of the base to melting point, permitting fast basal motion followed by massive discharge of icebergs. This is essentially a thermal mechanism of surging, an idea proposed by Robin (1955) long before the discovery of Heinrich Layers. It is a plausible mechanism, although it doesn't apply to any present-day surge. The first, and simplest, recent model is the binge-purge model of MacAyeal (1993). The climate is assumed to be constant throughout and the ice sheet stagnant during the quiescent period. The ice thickens by snow accumulation at the surface and the basal temperature rises because the geothermal gradient exceeds the atmospheric lapse rate. When the melting point is reached, water pressure builds up in the subglacial sediments and permits rapid deformation. This thins the ice and thereby increases the temperature gradient in it. The resulting increase in heat conduction cools the basal ice and, after 250 to 500 years, it refreezes. In this model, the period of the oscillation does not depend on accumulation rate. The predicted value is 7000 years if the mean annual air temperature at sea-level is $-10°$C. The uniform period stems from the unrealistic assumption of unchanging climate. In fact, the interval between events varies between 6000 and 16,000 years (Bond *et al.*, 1992). The predicted iceberg flux implies mean ice velocities of 20 to 40 km/yr, values I believe to be unrealistically high.

More sophisticated models (e.g. Verbitsky and Saltzman, 1994; Payne, 1995; Greve and MacAyeal, 1996; Saltzman and Verbitsky, 1996; Marshall and Clarke, 1997) take into account heat transfer by ice advection and heating by deformation and sliding friction. During the active phase, advection of cold ice from upstream helps to increase the temperature gradient in the lower layers.

The tidewater-glacier instability provides an alternative mechanism; advance of the Hudson Strait ice stream beyond the mouth of the Strait during a cold period triggered its collapse. This hypothesis could explain why each Heinrich Event occurred near the end of a cold period; the ice is expected to be at maximum extent then. This timing is surprising if an internal instability is the cause. Confirmation of simultaneous large iceberg discharges from other ice sheets would strengthen the case for the tidewater-glacier instability. On the other hand, a basal freeze-thaw cycle seems essential to explain the debris in the icebergs. The Antarctic ice shelves, for example, contain very little sediment because they are fed by ice streams whose fast motion generates enough heat to melt several tens of metres of basal ice, as they flow from the interior to the ice shelf. This removes virtually all debris. The upper layers contain no debris because there is no surface source of it (Bentley, 1994).

Alley and MacAyeal (1994) have addressed the problem of incorporating debris into basal ice. During the early stages of the discharge (purge), the frictional, deformational, and geothermal heat near the bed exceeds the heat conducted upwards through the ice. The excess heat melts basal ice. The ice thins as the purge proceeds. Upward heat conduction therefore increases, while the motion, and therefore the heat produced by deformation and friction, diminishes. In the later stages, conduction predominates and so the meltwater, which is assumed to have been stored in the sediments, refreezes and incorporates debris into the basal ice. When all the water-soaked sediment has been frozen, there is no more latent heat to balance the upward heat conduction. The bed refreezes and the rapid flow stops. Alley and MacAyeal believe that this mechanism can incorporate enough debris to account for the estimated volume in a Heinrich Layer.

An unanswered question is how the supply of sediment ("till") at the bed is maintained. Ice flow carries the till contained in it down-glacier and the till beneath may be deforming and moving downstream. This is believed to be happening under the West Antarctic ice streams where the till accumulates on the sea floor immediately downstream of where the ice goes afloat. New till could be generated by erosion but this would be possible only when the bed is unfrozen, that is during the short purge phase of the cycle. Alley and MacAyeal avoid this difficulty by assuming that the till layer is immobile and infinitely thick. Again, Andrews *et al.* (1993) have argued that because debris near the base of an iceberg could not be carried far across the Atlantic without melting out, the debris must have been dispersed throughout the ice column. It is difficult to envisage a mechanism for this. These problems deserve further study.

The difference between the hypotheses of an internal instability or a climatic trigger for the cause of massive calving may be smaller than appears at first sight. Once the ice stream reached the mouth of Hudson Strait, for whatever reason, there would be massive calving, rapid retreat, and thinning. To incorporate debris in the basal ice, a switch from basal melting to freezing seems necessary. This seems most likely to occur after retreat and thinning. Although it does not occur in present tidewater glaciers in Alaska, it might well have happened in Hudson Strait during the ice age. Perhaps the only difference is whether the ice advance was rapid as a result of an internal instability or slow as a result of climatic cooling. Andrews *et al.* (1994) found evidence of major pulses of meltwater, coinciding with Heinrich Events 1 and 2, in cores from the Labrador Sea. If it could be shown that these consisted largely of subglacial water, rather than surface melt or water from melting icebergs, this would support the hypothesis of a rapid advance.

The stability or otherwise of the present West Antarctic ice streams and of the ice-age ice sheets are two of the major problems in glaciology today.

Acknowledgements. Comments by an anonymous reviewer resulted in major improvements to the section on Heinrich Events. Discussions with Garry Clarke, Shawn Marshall, and graduate students at the University of British Columbia were also useful.

References

Alley, R. B., 1991. Sedimentary processes may cause fluctuations of tidewater glaciers, *Ann. Glaciol.*, **15**, 119–124.

Alley, R. B., and D. R. MacAyeal, 1994. Ice-rafted debris associated with binge/purge oscillations of the Laurentide Ice Sheet, *Paleoceanography*, **9** (4), 503–511.

Alley, R. B. and I. M. Whillans, 1991. Changes in the West Antarctic Ice Sheet, *Science*, **254**, 959–963.

Anandakrishnan, S. and R. B. Alley, 1997. Stagnation of Ice Stream C, West Antarctica, by water piracy, *Geophys. Res. Letters*, **24**, 265–268.

Andrews, J. T., and K. Tedesco, 1992. Detrital carbonate-rich sediments, northwestern Labrador Sea: Implications for ice-sheet dynamics and iceberg rafting (Heinrich) events in the North Atlantic, *Geology*, **20**, 1087–1090.

Andrews, J. T., K. Tedesco, and A. E. Jennings, 1993. Heinrich events: Chronology and processes, east-central Laurentide Ice Sheet and north-west Labrador Sea, In *Ice in the Climate System,* edited by W. R. Peltier, Springer-Verlag, New York, 167–186.

Andrews, J. T., H. Erlenkeuser, K. Tedesco, A. Aksu, and A. J. T. Jull, 1994. Late Quaternary (Stage 2 and 3) meltwater and Heinrich events, northwest Labrador Sea, *Quat. Res*, **41**, 26–34.

Bentley, C. R., 1994. Sedimentation by the Antarctic Ice Sheet from a glaciological perspective, *Terra Antarctica*, **1**, 249–250.

Blankenship, D. D., C. R. Bentley, S. T. Rooney, and R. B. Alley, 1986. Seismic measurements reveal a saturated porous layer beneath an active Antarctic ice stream, *Nature*, **322**, 54–57.

Bond, G. C., and R. Lotti, 1995. Iceberg discharges into the North Atlantic on millenial time scales during the last glaciation, *Science*, **267**, 1005–1010.

Bond, G. and 13 others, 1992. Evidence for massive discharges of icebergs into the North Atlantic Ocean during the last glacial period, *Nature*, **360**, 245–249.

Bond, G., W. Broecker, S. Johnsen, J. McManus, L. Labeyrie, J. Jouzel, and G. Bonani, 1993. Correlations between climate records from North Atlantic sediments and Greenland ice, *Nature*, **365** (6442), 143–147.

Boulton, G. S. and R. C. A. Hindmarsh, 1987. Sediment deformation beneath glaciers: rheology and geological consequences, *J. Geophys. Res.*, **92**, 9059–9082.

Broecker, W. S., 1994. Massive iceberg discharges as triggers for global climatic change, *Nature*, **372**, 421–424.

Broecker, W. S., G. Bond, J. McManus, M. Klas, and E. Clark, 1992. Origin of the North Atlantic's Heinrich events, *Clim. Dyn.*, **6**, 265–273.

Clarke, G. K. C., S. G. Collins and D. E. Thompson, 1984. Flow, thermal structure, and subglacial conditions of a surge-type glacier, *Can. J. Earth Sci.*, **21**, 232–240.

Cuffey, K. M. and 5 others, 1995. Large Arctic temperature change at the Wisconsinan-Holocene glacial transition, *Science*, **270**, 455–458.

Dahl-Jensen, D., 1989. Steady thermomechanical flow along two-dimensional flow lines in large grounded ice sheets, *J. Geophys. Res.*, **94**, 10355–10362.

Dansgaard, W. and 10 others, 1993. Evidence for general instability of past climate from a 250-kyr ice-core record, *Nature*, **364**, 218–220.

Engelhardt, H., N. Humphrey, B. Kamb, and M. Fahnestock, 1990. Physical conditions at the base of a fast moving Antarctic ice stream, *Science*, **248**, 57–59.

Fronval, T., E. Jansen, J. Bloemendal, and S. Johnsen, 1995. Oceanic evidence for coherent fluctuations in Fennoscandian and Laurentide ice sheets on millenium time scales, *Nature*, **374**, 443–447.

Greve, R. and D. R. MacAyeal, 1996. Dynamic/thermodynamic simulations of Laurentide ice-sheet instability, *Ann. Glaciol.*, **23**, 328–335.

Grousset, F. E., L. Labeyrie, J. A. Sinko, M. Cremer, G. Bond, J. Duprat, E. Cortijo and S. Huon, 1993. Patterns of ice-rafted detritus in the glacial North Atlantic, *Paleoceanography*, **8**, 175–192.

Heinrich, H., 1988. Origin and consequences of cyclic ice-rafting in the northeast Atlantic Ocean during the past 130,000 years, *Quat. Res.*, **29** (2), 141–152.

Johnsen, S. J., D. Dahl-Jensen, W. Dansgaard and N. S. Gundestrup, 1995. Greenland paleotemperatures derived from the GRIP bore hole temperature and ice core isotope profiles, *Tellus*, **47B**, 624–629.

Jouzel, J. and 16 others, 1993. Extending the Vostok ice-core record of palaeoclimate to the penultimate glacial period, *Nature*, **364**, 407–412.

Kamb, B., 1991. Rheological nonlinearity and flow instability in the deforming bed mechanism of ice stream motion, *J. Geophys. Res.*, **96**, 16,585–16,595.

Kamb, B., C. F. Raymond, W. D. Harrison, H. Engelhardt, K. A. Echelmeyer, N. Humphrey, M. M. Brugman and T. Pfeffer, 1985. Glacier surge mechanism: 1982–1983 surge of Variegated Glacier, Alaska, *Science*, **227**, 469–479.

MacAyeal, D. R., 1993. Binge/purge oscillations of the Laurentide Ice Sheet as a cause of the North Atlantic's Heinrich events, *Paleoceanography*, **8** (6), 775–784.

Marshall, S. J., and G. K. C. Clarke, 1997. A continuum mixture model of ice stream thermomechanics in the Laurentide Ice Sheet 2. Application to the Hudson Strait Ice Stream, *J. Geophys. Res.*, **102** (B9), 20,615–20,638.

Meier, M. F. and A. Post, 1987. Fast tidewater glaciers, *J. Geophys. Res.*, **92**, 9051–9058.

Paterson, W. S. B., 1992. Glaciology, *In Encyclopedia of Physical Science and Technology, 2nd Edition, Vol.* **7**, Academic Press, New York, 385–409.

Paterson, W. S. B., 1994. *The Physics of Glaciers, 3rd ed.*, 480 pp., Elsevier Sci., New York,

Payne, A. J., 1995. Limit cycles in the basal thermal regime of ice sheets, *J. Geophys. Res.*, **100** (B3), 4249–4263.

Revel, M., J. A. Sinko, F. E. Grousset, and P. E. Biscayne, 1996. Sr and Nd isotopes as tracers of North Atlantic lithic particles: Paleoclimatic implications, *Paleoceanography*, **11**, 95–113.

Robin, G. de Q., 1955. Ice movement and temperature distribution in glaciers and ice sheets, *J. Glaciol.*, **2** (18), 523–532.

Saltzman, B. and M. Ya. Verbitsky, 1996. Heinrich-scale surge oscillations as an internal property of ice sheets, *Ann. Glaciol.*, **23**, 348–351.

Verbitsky, M. Ya. and B. Saltzman, 1994. Heinrich type glacial surges in a low-order dynamical climate model, *Climate Dynamics*, **10**, 39–47.

Stable Isotope Records from Greenland Deep Ice Cores: The Climate Signal and the Role of Diffusion

Sigfús J. Johnsen,[1,4] H. B. Clausen,[1] Jean Jouzel,[2] Jakob Schwander,[3] Árny E. Sveinbjörnsdóttir,[4] and James White[5]

[1] The Niels Bohr Institute, Department of Geophysics, University of Copenhagen, Juliane Maries Vej 30, DK-2100, Copenhagen, Denmark

[2] Laboratoire des Sciences du Climat et de l'Environnement, UMR CEA-CNRS 1572, CEA Saclay, Orme des Merisiers 91191 Gif sur Yvette, France

[3] Physics Institute, University of Bern, Sidlerstrasse 5, CH-3012 Bern, Switzerland

[4] Science Institute, University of Iceland, Dunhaga 3, IS-107 Reykjavik., Iceland

[5] Institute of Alpine and Arctic Research and Department of Geological Sciences, University of Colorado, Boulder, USA

1 Introduction

Glaciers and ice caps exert a strong influence on Earth's environment and climate. They can grow and decay, change the sea level, control the albedo and shape the mountains and continents on which they reside. During the last glacial period the expanding glaciers surged frequently and brought an enormous number of icebergs into the North Atlantic which caused a whole suite of violent climatic oscillations (Heinrich events, Dansgaard-Oeschger cycles) (Bond et al., 1992; Broecker et al., 1985; Johnsen et al., 1972).

The most peculiar property of glaciers is perhaps in the ability to preserve, throughout their "third dimension", a record of past climatic and environmental changes. Each annual layer of snow deposited on the surface of a cold ice cap carries with it, as it sinks slowly towards to the bottom, a fingerprint of the impurities present in the atmosphere. Even the atmospheric gases are carried along, trapped in small bubbles when the snow is transformed into glacier ice.

The isotopic ratios $^{18}O/^{16}O$ or D/H (expressed in the following as the per mil deviations $\delta^{18}O$ or δD relative to that of Standard Mean Ocean Water, SMOW) of the snow reflect the temperature at which the cloud vapour was transformed into ice crystals or snow, which provides us with an extremely important proxy for past temperature changes.

The annual layers are nicely stacked in the ice cap until they arrive near the bottom strata where foldings and other stratigraphic disturbances from uneven bedrock undulations can disrupt the ordered layering. In Central Greenland this can happen after 150 thousand years or more and in smaller ice caps after 10 to 120 thousand years. By coring down to bedrock we can thus recover similarly old continuous records of the Earth's environment and climate.

NATO ASI Series, Vol. I 56
Ice Physics and the Natural Environment
Edited by John S. Wettlaufer, J. Gregory Dash and Norbert Untersteiner

The resolution of these records can be finer than a year up to 100 thousand years back in time, which allows the study of annual cycles as well as singular events in the Earth's history, like volcanic eruptions and sharp climatic shifts.

As an extra important bonus, high resolution records of selected constituents and visual core stratigraphy often allow precise dating of the cores by counting annual cycles several tens of thousands of years back in time (Alley et al., 1993; Hammer et al., 1997 (in press); Hammer et al., 1997; Meese et al., 1997). Such dating effort is a major undertaking which unveils the dynamics of environmental changes that are often found to proceed with catastrophic rapidity. It also gives a direct estimate of the past accumulation rates which is a very important climatic parameter.

In many cases the resolution of the records is impeded by diffusional processes in firn and ice that limit the usefulness of the data. On the other hand, as a compensation, they leave an imprint on the data that can be used to further constrain past environmental parameters like temperatures and accumulation rates. This is discussed in more detail in section 2.2 for the stable isotope data.

Refined and often continuous ice core analysis techniques allow several parameters to be obtained from the cores that help reveal past changes in the climate and the environment. The isotopic composition of the ice yields information about the temperature (Cuffey et al., 1995; Dansgaard, 1964; Johnsen et al., 1995a; Johnsen et al., 1989; Lorius, 1989) as stated earlier; the dust provides information about storminess and source efficiency (Hammer, 1977; Petit et al., 1981); the air bubbles tell us about the greenhouse gases in the atmosphere (Bender et al., 1994; Chappellaz et al., 1993; Oeschger et al., 1984; Raynaud et al., 1993; Sowers and Bender, 1995); the acidity holds clues about volcanic eruptions in both hemispheres (Clausen et al., 1995; Hammer et al., 1980; Hammer et al., 1997); chemical tracers teach us about various processes on land, in the sea and in the atmosphere (Clausen, 1995; Fuhrer et al., 1993; Legrand and Delmas, 1988; Mayewski et al., 1993), at the time of formation of the ice. Together, they give a detailed picture of the environment and climate of the past. All these parameters can be examined in great temporal detail and thereby illuminate various facets of the complicated mechanism of climate that create phase variations between individual parameters.

Several deep cores have been drilled and analyzed from Greenland and Antarctica. Figure 1 shows the geographic locations of the Greenland deep drilling sites discussed in this paper.

2 The Conversion of $\delta^{18}O$ to Temperature

The stable isotope ratios, $^{18}O/^{16}O$ or D/H, are the principal reference parameters in all ice core studies. If the cloud temperature at the moment of snow formation is described by the Rayleigh condensation process, then the stable isotope ratios are mainly determined by the water vapor volume faction and the fractionation constraints α_{18} and α_D This shows a direct climatic relevance of the stable isotope ratios, assuming unchanged temperature and humidity at the moisture source areas (Jouzel et

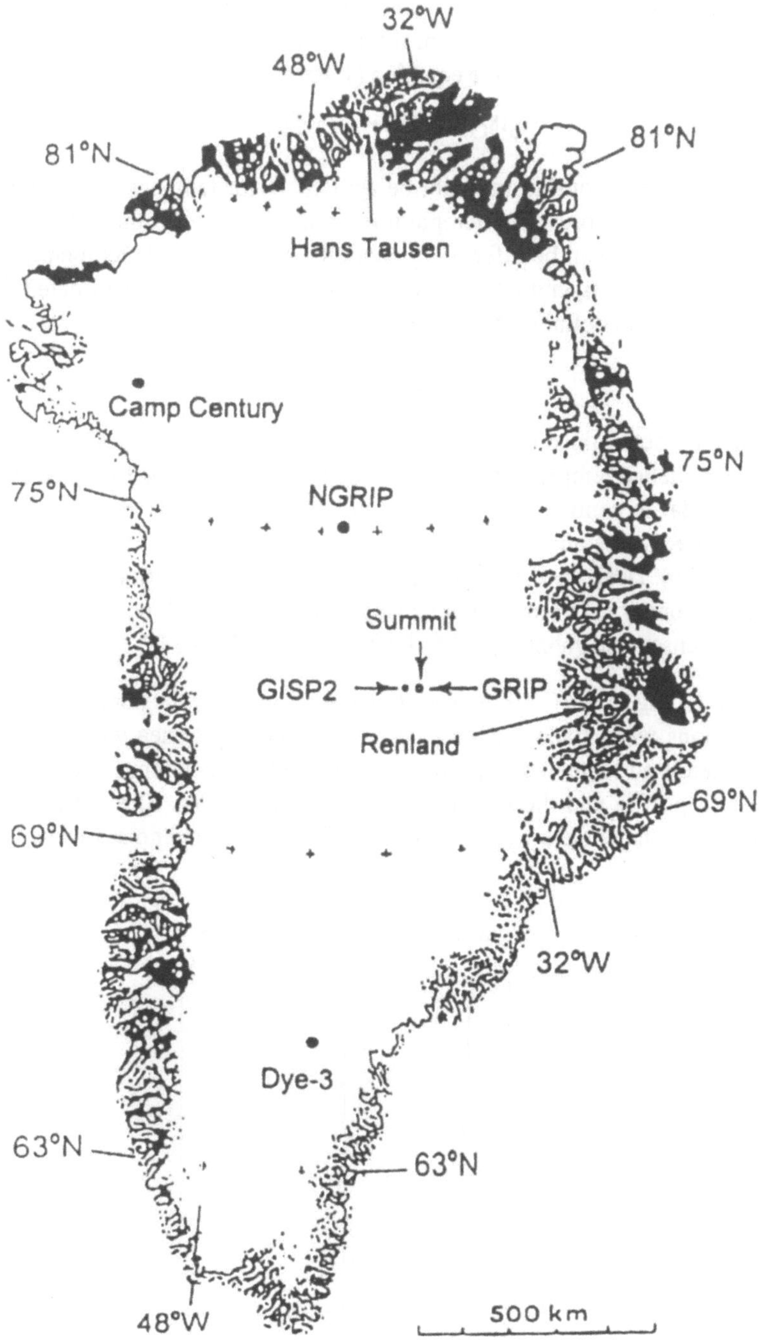

Fig. 1 Seven deep drilling sites in Greenland: Camp Century (U.S.A. CRREL, 1966); Dye 3 (GISP, 1981); Renland (Nordic Council of Ministers, 1988); Summit (GRIP, 1992); the USA camp (GISP2, 1993); Hans Tausen (Nordic Environmental Research Programme, 1995); a new deep drilling started at NGRIP in 1996.

al., 1997) and uniform deposition of snow throughout the year. On the Greenland ice sheet, the present mean annual $\delta^{18}O$ of the snow is closely related to the mean annual surface temperature, T(°C), by the equation (Johnsen et al., 1989)

$$\delta^{18}O = 0.67 \cdot T - 13.7‰. \quad (1)$$

However, this relationship may not be applicable for temporal changes measured along Greenland ice cores. The temporal relationship between the $\delta^{18}O$ and the surface temperature (T) can be inferred by modelling the bore hole temperature profile, based on a surface temperature history parameterized from the well-dated $\delta^{18}O$ history. The parameters defining the assumed $T - \delta^{18}O$ relationship

$$T = \alpha + \beta \cdot \delta^{18}O + \gamma \cdot (\delta^{18}O)^2 \quad (2)$$

α, β and γ are determined along with the geothermal heat flux by a least squares fitting of the modelled temperature profile to the measured temperature profile. Calibration for decadal to century long temperature changes can only be done for the past several hundred years and is obtained from shallow core data (Cuffey et al., 1994; Johnsen, 1977b). Deep ice core and bore hole data provide information for calibrating millennial or longer term temperature changes (Cuffey and Clow, 1997; Cuffey et al., 1995; Dahl-Jensen and Johnsen, 1986; Johnsen et al., 1995a).

In the modelling of the GRIP temperature profile we used the complete equation (2) and corrections were made for the past $\delta^{18}O$ variations of sea water. In Fig. 2 we show the measured temperature profile at GRIP and the difference between the modelled best fit profile and the measured temperature profiles. The GISP2 calibration assumed a linear dependency ($\gamma = 0$) between T and $\delta^{18}O$ and no sea water correction was applied.

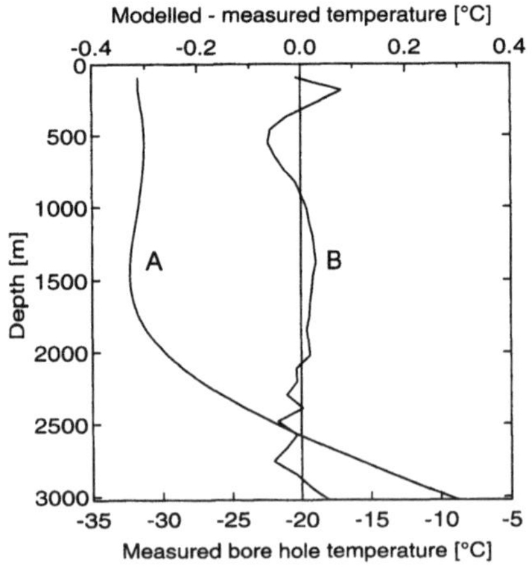

Fig. 2. A. The GRIP temperature profile and the modelled best fit profile. B. The deviations from the measured temperature profile are less than 0.1°C. Note the coherent deviations in the upper part indicating that the $\delta^{18}O$ history did not fully capture all the long term temperature shifts.

Figure 3 shows (i) the calculated cloud condensation temperature plotted against $\delta^{18}O$, based on a single-source precipitation model for Greenland (Johnsen et al., 1989); (ii) the surface (spatial) temperature to $\delta^{18}O$ relationship from equation (1) and the long-term bore hole temperature calibrations obtained from (iii) the GISP2 (Cuffey et al., 1995) and (iv) the GRIP (Johnsen et al., 1995a) deep bore holes.

The difference between the two GRIP and GISP2 calibration curves for the more negative $\delta^{18}O$ values is mainly due to the missing correction for the sea water $\delta^{18}O$ history in case of the GISP2 calibration.

Several points can be raised concerning the data in Fig. 3. First, we observe that the modelled cloud temperatures are generally higher than the surface temperatures, which accounts for the transport of heat by cyclones to the Greenland Ice Sheet (Loewe, 1936). Second, the slope of the cloud curve is only 1.04°C/‰ which indicates that precipitating clouds are not easily cooled off when lowering their $\delta^{18}O$ values, even during times of cold climate. Finally, the linear GISP2 calibration curve shows temperatures higher than the cloud temperatures at the upper end of the $\delta^{18}O$ scale, which is probably not a realistic scenario. This questions the validity of assuming a strict linear relationship between δ and T for millennial time scales.

The obvious difference in the various $\delta^{18}O$ – T relationships of Fig. 3 has generated a great deal of debate, and is discussed in Jouzel et al. (1997) where results from GCM models are included.

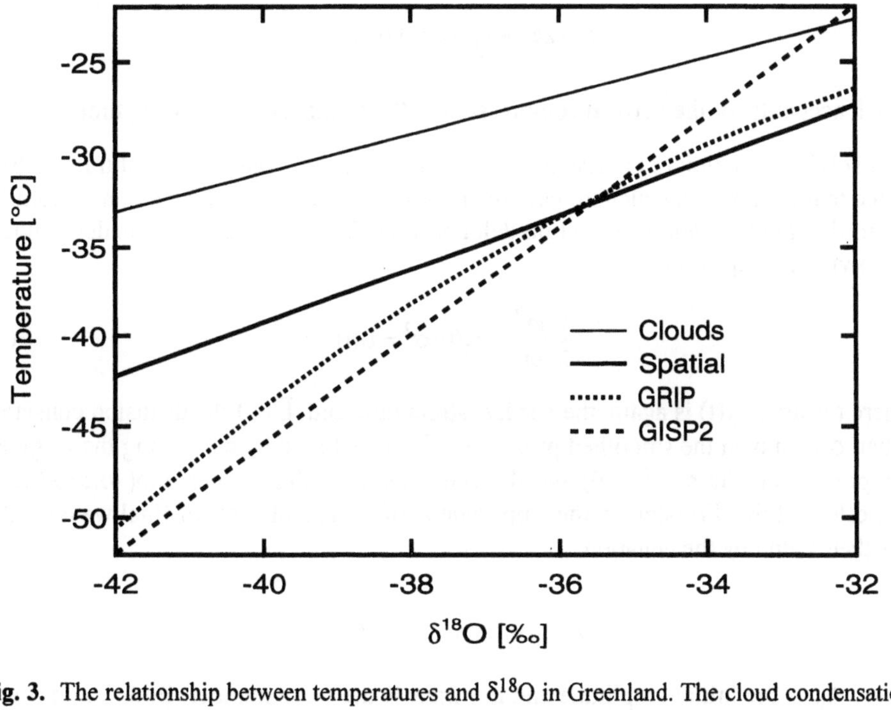

Fig. 3. The relationship between temperatures and $\delta^{18}O$ in Greenland. The cloud condensation temperature is shown as the thin line and the spatial relationship is shown as the thicker line. For long term temporal changes at Summit, the dotted curve shows the GRIP temperature profile calibration (Johnsen et al., 1995a) and the dashed line the GISP2 temperature profile calibration (Cuffey et al., 1995).

3 Diffusion of Stable Isotopes

3.1 Mathematical Formulation

In order to describe the smoothing of stable isotope data from ice cores we introduce a coordinate system with a vertical z-axis and an origin that follows a horizontal layer from the surface as it sinks into the firn and ice matrix. The medium in the new coordinate system is assumed to have uniform vertical strain rate $\dot{\varepsilon}_z(t)$ and a diffusion constant for each isotopic species D(t) that is dependent on the temperature and the firn density through the age parameter t. The age t is known as a function of depth from a deformation model valid for the particular coring site that mainly depends on the temperature and the accumulation rate. The diffusion equation for the smoothing of an initial stable isotope profile $\delta = \delta(z,0)$ then becomes

$$\frac{\partial \delta}{\partial t} = D(t)\frac{\partial^2 \delta}{\partial z^2} - \dot{\varepsilon}_z(t)\cdot z\frac{\partial \delta}{\partial z} \tag{3}$$

where $\delta = \delta(z',t)$ is the resulting smoothed delta profile after time t and z' and is found from

$$z' = z\exp(-\int_0^t \dot{\varepsilon}_z(\tau')d\tau'), \tag{4}$$

which accounts for the vertical compression of the original profile during time t.

In solving equation (3) a new parameter, the diffusion length σ, is naturally introduced that is most convenient to use when discussing isotopic diffusion in a uniformly deforming medium and its effect on delta profiles. This new parameter is determined by solving the equation

$$\frac{1}{2}\frac{d\sigma^2}{dt} - \dot{\varepsilon}_z(t)\cdot\sigma^2 = D(t) \tag{5}$$

where t is age, $\dot{\varepsilon}_z(t)$ is again, the vertical strain rate, and $D(t)$ the diffusion constant. When σ is known the smoothed profile $\delta = \delta(z',t)$ is found by combining the original compressed profile $\delta = \delta(z',0)$ and the Gaussian normalized clock $\exp(-0.5\cdot z^2/\sigma^2)$. The effect of the diffusion on the amplitude A of an annual cycle of final wavelength λ is then found by the equation

$$A = A_0 \exp(-\tfrac{1}{2}k^2\cdot\sigma^2) \tag{6}$$

where A_0 is the initial amplitude and $k = 2\pi/\lambda$ is the wave number. Both λ and σ are known as a function of depth from the age/deformation model used to solve equation (3). The effect of diffusion on the power spectral densities P(k) of the smoothed data is similarly found from

$$P(k) = P_0(k)\exp(-k^2 \cdot \sigma^2), \tag{7}$$

where $P_0(k)$ is the spectral density of the original undiffused profile. Equations (6) and (7) can be used to estimate the diffusion length σ from estimates of the amplitudes of the annual cycle and the spectral densities found in the data series. The latter method seems to work since the high wave number part of the original spectral density $P_0(k)$ is almost white, or constant, due to the strong depositional noise (Fisher et al., 1985; Johnsen et al., 1997) normally encountered in high resolution ice core δ series.

The above formulation of the diffusion process in moving firn and ice is valid for any other diffusing constituent, we only need to know the appropriate diffusion constant to be used in equation (5).

3.2 Firn Diffusion

After deposition on the surface several of the constituents in the ice may undergo changes related mostly to diffusional processes. These changes may not be so important for the longer term climatic or environmental information contained in the cores but some of the important annual cycles used for dating may become weakened and will eventually vanish at depth. The smoothing of stable isotope profiles was first observed and discussed by C.C. Langway Jr. in his pioneering work on the NW-Greenland Site II core (Langway, 1967).

During the densification process, in which the firn density is increased with depth, a substantial intergranular exchange of water molecules takes place through the vapor phase in the open interconnected pores. The effective firn diffusion constant D_{eff}, to be used in equation (5), for water molecules in the firn matrix is proportional to the vapor pressure and thus highly dependent on temperature (Johnsen, 1977a; Whillians and Grootes, 1985). It is also inversely proportional to both the tortuosity τ, which accounts for the shape of the open pores, and the fractionation constants α_{18} or α_D, which accounts for the partial pressure of the isotopic species considered (Johnsen et al. in preparation). The full expression for D_{eff}, including the other parameters, then becomes

$$D_{eff} = \frac{m \cdot p \cdot D_i}{R \cdot T \cdot \alpha_i \cdot \tau}\left(\frac{1}{\rho} - \frac{1}{\rho_{ice}}\right) \tag{8}$$

where m is the molar weight of H_2O, p the saturation vapor pressure over ice at the temperature T(k), R the gas constant and D_i the diffusion constant for the particular isotopic species in air at the ambient pressure and temperature. The tortuosity τ is a function of the firn density ρ (Schwander, 1996; Schwander et al., 1988) and ρ_{ice} is the density of glacier ice. Of the two fractionation factors, α_D is dominates and is highly temperature dependent. This implies that comparison of detailed $\delta^{18}O$ and δD profiles measured on the same core increments would help constrain the initial temperatures during deposition of the layers. The strain rates to be used in solving equation (5) can be obtained from the Herron-Langway densification model (Herron and Langway, 1980).

Figure 4 shows the measured $\delta^{18}O$ signal based on 2.5 cm samples for the GRIP core on Summit Greenland and the annual $\delta^{18}O$ amplitudes from 0 to 100 m depth. The amplitudes of the annual cycle are determined in 5 meter long core increments using the MEM spectral analysis technique by integrating analytically the annual peak (Johnsen and Andersen, 1978) in the spectrum. The initial amplitude is close to 5‰ at the surface and is rapidly reduced to 0.4‰ at a depth of 55 meters where the open pores have almost been closed off. This corresponds to a diffusion length of slightly over 8 cm of ice equivalent and implies that each water molecule has on the average been displaced vertically a distance of 8 cm from its position in the original layer.

The firn diffusion lengths found for most Greenland sites (Dansgaard et al., 1973; Hammer et al., 1978; Johnsen, 1977a) are all close to 8 cm, which is due to a general positive correlation between temperature and accumulation in Greenland. These findings are in agreement with recent results of modelling the firn diffusion lengths as a function of temperature and accumulation (Cuffey and Steig, in press). Generally the firn diffusion lengths are a function of both the initial accumulation rate and temperature and can, under favorable conditions, be considered as a length scale frozen into the ice matrix for a very long time. When the diffusion lengths have been determined in a core segment for both $\delta^{18}O$ and δD under such conditions, they can be used to constrain the initial temperature and accumulation rate along with the total thinning of the segment.

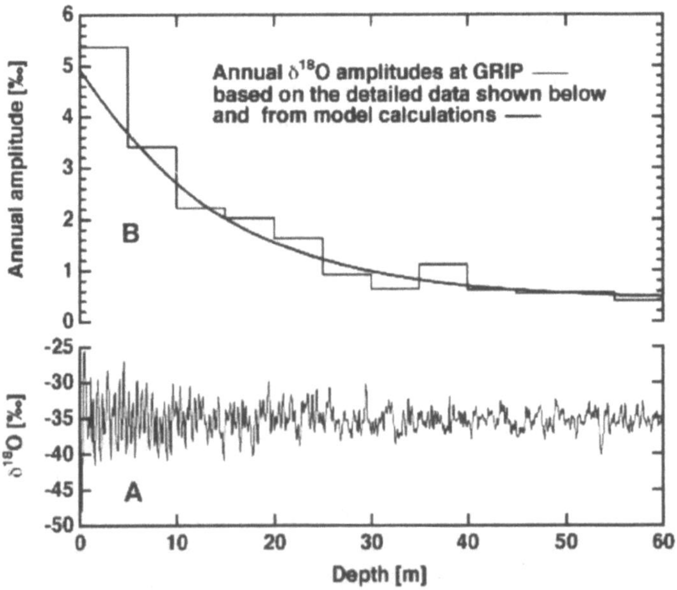

Fig. 4. A. The detailed $\delta^{18}O$ profile from the upper 60 meters from the GRIP Summit deep core. B. The estimated amplitudes of the $\delta^{18}O$ annual cycle for 5 meter core increments. Note the rapid decay of the annual cycle in the firn.

3.3 Ice Diffusion

Below the firn layers normal self diffusion in ice single crystals is expected to take over. In order to solve equation (5) at greater depths we follow a layer from its transformation to ice and use a coupled temperature ice flow model (Johnsen et al., 1995a) to obtain the temperature and strain rate history $\dot{\varepsilon}_z(t)$ of the layer together with the self diffusion constants measured by Ramseier (Ramseier, 1967) from the modelled in-situ temperature.

The model calculations show that the diffusion length after firnification is only affected by vertical thinning in the Holocene part of the core. This implies that the amplitudes for the annual $\delta^{18}O$ cycle should be constant for this period.

MEM power spectra of detailed $\delta^{18}O$ profiles at five depth levels in the Holocene ice are shown in Fig. 5 The damping of the $\delta^{18}O$ signal due to diffusion lowers the spectral densities at higher frequencies down to the flat white noise level of the measuring errors. The spectral peaks of the annual cycles are clearly discernible on the spectral ramp at 1 cycle/yr and are obviously being increasingly attenuated by diffusion with increasing depth, which is at odds with model predictions. Analytical integration of the annual spectral peaks (Johnsen and Andersen, 1978) shows the mean amplitude of the annual $\delta^{18}O$ cycle to be about 5.0‰ at the surface, 0.4‰ at 130 m, 0.24‰ at 725 m and 0.13‰ at 1540 m.

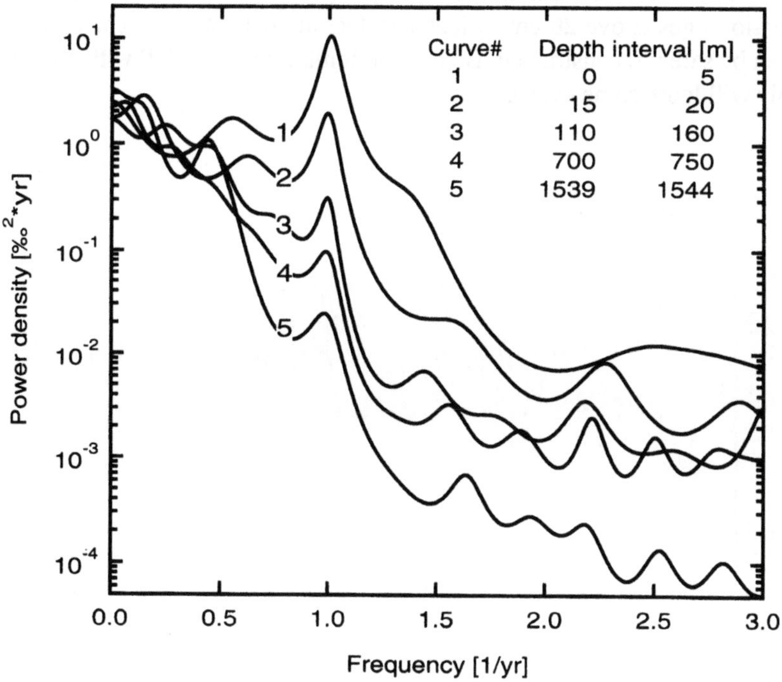

Fig. 5. MEM power spectra for detailed $\delta^{18}O$ profiles demonstrating, for five different levels in the GRIP deep core, the presence of the weak annual $\delta^{18}O$ cycle by the spectral peaks observed at 1 cycle/yr. The effect of smoothing by diffusion is manifested in the drastic lowering of the power densities at frequencies close to 1 cycle/yr.

These results are consistent with a diffusion constant an order of magnitude higher than expected from the work of Ramseier on single crystal diffusion at the low (–32°C) temperature in the GRIP ice. We thus have to conclude that processes other than single crystal diffusion are active in the ice matrix of natural polycrystalline ice. The processes responsible for the observed "excess diffusion", are believed to be a form of crystal boundary ("CB") diffusion (Johnsen et al., in preparation) combined with ongoing recrystallization with crystal rotation and diagonalisation in the ice matrix (Thorsteinsson et al., 1997). Water filled veins at crystal triple junctions are also expected to play an important role in this process (John Nye, personal communication). The vein diffusion process could also expedite diffusion of ionic impurities and help induce chemical reactions as observed by J.P. Steffensen in the GRIP Eemian ice (Steffensen et al., 1997).

In order to make full use of the isotopic diffusion in ice core studies we need to know to what extent the diffusion constants for isotopic diffusion in glacier ice are independent of the isotopic species.

In spite of the strong damping of the annual $\delta^{18}O$ cycle in the Holocene ice of the GRIP core, it is still possible to reconstruct to a sufficient degree the original cycle by using deconvolution techniques based on the inverse of equation (6) (Johnsen, 1977a). Figure 6 shows an example of such reconstruction. The annual cycle is hardly discernible in the raw data, but in the deconvoluted series the annual cycles are clearly visible and can be used for dating the core. The general rule for Greenland is that for accumulation rates above 20 cm of ice/year, like at GRIP with 23 cm/yr, the annual cycles can be safely reconstructed. Below that limit, like at NGRIP with 19 cm/yr, we inevitably will loose some cycles.

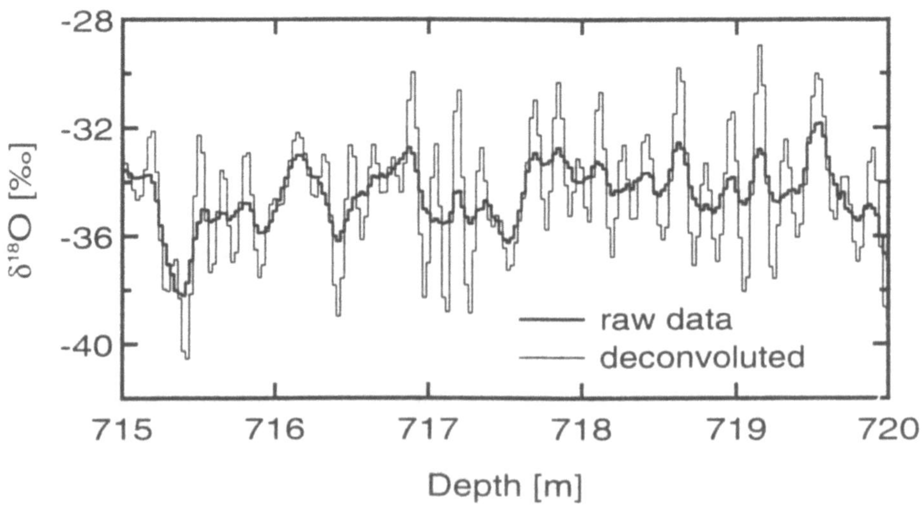

Fig. 6. A detailed $\delta^{18}O$ profile from 715 to 720 m depth is shown as the thick curve. The corresponding deconvoluted profile (thin curve), shows the restored annual $\delta^{18}O$ cycle, which is used for counting annual layers.

4 The Greenland Deep $\delta^{18}O$ Profiles

Continuous $\delta^{18}O$ profiles along sections of four Greenland deep ice cores from Camp Century (Northwest), Dye-3 (Southeast), GRIP (Central), and Renland (East Greenland), spanning nearly the same time interval including the last glacial period, are shown in Fig. 7. The GISP2 profile (Grootes et al., 1993), which is almost identical with the GRIP profile except for the lowest 100 m shown, is not included in this comparison. The four records are all plotted on linear depth scales and extend down to bedrock except for the GRIP record. The silty ice layers sitting immediately above bedrock are indicated for both the Camp Century and Dye-3 profiles.

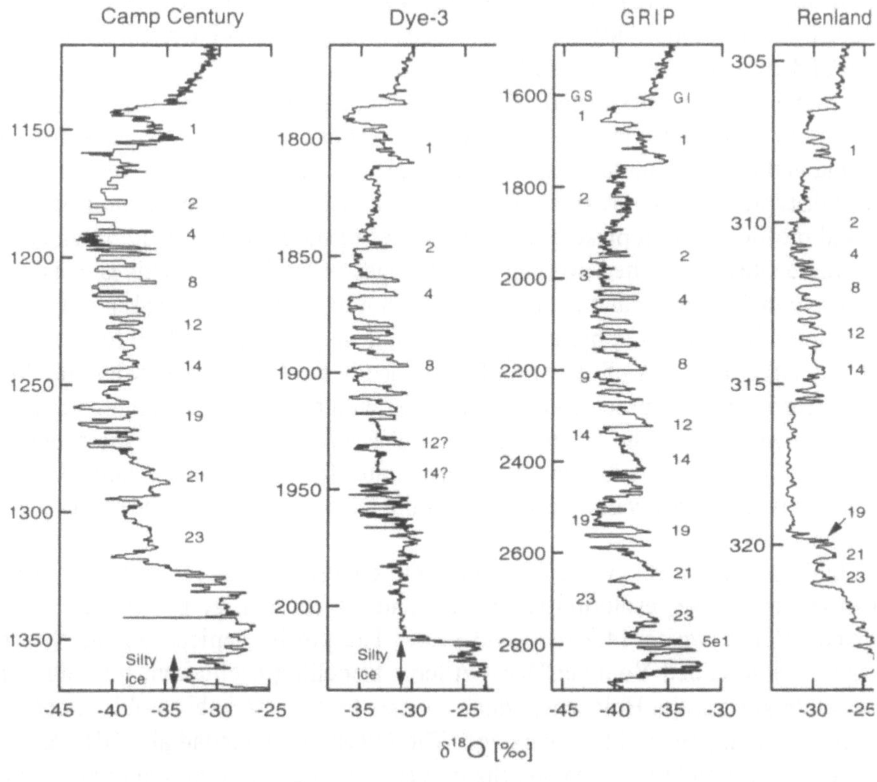

Fig. 7. Continuous $\delta^{18}O$ profiles along sections of four Greenland deep ice cores (located in Fig. 1) from Camp Century (Northwest), Dye-3 (Southeast), GRIP/Summit (Central), and Renland (East Greenland), spanning nearly the same time interval including the last glacial period. The GISP2 profile, which is almost identical with the Summit profile except for the lowest 100 m shown, is not included. The four records are all plotted on linear depth scales and extend down to bedrock except for the GRIP record. The silty ice layers are indicated for both the Camp Century and the Dye-3 core. Some of the warm Greenland Interstadials (GI), as defined from the GRIP $\delta^{18}O$ record are shown by the numbers adjacent to the $\delta^{18}O$ profiles and serve to highlight the correlation between all four records. Some of the cold Greenland Stadials (GS) are also numbered to the left of the GRIP record.

Some of the warm GRIP Interstadials (GI), as defined from the GRIP $\delta^{18}O$ record (Dansgaard et al., 1993), are shown by the numbers adjacent to the $\delta^{18}O$ profiles and serve to highlight the correlation among all four records. Judging from the correlation with the GRIP record the Dye-3 record seems to deteriorate below GI 8 manifesting the effect of the rough bedrock topography in the Dye-3 area (Overgaard and Gundestrup, 1985; Whillans et al., 1984) on the layering in the Dye-3 core. The Camp Century record shows the highest, and the Renland record the lowest, degree of glacial instability, suggesting different source regions for the precipitation falling at these two sites.

The last Renland glacial record (Hansson, 1994; Johnsen et al., 1992), which spans less than 20 m of ice, shows a surprisingly accurate correlation with the more than 1200 m long glacial GRIP record when the 4 m low $\delta^{18}O$ interval above GI 19 is excluded. The low amplitude and blocky shape of the Renland interstadials are still a matter of debate (Johnsen et al., 1992).

The high $\delta^{18}O$ sections situated below GI 23 in the profiles of Fig. 7 are assumed to represent parts of the last interglacial period, the Eemian. Evidently no consensus is to be found in the data about the shape of the Eemian in Greenland. The Dye-3 profile is clearly disrupted and can be ruled out in this discussion. The Camp Century and the Renland profiles could represent the latest stages of the Eemian period provided the ice at these sites had melted away in earlier stages of the Eemian. This is consistent with heavy melt features and an associated high dust content observed in the lowest meter of the Renland core pointing to a rebuilding stage of the Renland Ice Cap. The relatively large thickness of the warm Eemian ice in the Camp Century and the Renland ice cores seems to preclude any correlation with the GRIP Eemian record. If we assume, however, that the Camp Century and Renland Eemian ice was deposited during a build-up phase with low initial strain rates for thinning the Eemian layers a correlation with the last warm phase (5e1) of the GRIP Eemian, as indicated in Fig. 7, seems possible.

A highly detailed $\delta^{18}O$ profile from the GRIP core for the last 20 thousand years is shown in Fig. 8. The resolution is 20 years and the chronology is based on annual layer counting down to 14.7 thousand years. The profile depicts violent climatic changes, the so called Younger Dryas-Alleroed/Boelling oscillation, preceding the present interglacial, the Holocene, which appears to be remarkably stable except for the strong cooling event 8210 years ago. The Greenland Interstadials (GI) are indicated to the right and the Greenland Stadials (GS) to the left in the figure along with obvious substages marked as a,b,c etc.; GS 1 corresponds to the Younger Dryas cold period and GI 1 to the Alleroed/Boelling warm period. Therefore, we propose that the GRIP isotope stadials should be assigned the same number as the preceding interstadial defined in Dansgaard et al. (1993).

The GRIP $\delta^{18}O$ profile has been converted to a climatic temperature signal by using the calibration from Fig. 3. The resulting temperature history is shown in Fig. 9 as deviations from the present Summit temperature corrected for past elevation changes. The correlation to the marine isotope stages (MIS) is marked to the right.

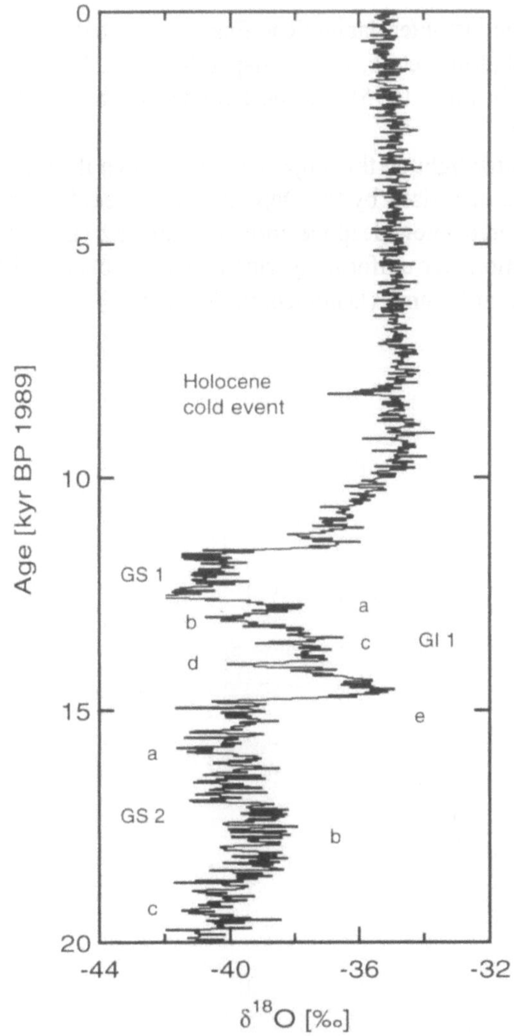

Fig. 8. High resolution $\delta^{18}O$ profile from the GRIP core reaching 20 kyr back in time. This includes the transition from the last glacial (Wisconsin/Weischel) to the present interglacial (Holocene) and illustrates the violent environmental changes taking part in the transition known as the Younger Dryas-Alleroed/Boelling oscillation.

The Holocene climate shows a 3°C cooling over the past 10 thousand years and should not be considered a stable climatic period. This is in good agreement with early Holocene melt feature observations in cores drilled in the Agassiz Ice Cap, NE Canada (Fisher et al., 1995). During the last glacial maximum the Greenland temperatures are found to be 20°C colder than at present and the violent Dansgaard-Oeschger cycles depict up to 15°C temperature shifts. In general it seems that the Greenland climate is unstable on all time scales.

The integrity of the last interglacial (The Eemian denoted as Eem) shown in Fig. 9 is still being debated (Alley et al., 1995; Chappellaz et al., 1997; Johnsen et al., 1997; Johnsen et al., 1995b; Peel, 1995) and the Eemian part of the GRIP core may need further interpretation.

We hope to shed new light on this important issue when the recently started NGRIP deep drilling project, organised by the Department of Geophysics, University of Copenhagen, has recovered a new deep ice core in an area 320 km NNW of Summit (see Fig. 1), where conditions for uniform layering at great depths are believed to be much more favourable than at Summit (Dahl-Jensen et al., 1997).

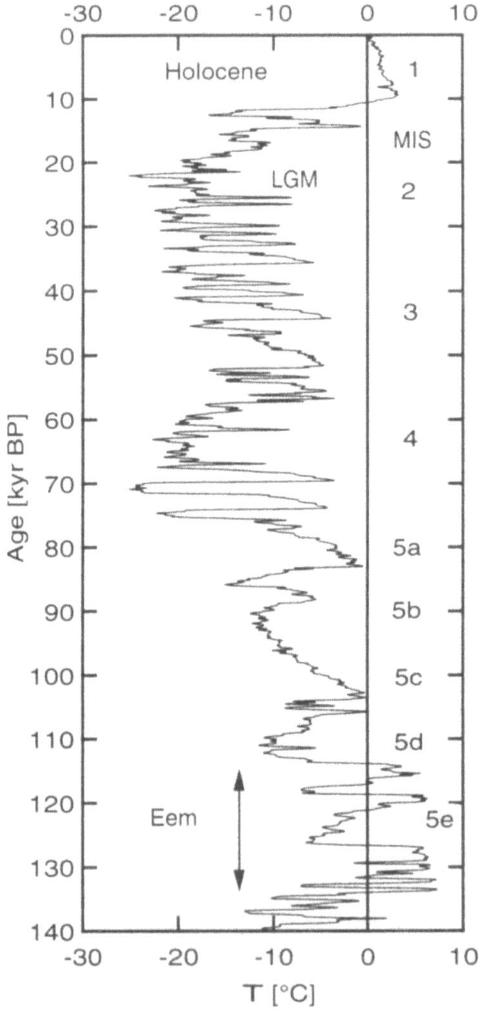

Figure 9. Past temperatures for Greenland based on the GRIP $\delta^{18}O$ and the borehole thermometry calibration discussed in section 2. The Eemian part of the record between 115 kyr and 135 kyr is still being debated.

4 Acknowledgements

This work is a contribution to the Greenland Ice-core Project (GRIP) organised by the European Science Foundation. We thank the GRIP scientists, students and logistic supporters for their co-operative effort. We also thank the national science foundations in Belgium, Denmark, France, Germany, Iceland, Italy, Switzerland and the United Kingdom, as well as the XII Directorate of CEC, the Carlsberg Foundation, the Commission for Scientific Research in Greenland and the University of Iceland Research Foundation for financial support. One of us (JW) was supported by the US National Science Foundation, Office of Polar Programs.

References

Alley, R.B., A.J. Gow, S.J. Johnsen, J. Kipfstuhl, D.A. Meese, and T. Thorsteinsson, (1995), Comparison of deep ice cores, Nature, 373: 393–394.

Alley, R.B., D.A. Meese, C.A. Shuman, A.J. Gow, K.C. Taylor, P.M. Grootes, J.W.C. White, M. Ram, E.D. Waddington, P.A. Mayewski, and G.A. Zielinski, (1993), Abrupt increase in Greenland snow accumulation at the end of the Younger Dryas event, Nature, 362 (6420): 527–529.

Bender, M., T. Sowers, M.-L. Dickson, J. Orchardo, P. Grootes, P.A. Mayewski, and D.A. Meese, (1994), Climate connections between Greenland and Antarctica during the last 100,000 years, Nature, 372: 663–666.

Bond, G., H. Heinrich, W. Broecker, L. Labeyrie, J. McManus, J. Andrews, S. Huon, R. Jantschik, S. Clasen, C. Simet, K. Tedesco, M. Klas, G. Bonani, and S. Ivy, (1992), Evidence for massive discharges of icebergs into the North Atlantic ocean during the last glacial period, Nature, 360: 245–249.

Broecker, W.S., D.M. Peteet, and D. Rind, (1985), Does the ocean-atmosphere system have more than one stable mode of operation?, Nature, 315: 21–26.

Chappellaz, J., T. Blunier, D. Raynaud, J.M. Barnola, J. Schwander, and B. Stauffer, (1993), Synchronous changes in atmospheric CH_4 and Greenland climate between 40 and 8 kyr BP, Nature, 366: 443–445.

Chappellaz, J., E. Brook, T. Blunier, and B. Malaizé, (1997), CH_4 and $\delta^{18}O$ of O_2 records from Antarctic and Greenland ice: A clue for stratigraphic disturbance in the bottom part of the Greenland Ice Core Project and Greeland Ice Sheet Project 2 ice cores, Journal of Geophysical Research, 102 (C12): 26,547–26,557.

Clausen, H.B., (1995), Group Meeting on Nitrate Sources in Antarctica and Greenland, in Ice Core Studies of Global Biogeochemical Cycles, edited by R.J. Delmas, pp. 247–248, Springer-Verlag Berlin Heidelberg, Annecy, France.

Clausen, H.B., C.U. Hammer, J. Christensen, C. Schøtt Hvidberg, D. Dahl-Jensen, M. Legrand, and J.P. Steffensen, (1995), 1250 Years of Global Volcanism as Revealed by Central Greenland Ice Cores, in Ice Core Studies of Global Biogeochemical Cycles, edited by R.J. Delmas, pp. 175–194, Springer-Verlag Berlin Heidelberg, Annecy, France.

Cuffey, K.M., R.B. Alley, P.M. Grootes, J.M. Bolzan, and S. Anandakrishnan, (1994), Calibration of the $\delta^{18}O$ isotopic paleothermometer for central Greenland, using borehole temperatures, Journal of Glaciology, 40 (135): 341–349.

Cuffey, K.M., and G.D. Clow, (1997), Temperature, accumulation and ice sheet elevation in central Greenland through the last deglacial transition, Journal of Geophysical Research, 102 (C12): 26,383–26,396.

Cuffey, K.M., G.D. Clow, R.B. Alley, M. Stuiver, E.D. Waddington, and R.W. Saltus, (1995), Large Arctic Temperature Change at the Wisconsin-Holocene Glacial Transition, Science, 270: 455–458.

Cuffey, K.M., and E.J. Steig, (in press), Isotopic diffusion in polar firn: implications for interpretation of seasonal climate parameters in ice core records, with emphasis on central Greenland, Journal of Glaciology: .

Dahl-Jensen, D., N.S. Gundestrup, K.R. Keller, S.J. Johnsen, S.P. Gogineni, C.T. Allen, T.S. Chuah, H. Miller, J. Kipfstuhl, and E.D. Waddington, (1997), A search in north Greenland for a new ice-core drill site, Journal of Glaciology, 43 (144): 300–306.

Dahl-Jensen, D., and S.J. Johnsen, (1986), Palaeotemperatures still exist in the Greenland ice sheet, Nature, 320 (6059): 250–252.

Dansgaard, W., (1964), Stable isotopes in precipitation, Tellus, 16: 436–468.

Dansgaard, W., S.J. Johnsen, H.B. Clausen, D. Dahl-Jensen, N.S. Gundestrup, C.U. Hammer, C.S. Hvidberg, J.P. Steffensen, A.E. Sveinbjörnsdottir, J. Jouzel, and G. Bond, (1993), Evidence for general instability of past climate from a 250-kyr ice-core record, Nature, 364 (6434): 218–220.

Dansgaard, W., S.J. Johnsen, H.B. Clausen, and N. Gundestrup, (1973), Stable isotope glaciology, pp. 1–53.

Fisher, D.A., R.M. Koerner, and N. Reeh, (1995), Holocene climatic records from Agassiz Ice Cap, Ellesmere Island, NWT, Canada, The Holocene, 5 (1): 19–24.

Fisher, D.A., N. Reeh, and H.B. Clausen, (1985), Stratigraphic noise in time series derived from ice cores, Annals of Glaciology, 7: 76–83.

Fuhrer, K., A. Neftel, M. Anklin, and V. Maggi, (1993), Continuous measurements of hydrogen peroxide, formaldehyde, calcium and ammonium concentrations along the new GRIP ice core from Summit, Central Greenland, Atmospheric Environment, 27A: 1873–1880.

Grootes, P.M., M. Stuiver, J.W.C. White, S.J. Johnsen, and J. Jouzel, (1993), Comparison of oxygen isotope records from the GISP2 and GRIP Greenland ice cores, Nature, 366 (6455): 552–554.

Hammer, C.U., (1977), Dust studies on Greenland ice cores, in Proc. of Symp. on Isotopes and Impurities in Snow and Ice, Int. Ass. of Hydrol. Sci., Commission of Snow and Ice, I.U.G.G. XVI, General Assembly, Grenoble Aug.–Sept. 1975, pp. 365–370, IAHS-AISH Publ.

Hammer, C.U., K.K. Andersen, H.B. Clausen, D. Dahl-Jensen, C. Schøtt Hvidberg, and P. Iversen, (1997 (in press)), The Stratigraphic Dating of the GRIP Ice Core, Department of Geophysics, Niels Bohr Institute of Astronomy, Physics and Geophysics, University of Copenhagen, Copenhagen.

Hammer, C.U., H.B. Clausen, and W. Dansgaard, (1980), Greenland ice sheet evidence of post-glacial volcanism and its climatic impact, Nature, **288**: 230–235.

Hammer, C.U., H.B. Clausen, W. Dansgaard, N. Gundestrup, S.J. Johnsen, and N. Reeh, (1978), Dating of Greenland ice cores by flow models, isotopes, volcanic debris, and continental dust, Journal of Glaciology, 20: 3–26.

Hammer, C.U., H.B. Clausen, and C.C. Langway, Jr., (1997), 50,000 Years of recorded global volcanism, Climatic Change **35**: 1–15.

Hansson, M.E., (1994), The Renland ice core. A Northern Hemisphere record of aerosol composition over 120,000 years, Tellus, 46B: 390–418.

Herron, M.M., and C.C. Langway, Jr., (1980), Firn densification: an empirical model, Journal of Glaciology, 25 (93): 373–385.

Johnsen, S., D. Dahl-Jensen, W. Dansgaard, and N. Gundestrup, (1995a), Greenland palaeotemperatures derived from GRIP bore hole temperature and ice core isotope profiles, Tellus, 47B: 624–629.

Johnsen, S.J., (1977a), Stable isotope homogenization of polar firn and ice, in Proc. of Symp. on Isotopes and Impurities in Snow and Ice, Int. Ass. of Hydrol. Sci., Commission of Snow and Ice, I.U.G.G. XVI, General Assembly, Grenoble Aug.–Sept., 1975, pp. 210–219, IAHS-AISH Publ.

Johnsen, S.J., (1977b), Stable isotope profiles compared with temperature profiles in firn and with historical temperature records, in Proc. of Symp. on Isotopes and Impurities in Snow and Ice, Int. Ass. of Hydrol. Sci., Commission of Snow and Ice, I.U.G.G. XVI, General Assembly, Grenoble Aug.–Sept., 1975, pp. 388–392, IAHS-AISH Publ.

Johnsen, S.J., and N. Andersen, (1978), On power estimation in maximum entropy spectral analysis, Geophysics, 43: 681–690.

Johnsen, S.J., H.B. Clausen, W. Dansgaard, N.S. Gundestrup, C.U. Hammer, U. Andersen, K.K. Andersen, C.S. Hvidberg, D. Dahl-Jensen, J.P. Steffensen, H. Shoji, Á.E. Sveinbjörnsdóttir, J. White, J. Jouzel, and D. Fisher, (1997), The $\delta^{18}O$ record along the Greenland Ice core Project deep ice core and the problem of possible Eemian climatic instability, Journal of Geophysical Research, 102 (C12): 26,397–26,410.

Johnsen, S.J., H.B. Clausen, W. Dansgaard, N.S. Gundestrup, C.U. Hammer, and H. Tauber, (1995b), The Eem Stable Isotope Record along the GRIP Ice Core and Its Interpretation, Quaternary Research, 43: 117–124.

Johnsen, S.J., H.B. Clausen, W. Dansgaard, N.S. Gundestrup, M. Hanson, P. Jonson, J.P. Steffensen, and A.E. Sveinbjörnsdottir, (1992), A "deep" ice core from East Greenland, pp. 22.

Johnsen, S.J., W. Dansgaard, H.B. Clausen, and C.C. Langway, Jr., (1972), Oxygen isotope profiles through the Antarctic and Greenland ice sheets, Nature, 235 (5339): 429–434.

Johnsen, S.J., W. Dansgaard, and J.W.C. White, (1989), The origin of Arctic precipitation under present and glacial conditions, Tellus, 41B: 452–468.

Jouzel, J., R.B. Alley, K.M. Cuffey, W. Dansgaard, P. Grootes, G. Hoffmann, S.J. Johnsen, R.D. Koster, D. Peel, C.A. Shuman, M. Stievenard, M. Stuiver, and J.

White, (1997), Validity of the temperature reconstruction from water isotopes in ice cores, Journal of Geophysical Research, 102 (C12): 26,471–26,487.
Langway, C.C., Jr., (1967), Stratigraphic analysis of a deep ice core from Greenland, : pp. 130.
Legrand, M.R., and R.J. Delmas, (1988), Soluble impurities in four Antarctic ice cores over the last 30 000 years, Annals of Glaciology, 10: 1–5.
Loewe, F., (1936), The Greenland ice cap as seen by a meteorologist, Q. J. R. Met. Soc., 62: 359–377.
Lorius, C.J., (1989), Polar ice cores and climate, in Climate and Geo-Sciences, edited by A. Berger, and et al., pp. 77–103, Kluwer Academic Publishers.
Mayewski, P.A., L.D. Meeker, M.C. Morrison, M.S. Twickler, S.I. Whitlow, K.K. Ferland, D.A. Meese, M. Legrand, and J.P. Steffensen, (1993), Greenland Ice Core "Signal" Characteristics: An Expanded View of Climate Change, Journal of Geophysical Research, 98 (D7): 12,839–12,847.
Meese, D.A., A.J. Gow, R.B. Alley, G.A. Zielinski, P.M. Grootes, M. Ram, K.C. Taylor, P.A. Mayewski, and J.F. Bolzan, (1997), The Greenland Ice Sheet Project 2 depth-age scale: Methods and results, Journal of Geophysical Research, 102 (C12): 26,411–26,423.
Oeschger, H., J. Beer, U. Siegenthaler, B. Stauffer, W. Dansgaard, and C.C. Langway, Jr., (1984), Late glacial climate history from ice cores, in Climate Processes and Climate Sensitivity. Geophysical monograph 29, Maurice Ewing, edited by J.E. Hansen, and T. Takahashi, pp. 299–306.
Overgaard, S., and N. Gundestrup, (1985), Bedrock topography of the Greenland ice sheet in the Dye 3 area, in Greenland Ice Cores: Geophysics, Geochemistry and Environment, edited by C.C. Langway, Jr., H. Oeschger, and W. Dansgaard, pp. 49–56, American Geophysical Union (AGU) Monograph.
Peel, D.A., (1995), Profiles of the past, Nature, 378: 234–235.
Petit, J.-R., M. Briat, and A. Royer, (1981), Ice age aerosol content from East Antarctic ice core samples and past wind strength, Nature, 293 (5831): 391–394.
Ramseier, R.O., (1967), Self-diffusion of tritium in natural and synthetic ice monocrystals, Journal of Applied Physics, 38: 2553–2556.
Raynaud, D., J. Jouzel, J.M. Barnola, J. Chappellaz, R.J. Delmas, and C. Lorius, (1993), The Ice Record of Greenhouse Gases, Science, 259: 926–934.
Schwander, J., (1996), Gas diffusion in firn, in Chemical Exchange Between the Atmosphere and Polar Snow, edited by E.W. Wolff, and R.C. Bales, pp. 527–540, Springer-Verlag Berlin Heidelberg, Il Ciocco, Italy.
Schwander, J., B. Stauffer, and A. Sigg, (1988), Air mixing in firn and the age of the air at pore close-off, Annals of Glaciology, 10: 141–145.
Sowers, T., and M. Bender, (1995), Climate Records Covering the Last Deglaciation, Science, 269: 210–214.
Steffensen, J.P., H.B. Clausen, C.U. Hammer, M. Legrand, and M. De Angelis, (1997), The chemical composition of cold events within the Eemian section of the Greenland Ice Core Project ice core from Summit, Greenland., Journal of Geophysical Research, 102 (C12): 26,747–26,754.

Thorsteinsson, T., J. Kipfstuhl, and H. Miller, (1997), Textures and fabrics in the GRIP ice core, Journal of Geophysical Research, 102 (C12): 26,583–26,599.

Whillans, I.M., K.C. Jezek, A.R. Drew, and N. Gundestrup, (1984), Ice flow leading to the deep core hole at Dye 3, Greenland, Annals of Glaciology, 5: 190–195.

Whillians, I.M., and P.M. Grootes, (1985), Isotopic Diffusion in Cold Snow and Firn, Journal of Geophysical Research, 90 (D2): 3910–3918.

Present and Past Glaciations: A Geological Perspective

Krzysztof Birkenmajer

Institute of Geological Sciences, Polish Academy of Sciences,
Senacka 3, 31-002 Kraków, Poland

Abstract. The great global glaciations were a cyclic though rather infrequent phenomenon in the history of the Earth. The Late Palaeozoic (330-270 Ma), the Early Palaeozoic (about 450 Ma), probably also the Late Proterozoic (Vendian: about 600 Ma) glaciations, affected large continents in high latitude south polar position. Asynchronous glaciation affected the southern and the northern continents during the Cenozoic. In Antarctica, it started during Eocene (at about 50 Ma), its ice-cap at sea-level developed in Early Oligocene (at 32-30 Ma), and became a semi-permanent feature of this continent since Early Miocene (22-20 Ma). In the Arctic, the glaciation started much later, during Pliocene (at about 3.4 Ma). Its Pleistocene (2-0.1 Ma) ice-sheets developed around a relatively small Arctic Ocean in northern North America and Eurasia, and in Greenland. Presently, the Arctic glaciation is reduced to a few much reduced ice-caps, the Greenland and the Svalbard ones being the largest. There is a direct correlation between global glaciations and world-ocean level recognizable in Cenozoic marine geological record: low stands of sea level correspond to glacial epochs, while high stands to interglacials. Global-scale glaciations have both terrestrial and extraterrestrial causes.

Keywords. Glaciations, Late Precambrian to Recent, global impact.

1 Recent Glaciation

The recent glaciation of the Earth includes the Antarctic and the Greenland ice-caps located in high latitude position, the ice-shelves (mainly in Antarctica), a number of smaller ice-caps and ice-domes, and a variety of mountain glaciers. Total area covered by glacier-ice is calculated at between 15.9 and 16.3 million square kms, i.e., about 11% of the total land area. Of this, the Antarctic ice-cap and ice-shelves represent 85.7%, and the Greenland ice-cap - 10.9% of the total glaciated area (Jania, 1993; Hambrey & Alean, 1992).

The role of glaciers in reshaping the Earth's surface depends mainly on their temperature - cold *versus* temperate glaciers, and availability of running water during the warmer season. Effects of erosion by glaciers on modifying and deepening pre-existing river-valley systems, and on sharpening mountain reliefs, are apparent in

NATO ASI Series, Vol. I 56
Ice Physics and the Natural Environment
Edited by John S. Wettlaufer, J. Gregory Dash and Norbert Untersteiner

every high mountain range of the world. Levelling of the mountains is a characteristic feature of erosional activity by ice-caps.

There are substantial differences between glaciation of the southern and the northern polar regions. In the south, Antarctica is a sole continent in polar position. It is surrounded by large oceans, and is screened from warm equatorial currents at about 60°S by the Antarctic Convergence. In the north, a relatively small but deep Arctic Ocean is surrounded by continents, its waters communicating with the Atlantic and the Pacific oceans by narrow straits, the Nansen Strait and the Bering Strait, respectively.

There are no large rivers on land in the Antarctic. However, subglacial drainage of its enormous ice shelves (e.g., Ronne-Filchner; Ross) supplies large amount of cold meltwater to adjoining seas. In the Arctic, there is a steady flow of freshwater supplied to the surface of the Arctic Ocean by large Siberian and Canadian rivers.

1.1 Mountain Glaciers

Glaciation of mountain ranges depends mainly on their altitude, geographical position and patterns of atmospheric circulation (availability of precipitation). About two-thirds of the Earth's small glaciers are located in the Arctic. Owing to their quick response to climatic warming or cooling, their status may be used as a reliable indicator of regional and global climatic change (Jania & Hagen, 1996).

Though mountain glaciers represent only 1.3% of the total glaciated area on the Earth, their role in reshaping its surface is obvious to everybody entering high mountain ranges at nearly all latitudes. Deep U-shaped valleys, rimmed by lateral moraine ridges, arcuate marginal moraines, glacial lakes and roches moutonnées, are among characteristic signs of past extension, erosional activity and accumulation by the glaciers.

1.2 Ice-caps and Ice-shelves

Large ice-caps are located only in high-latitude polar positions. The Antarctic ice-cap and its ice-shelves occupy an area of about 13.9 million km^2 (85.7% of the total glaciated area of the Earth). The Greenland ice-cap occupies about 1.8 million km^2 (10.9% of the total glaciated area of the Earth). Smaller ice-caps and ice-domes (mainly in the Arctic) cover about 2.1% of the total glaciated area (Jania, 1993).

About 98-99% of the Antarctic continent is covered by ice, its mean thickness being 2000-2500 m, with maxima 4800 m in East Antarctica, and over 4250 m in West Antarctica. The ice-cap forms a gently sloping plateau at 2000, rising to about 4200 m a.s.l. It is dissected by high mountain ranges (e.g., the Transantarctic Mts, the Ellsworth Mts and the Antarctic Peninsula ranges) and by numerous nunataks. Large glaciers and icestreams coalesce to form grounded or floating ice shelves (e.g., Ronne-Filchner; Ross; Amery; Larsen) 200-1300 m thick. The yearly production of icebergs in Antarctica amounts to 57 x 10^{10} tonnes of freshwater which returns to the sea (Bentley, 1964; Lovering & Prescottt, 1979). Mean thickness of the Greenland ice-cap amounts to 1530 m, its volume amounts to 2.6 x 10^6 km^3 (Jania, 1993).

2 Tertiary to Quaternary Glaciations

2.1 Arctic

The present glaciation of the Arctic is small compared with that during the Pleistocene epoch (2-0.01 Ma). During that epoch, four major - or six according to some authors - ice-sheets invaded Europe and North America, their deposits being most readily studied at the southern peripheries of the glaciated areas.

The Scandinavian ice-cap was connected northward across Barents Sea (ice-shelf or ice-sheet) with smaller ice-caps of Svalbard and Franz-Joseph Lands (e.g., Mangerud *et al.*, 1996). Southward, the Scandinavian ice-cap expanded across North Sea to cover British isles, and across the Baltic Sea over a large area of continental Europe. The oldest Scandinavian ice-sheet reached in foothills of the Carpathians and the Sudeten mountains in Poland, the three younger glaciations (Elsterian, Saalian and Vistulian) had successively smaller and smaller extensions (Lindner *et al.*, 1983).

2.2 Antarctica

Paradoxically enough, in Antarctica, there is much better evidence for the history and extent of the Tertiary, than the Pleistocene glaciations. Most of geological evidence for past glaciations is hidden under the enormous Antarctic ice-cap, and only 1-2% of the continent is exposed to direct geological study. Evidence for past glaciations is derived mainly from continental-glacial and glacio-marine deposits exposed on offshore islands, and ice-free margins of the continent, and in some mountain ranges and nunataks emerging from the ice (Birkenmajer, 1990; Moriwaki *et al.*, 1992).

Subglacial and submarine evidence for Cenozoic climatic changes in Antarctica has so far been available only from scarce drillholes in continental and shelf areas (Barrett *et al.*, 1989; Hambrey & Barrett, 1993). The DSDP and the ODP programmes (Kennett, 1977; Kennett & Barker, 1990), and seismoacoustic studies (Cooper *et al.*, 1995), yielded data on geological history of seas surrounding the Antarctic continent. The latter data, in most cases, still need better geological time calibration by palaeontological and radiometric dating.

An example of rather continuous record of the Tertiary glaciations in West Antarctica is provided by coastal exposures on King George Island (South Shetland Islands), at about 62°S. Four cold (glacial) and three warm (interglacial) climatic epochs, spanning some 30 Ma from Eocene through Miocene have been recognized (Birkenmajer, 1990, 1996a, b).

(1) The oldest Kraków Glaciation (Early/Middle Eocene, about 50 Ma) is evidenced by fossiliferous shallow-water glacio-marine deposits. The glacial centre with valley glaciers descending to the sea was probably located in a southern part of Antarctic Peninsula (about 70°S) at about 2000 m a.s.l.

(2) During the Arctowski Interglacial (Middle Eocene - Early Oligocene, 50-32 Ma), rich warm and moist climate *Nothofagus*-pteriodophyte vegetation developed on King George Island. Ice-caps covered only tops of higher volcanoes.

(3) The Polonez Glaciation (late Early Oligocene, 32-30 Ma) was the largest Cenozoic glacial epoch in Antarctica. The West Antarctic ice-cap crossed Bransfield Strait and reached King George Island. Lodgement till deposition was followed by submarine diamictites and by shallow-marine, glacially-controlled fossiliferous strata with strong contemporaneous volcanic contribution.

(4) During the Wesele Interglacial (mid-Oligocene, at about 30 Ma), the continental ice-cap retreated to Antarctic Peninsula. Bransfield Strait and King George Island were dry land devoid of ice cover, being dissected by a primitive river system.

(5) The Legru Glaciation (Late Oligocene, 30-26 Ma) was a local glaciation of King George Island, disconnected from continental ice-sheet. Laharic agglomerates alternating with andesitic/basaltic lavas were followed by valley glacier tills.

(6) The Wawel Interglacial (Oligocene/Miocene transition, 26-22 Ma) began with recolonization of ice-free land area by temperate (cool and warm) *Nothofagus* rain forests. A shallow marine incursion followed.

(7) During the Melville Glaciation (Early Miocene, at 22-20 Ma), glacially-controlled fossiliferous marine deposits, corresponding to outer shelf/upper slope of a marginal sea, were laid down.

The oldest direct evidence for glaciation in continental Antarctica is from the Lambert Glacier/Amery Ice Shelf (Prydz Bay) area. This glaciation is tentatively dated at between Late-Middle Eocene (Hambrey *et al.*, 1989) and about 40 Ma (Moriwaki *et al.*, 1992). In the first case, it could correlate with the Kraków Glaciation, in the second - with a cold period not yet recognized within the Arctowski Interglacial.

The Early/Late Oligocene transition, about 30-32 Ma, was the time of ice-sheet development in West and East Antarctica (Barrett *et al.*, 1989; Barrett, 1991; Moriwaki *et al.*, 1992). Significant cooling of seawater at about 30 Ma recorded on oxygen-isotope curves obtained from deep-sea cores in south Pacific (Miller *et al.*, 1987), and at the beginning of Oligocene in Weddell Sea (Kennett & Barker, 1990), correlates with the Polonez Glaciation on King George Island (Birkenmajer, 1996a, b). In Ross Embayment, marine marine drillholes recorded Early Oligocene (about 36 Ma) marine strata containing debris from glacier ice calving at sea level (Barrett *et al.*, 1989).

Between 30 and 20 Ma, a major ice build-up is postulated for the Weddell Sea and the Ross Sea regions, and a glacial maximum for the Prydz Bay-Prince Charles Mountains region in East Antarctica. Several interglacial-type epochs of deglaciation or warm condition, were separated by glacial maxima (Barrett, 1991; Moriwaki *et al.*, 1992); they might correlate with mid- to late Oligocene glacial/interglacial epochs on King George Island.

Since Early Miocene (22-20 Ma), the ice-sheet in continental Antarctica became a semi-permanent feature. There are evidences for glacier advances and retreats, even for interglacial-type warming epochs (during Pliocene). However, the succession and ages of the Middle Miocene through Pleistocene events are still inadequately known.

3 Global Impact of Continental Glaciations

Studies of Tertiary glacially-controlled sedimentary sequences in Antarctica provide one of the best means for assessing global impact of continental glaciation (Webb, 1990; Barrett, 1991). The impact is manifold. It affects climate and atmospheric circulation; temperature, salinity and circulation of marine water; distribution and evolution of marine and terrestrial biota. It causes fluctuation of world ocean water level. Loading/unloading of continental crust by ice-cap causes isostatic down-/upwarping of glaciated areas, etc.

3.1 Circum-Antarctic Current

According to Kennett (1977), the Circum-Antarctic Current which presently screens Antarctica from warm equatorial currents, at about 60°S, had formed as a result of continental dispersion and opening of the Scotia Seaway at about 25 Ma. The history of climatic changes recorded from King George Island suggests a younger date, 22-20 Ma, as based on the Early Miocene Melville Glaciation. At that time, the Scotia Sea was already wide and deep enough to serve as a way for intermixing on a large scale the Atlantic and Pacific (Birkenmajer, 1996a). Since 22-20 Ma, the Circum-Antarctic Current became a semi-permanent feature of the southern oceans, disappearing during the interglacial, and reappearing during the glacial epochs.

3.2 Glaciation and Sea-level Changes

Accumulation of ice in form of ice-sheets/caps and ice-shelves during global-scale glaciations results in lowering of world-ocean level. Melting of ice-caps at the end of a glaciation causes rising of sea level and marine transgression. The ocean both forces and responds to ice-sheet evolution. From palaeoclimate records, it is known that global sea level has risen at an average rate of ca 6 mm/yr over the past 20,000 years. The average rate of global sea-level rise over the past century has been 1-2 mm/yr, with higher values (2.3-2.4 mm/yr) over the past 50 years (Jacobs, 1990). The Antarctic ice-sheet, with its volume of some 24.5 million km^3, is the largest storehouse of freshwater on the Earth, representing approximately 90% of the world's freshwater reserves. If it melted completely, the world-ocean level would rise by about 60 m (Lovering & Prescott, 1979).

In West Antarctica, there is a good correlation between the Eocene (Kraków) and the Early Oligocene (Polonez) glaciations and low stands of world ocean level as recorded on eustatic curve of Haq *et al.* (1987). The character and succession of glaciomarine strata formed during the Polonez Glaciation (30-32 Ma), indicate that sea-level rapidly fell at the beginning of the glaciation when the continental ice-sheet crossed Bransfield Strait and reached King George Island. It had risen by about 20 m at the end of the glaciation, when the ice-sheet receded back to the continent. There is no good correlation between low stands of world ocean level and the Late Oligocene

(Legru) and the Early Miocene (Melville) glaciations. This was the time of opening of the back-arc Bransfield Rift, and of up- and down-warping of tectonic blocks in subduction-controlled magmatic arc of the South Shetland Islands in West Antarctica (Birkenmajer, 1996a, b).

3.3 Glacio-isostasy

It is supposed that bedrock surface of the Antarctic ice-cap was isostatically lowered down by about 600 m due to ice load. The continental shelves of Antarctica are lowered down to about 500 m below the sea, a value twice that of the continental shelf depths elsewhere in the world. The Holocene land uplift in coastal Antarctica as a result of ice-volume reduction, was small. It is calculated at between 54 m (John & Sugden, 1971) and 30 m (Pallàs *et al.*, in press). The total melting of Antarctic ice-cap would raise the bedrock surface by about 960 m (Drewry, 1983).

Glacio-isostatic models of Holocene land uplift following decay of the young Pleistocene ice-caps, are known from Canada, Svalbard and Europe, the Fennoscandian model being the classic one (e.g., Andrews, 1970; Andrews, ed., 1974; Embleton & King, 1975). The present rate of land-uplift in Fennoscandia suggests that still more than 200 m of uplift in the Bothnia Bay (Baltic Sea) would be needed to reach the isostatic equilibrium. The Late Pleistocene down-warping of the area is calculated at 700-750 m that correlates with ice-cap thickness of at least 2300 m.

3.4 Climate

Past variations of temperature and chemistry of the atmosphere are most readily reconstructed from ice-records extending for tens and even hundreds of thousands of years in ice cores recovered from large ice-caps, such as Greenland and Antarctica.

The major cycles thought to affect Earth's climates are related to: (i) the length of the cosmic year and to the position of the solar system relative to the Magellanic clouds (period between ca 1.5 and 5 x 10^8 yrs); (ii) eccentricity of earth's orbit around the Sun (period ca 9.3 x 10^4 yrs); (iii) precession of the equinoxes (period ca 2.1 x 10^4 yrs); (iv) obliquity of the ecliptic of the Earth (period ca 4.1 x 10^4 yrs). The latter cycle is very significant for high latitude zones, including Antarctica (Frakes, 1982). These cycles are, however, usually very difficult to identify in geological record, particularly in the pre-Quaternary ones.

4 Pre-Cenozoic Glaciations

There is a good geological record of pre-Cenozoic glaciations, expressed, among others, by marks left by moving glaciers/ice-caps on the bedrock, by terrestrial glacial and glacio-marine deposits, and by characteristic fauna and flora typically associated with cold climates (Schwarzbach 1974; Frakes, 1979; Chumakov, 1985). In the

Earth's history, the great global glaciations were a cyclic though rather infrequent phenomenon. There are evidences for large-scale glaciations since Precambrian, with the most important ones recognized in Late Proterozoic (Vendian or Varangian glaciation), in Early Palaeozoic (Late Ordovician-Early Silurian), and in Late Palaeozoic (Late Carboniferous-Early Permian) deposits.

4.1 Late Palaeozoic Glaciation of Gondwanaland

The great Late Palaeozoic glaciation covered the supercontinent of Gondwanaland prior to its Mesozoic-Cenozoic break-up and dispersion. Palaeomagnetic evidence locates centres of glaciation in a south polar position, similar to that of the recent Antarctica. Enormous area of Gondwanaland combined the present South American, African, Antarctic and Australian continents, the Indian subcontinent, and a number of smaller lands. The landmass extended far to the north reaching temperate and equatorial areas of the Late Palaeozoic Earth. There was no Antarctic-type convergence, and the cold polar waters mixed freely with warm equatorial ones.

During a minimum of 60 million years duration of the Gondwanaland glaciation (Late Carboniferous-Early Permian, 330-270 Ma), there were no less than nine glacier advances/glaciations separated by warm epochs (interglacials). The growth and decay of south-polar ice-sheets caused fluctuations of world ocean level (Wanless & Shepard, 1936; King, 1961), probably of the order of about several hundred metres (Crowell, 1978). These fluctuations were recorded in coeval shallow marine sequences as far north as Svalbard (presently between 75° and 80°N) whose Early Permian position was 30-35°N (Birkenmajer, 1964). The impact of the Gondwanaland glaciation on marine and atmospheric circulation, and on dispersion of terrestrial and marine biota was without parallel in the Earth's history.

4.2 Early Palaeozoic Glaciation

Geological evidence for a Late Ordovician-Early Silurian glaciation (at about 450 Ma) was first found in North Africa. Other evidences were supplied mainly from South Africa and South America (Hambrey & Harland, eds, 1981). In plate-tectonic reconstruction, the African and South American continents are positioned south of latitude 60°S (Chumakov, 1985),

4.2 Late Proterozoic Vendian Glaciation

The Late Proterozoic Vendian (Varangian) glaciation, dated at about 600 Ma, covered an area of central and northern Europe, Svalbard and East Greenland (Hambrey & Harland, eds, 1981; Harland *et al.,* 1993). This part of the present Eurasian continent was then probably also located in a south-polar position.

5 Concluding Remarks

1. Glaciers are a powerful factor in reshaping the Earth's surface, remodelling its atmosphere and hydrosphere, and profoundly affecting the biosphere.

2. Glaciers of all types are among the most sensitive instruments of Nature immediately responding to and registering regional and global climatic change, its cooling or warming. For the International Geosphere-Biosphere Project Global Change, they naturally become one of the major fields of coordinated scientific research in an attempt to construct models of present and future climatic trends on the Earth.

3. Past variations of temperature and chemistry of the atmosphere are most readily reconstructed from ice-records extending for tens and even hundreds of thousands of years in ice cores recovered from large ice-caps, such as Greenland and Antarctica.

4. Global-scale glaciations, represented by continental ice-sheets, ice-caps, and marine ice-shelves, have both terrestrial and extraterrestrial causes. An important prerequisite is presence of a large landmass(es) in polar or high latitude position.

5. The great global glaciations were a cyclic though rather infrequent phenomenon in the history of the Earth, the oldest one recognized already at the dawn of life on our planet (Late Proterozoic Vendian Glaciation) about 600 million years ago.

6. The great Carboniferous-Permian glaciation (330-270 Ma) covered the supercontinent of Gondwanaland prior to its Mesozoic-Cenozoic dispersion.

7. The Cenozoic glaciation started in Antarctica already during Eocene at c. 50 Ma, and the ice-cap at sea-level developed there in Early Oligocene, at 32-30 Ma. There is a good evidence for instability of Antarctic ice-sheet, with glacial and interglacial epochs alternating, during the Tertiary. As yet, there is a poor terrestrial record of ice-cap fluctuation on Antarctic continent during the Quaternary (2-0 Ma).

8. The Cenozoic glaciation in the Arctic started much later, during late Pliocene (at 3.4 Ma). Its full development during the Pleistocene (2-0.01 Ma), with cyclically repeating glaciations and interglacials, is well recorded from Eurasia and North America. The present stage (Holocene, 0.01 Ma duration thus far) of glaciation in the Arctic might well represent an interglacial.

9. There is a direct correlation between global glaciation and world-ocean level. Low stands of ocean level during the Tertiary, as recognized in Antarctica, correlate with epochs of glaciation, while high-stands of ocean-level - with interglacial epochs of ice-cap decay on the continent.

10. A similar correlation exists between Quaternary glacial/interglacial epochs and world-ocean level in the Northern Hemisphere.

11. Models for isostatic response of land masses to glaciation and deglaciation, based on distribution of raised marine strandlines/beaches, and on long-term geodaetic recording, are best known from the areas of Europe (Scandinavia) and North America which were covered by ice-caps during the last Pleistocene glaciation. Glacial-isostatic studies in the Antarctic continent, the ice-cap of which has considerably been reduced during the past 8-10 thousand years, are yet far from being completed.

12. The global glaciations, such as the Late Palaeozoic (Gondwanian), and the Cenozoic (Tertiary and Quaternary in Antarctica; Quaternary in the Northern Hemisphere) ones, were a powerful factor in influencing evolution and dispersal of biota on our planet.

13. During the past 2 Ma, there was a profound impact of the glaciations and related climatic and sea-level changes on the evolution, culture, and dispersal of Man and its civilisation.

References

Andrews, J. T., 1970. A geomorphological study of post-glacial uplift with particular reference to Arctic Canada. - Inst. Brit. Geogr., Spec. Publ., **2**: 156 pp.

Andrews, J. T., ed., 1974. Glacial isostasy. Benchmark Papers in Geology. Dowden, Hutchinson & Ross Inc. Stroudsburg, Penn., 491 pp.

Barrett, P. J. (ed.) *et al.,* 1989 Antarctic Cenozoic history from the CIROS-1 drillhole, McMurdo Sound. - DSIR Bull., Wellington, **245**: 5-250.

Barrett, P. J., 1991. Antarctica and global climatic change: a geological perspective. *In*: C. Harris & B. Stonehouse, Antarctica and Global Climate Change. Scott Polar Res. Inst. & Belhaven Press, Cambridge: 35-50.

Bentley, Ch. R., 1964. The structure of Antarctica and its ice cover. *In:* H. Odishaw (ed.), Research in Geophysics. 2. - MIT, Cambridge, Mass.: 335-389.

Birkenmajer, K., 1964. Devonian, Carboniferous and Permian formations of Hornsund, Vestspitsbergen. - Stud. Geol. Polon., **11**: 47-123.

Birkenmajer, K., 1990. Geochronology and climatostratigraphy of Tertiary glacial and interglacial successions on King George Island, South Shetland Islands (West Antarctica). - Ztrbl. Geol. Paläont. (Stuttgart), **1**: 127-140.

Birkenmajer, K., 1996a. Tertiary glaciation and sea-level changes: record from King George Island (South Shetland Islands), West Antarctica. - Korean Jrl Polar Res. (Seoul), **7** (1/2): 1-10.

Birkenmajer, K., 1996b. Tertiary glacial/interglacial palaeoenvironments and sea-level changes, King George Island, West Antarctica. An overview. - Bull. Pol. Acad. Sci., Earth Sci., **44** (3): 157-181.

Chumakov, N. M., 1985. Glacial events of the past and their geological significance. - Palaeogeogr., Palaeoclimatol., Palaeoecol., **51**: 319-346.

Cooper, A. K., Barker, P. F. & Brancolini, G. (eds), 1995. Geology and seismic stratigraphy of the Antarctic margin. - Am. Geophys. Un., Antarctic Res. Ser., **68**: 303 pp. & maps.

Crowell, J. C., 1978. Gondwanan glaciation, cyclothems, continental positioning, and climate change. - Am. Jrl Sci., **278**: 1345-1372.

Drewry, D. J., 1983. Antarctica: glaciological and geophysical folio. Scott Polar Res. Inst., Cambridge.

Embleton, C. & King, C. A. M., 1975. FlXIl fwomoepholofy. E. Arnold, London, 573 pp.

Frakes, L. A., 1979. Climates throughout geologic time. Elsevier Sci. Publ. Co., Amsterdam etc., 310 pp.

Frakes, L. A., 1982. The history of the Antarctic climates over 'geologic' time-scales. - Aust. Met. Mag., **30**: 175-179.

Hambrey, M. J. & Alean, J., 1992. Glaciers. Cambridge Univ. Press, 208 pp.

Hambrey, M. J. & Barrett, P. J., 1993. Cenozoic sedimentary and climatic record, Ross Sea region, Antarctica. - Am. Geophys. Un., Antarctic Res. Ser., **60**: 91-124.

Hambrey, M. J. & Harland, W. B. (eds), 1981. Earth's pre-Pleistocene glacial record (IGCP Project 38). Cambridge Univ. Press, 1004 pp.

Hambrey, M. J., Larsen, B., Ehrmann, W. U. *et al.*, 1989. Forty million years of Antarctic glacial history yielded by Leg 119 of the Ocean Drilling Program - Polar Rec., **25**: 99-106.

Haq, B. U., Hardenbol, J. & Vail, P. R., 1987. Chronology of fluctuating sea levels since the Triassic. - Science, **235**: 1156-1165.

Harland, W. B., Hambrey, M. J. & Waddams, P., 1993. Vendian geology of Svalbard. - Norsk Polarinst. Skr., **193**: 150 pp.

Jacobs, S. S., 1990. Sea-level response to ice-sheet evolution: an ocean perspective. *In*: R. A. Bindschadler (ed.), West Antarctic Ice Sheet Initiative. - NASA Confer. Publ., **3115** (2): 23-47.

Jania, J., 1993. Glaciology (in Polish). Polish Sci. Publ., Warszawa, 359 pp.

Jania, J. & Hagen, J. O., 1996. Mass balance of Arctic Glaciers. - Int. Arctic Sci. Cttee (W.G. on Arctic Glaciology), Rept **5**: 62 pp.

John, B. S. & Sugden, D. E., 1971. Raised marine features and phases of glaciation in the South Shetland Islands. - Brit. Antarct. Surv., Bull., **24**: 45-111.

Kennett, J. P., 1977. Cenozoic evolution of Antarctic glaciation, the circum-Antarctic ocean, and their impact on global paleoceanography. - Jrl Geophys. Res., **82**, 3843-386.

Kennett, J.P. & Barker, P. F., 1990. Latest Cretaceous to Cenozoic climate and oceanographic developments in the Weddell Sea, Antarctica: an ocean-drilling perspective. - Proc. ODP Sci. Res., **113**: 937-960.

King, L. C., 1961. The palaeoclimatology of Gondwanaland during the Palaeozoic and Mesozoic eras. *In*: A. E. M. Nairn (ed.) Descriptive Palaeoclimatology. Intersci. Publ., New York-London: 307-331.

Lindner, L., Marks, L. & Pekala, K., 1983. Quaternary glaciations of South Spitsbergen and their correlation with Scandinavian glaciations in Poland. - Acta Geol. Polon., **33** (1-4): 169-182.

Lovering, J. F. & Prescott, J. R. V., 1979. Last of Lands ... Antarctica. Melbourne Univ. Press, 212 pp.

Mangerud, J., Jansen, E. & Landvik, J. Y., 1996. Late Cenozoic history of the Scandinavian and Barents Sea ice sheets. - Global & Planet. Change, **12**: 11-26.

Miller, K. G., Fairbanks, R. G. & Mountain, G. S., 1987. Tertiary oxygen isotope systnesis, sea leavel history, and continental margin erosion. - Palaeoceanography, **2**: 1-19.

Moriwaki, K., Yoshida, Y. & Harwood, D.M., 1992. Cenozoic glacial history of Antarctica - a correlative synthesis. *In:* Y. Yoshida, K. Kaminuma & K. Shiraishi (eds), Recent Progress in Antarctic Earth Science. Terra Sci. Publ., Tokyo: 733-780.

Pallàs, R., James, T. S., Sàbat, F., Vilaplana, J. M. & Grant, D. R. (in press). Holocene uplift in the South Shetland Islands: evaluation of tectonics and glacio-isostasy. - Proceed. VII Int. Sympos. Antarctic Earth Sci., 1995, Siena.

Schwarzbach, M., 1974. Das Klima der Vorzeit. Eine Einführung in die Paläoklimatologie. F. Enke Verl., Stuttgart, 380 pp.

Wanless, H. R. & Shepard, F. P., 1936. Sea level and climatic changes related to late Paleozoic cycles. - Bull. Geol. Soc. Am., **47**: 1177-1206.

Webb, P. N., 1990. Review: The Cenozoic history of Antarctica and its global impact. - Antarctic Science, **2** (1): 3-21.

Ice in the Troposphere

M. B. Baker

Department of Geophysics and Department of Atmospheric Sciences,
University of Washington, Seattle, Washington 98195, USA

1 Introduction

Ice particles play important roles in modulating radiative and hydrological fluxes and chemical composition in the troposphere, and in creation of the large in-cloud electric fields that give rise to lightning. These roles depend on the numbers, shapes, sizes, surface properties and spatial distributions of the particles. In this paper we discuss current understanding of the important processes involving tropospheric ice particles, with particular focus on the roles of atmospheric impurities (i.e., nonwater substance) of both natural and anthropogenic origins. For additional background material see [1, 2, 3, 4].

Figure 1 shows the mean annual distribution of the fractional area covered by clouds whose tops are above the 440 mb pressure level. All these clouds are likely to contain ice. The total atmospheric burden of ice is not known, but very crude estimates suggest it is around 10^{13} to 10^{14} kg. This mass is small compared with that (over 10^{16} kg [5]) in surface ice and snow. Despite its relatively small abundance, atmospheric ice has considerable impact on global climate, the chemical composition of the atmosphere and of precipitation, and on thunderstorm electrification. Figure 2 represents the ice-related tropospheric processes to be discussed in this paper, and we will refer to this figure in succeeding sections.

We begin with an examination of ice formation and evolution in clouds. We first discuss ice in the 'mixed phase', where liquid and ice particles coexist, and then ice in fully glaciated clouds at lower temperatures.

NATO ASI Series, Vol. I 56
Ice Physics and the Natural Environment
Edited by John S. Wettlaufer, J. Gregory Dash and Norbert Untersteiner

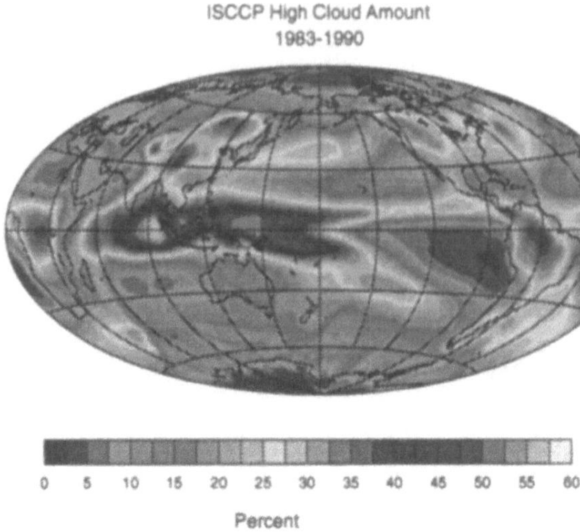

Figure 1: Annual average cloud fractional area coverage, for clouds with tops higher than the 440 mb level, estimated from satellite data under the International Satellite Cloud Climatology Project. Data kindly provided by D. Hartmann.

2 Mixed-Phase Clouds

2.1 Formation of Ice Particles

Ice formation in mixed-phase clouds is often preceded by the formation of drops at temperatures $T > 0°C$ as parcels of humid air ascend, expand and cool. When the relative humidity (RH) in such a parcel rises to just over 100%, water begins to condense on small ($0.01 - 0.1$ μm) aerosol particles known as Cloud Condensation Nuclei (CCN). These consist largely of sulfates and organic materials, of both natural and anthropogenic origin. Approximately $10^8 - 10^9/\text{m}^3$ droplets are thus formed in continental convective clouds, and approximately an order of magnitude fewer over the remote oceans [3]. Condensation of vapor on the small droplets increases their sizes and liberates latent heat which stimulates cloud growth. Those drops that become large enough to fall against the rising cloud can collide with and collect smaller droplets, forming warm rain (Fig. 2).

As air parcels rise through the 0°C isotherm the droplets remaining become supercooled and ice begins to form at slightly lower temperatures. Laboratory and field studies show that at temperatures higher than $\approx -35°C$ most atmospheric ice crystals form primarily via heterogeneous freezing of supercooled

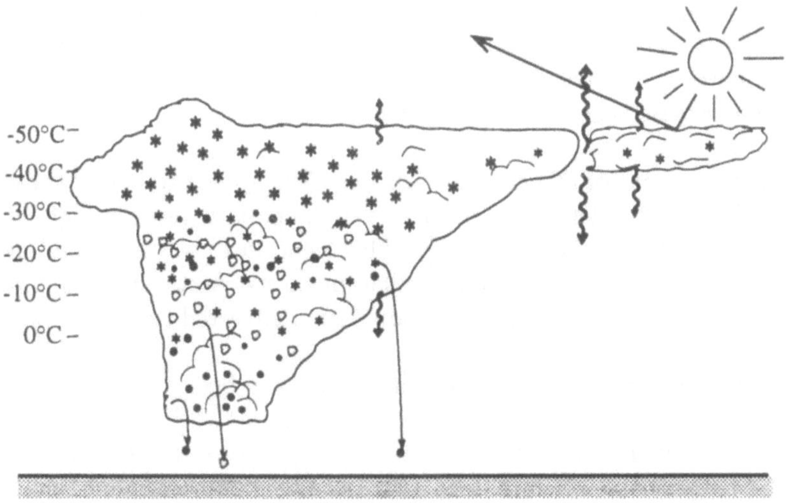

Figure 2: Schematic of a mesoscale convective cloud system and trailing cirrus cloud. (∘): liquid drops, (*): unrimed ice particles, (irregular shaded objects): rimed ice particles. ($\longrightarrow$): SW radiative fluxes; wavy arrows show LW radiative fluxes.

drops containing particles known as ice nuclei (IN), and/or via condensation of supercooled water onto IN directly from the vapor, followed by freezing. For detailed discussion of atmospheric ice nucleation see reviews [6, 7, 8, 9, 10, 11].

Most natural atmospheric ice nuclei appear to originate at the earth's surface, over both continents and regions of high biological productivity in the oceans [6, 8, 12, 13, 14], where the primary source is gas-to-particle conversion from the gases emitted into the atmosphere by the ocean surface. There are typically $\approx 10^2$ to 5×10^5 m^{-3} IN active at $T \approx -15$ to -40°C [15]; their concentrations increase rapidly with supercooling and with supersaturation with respect to ice.

Biological materials, such as leaf mold or bits of bacterial DNA or proteins, appear to participate in atmospheric ice nucleation, particularly over plant canopies [7, 11] at temperatures as high as $T \approx -4°C$. Certain clays, silicates and metal oxides [8, 12, 13, 14, 16] appear to be implicated in ice nucleation although it is possible that these act only to transport the biological molecule

Anthropogenic factors can either increase or decrease IN activity. Both sulfates and organics are usually found in atmospheric IN [17]; urban production of both types of particles and vapors probably increases the IN populations, while deactivation of the nucleating sites by other gases or by coagulation with other particles can decrease the ice nucleating efficiency of a particle population [3, 11, 12]. Pyrolysis of biomass tends to produce molecules that have both organic ring structures and polar chains [18]. These may condense with water and sulfates to form IN in situ, and may play important roles in cloud microphysical processes, due to their partial solubility in aqueous solutions and their surface active properties. The particles in jet contrails may also be transformed into IN by in-plume physico-chemical transformations [19].

The concentrations of IN active at -20 to $-15°C$ appear to have decreased (at the ground) by 1–2 orders of magnitude over the past 30 years at high latitudes in the Southern Hemisphere ([11, 12]). This trend has not been found over the Arctic [14], but the large seasonal oscillation in the Arctic freezing level may indicate seasonal variability of the IN population advected from lower latitudes.

A major puzzle in atmospheric cloud physics has been the finding that the concentrations of small, pristine ice crystals in clouds at $T > -12°C$ are often many orders of magnitude higher, and the temperature dependence of the ice particle concentration is much less, than would be expected on the basis of nucleation alone ([20]. It now appears [21, 22] that most, but not all ([20]) the observations of high ice concentrations are consistent with laboratory observations on splintering accompanying riming ([23]). Other possibilities are splintering during drop freezing, seeding by overlying clouds, freezing of evaporating droplets, perhaps induced by impurities or induced charges in electrified clouds, and breakup of crystals as they sublimate [24]. Observations indicate correlations of upper tropospheric ice with solar activity; however, attempts to find a mechanism of ice nucleation modulated by the solar wind have so far not been successful [25].

2.2 Growth and Sublimation of Ice Particles

The concentrations of ice particles are typically ≈ 100 m^{-3} at $T \approx -10°C$; far smaller than drop concentrations at these levels. Since the equilibrium vapor pressure over ice is less than that over supercooled water, the ice particles begin to grow in this temperature range and the drops evaporate. The growing crystals have hexagonal symmetry and shapes defined by their aspect ratios $\Gamma \equiv c/a$, where $c(a)$ is the crystal length perpendicular to the basal (prism) facet. Γ depends on local temperature and relative humidity [26], crystal size and rate of growth [27, 28] and, in the atmosphere, on sedimentation velocity [36], crystal defects and particle history. At relatively low supersaturations (with respect to ice) the crystals are compact solids whose aspect ratios take on values just below to just above 1 as the temperature is lowered from 0 to $-30°C$; the variations in habit appear to occur very abruptly at certain well-defined temperatures [29, 30] under certain humidity conditions. In mixed phase clouds, vapor growth takes place at or near water saturation, and can produce elaborate dendritic shapes, particularly around $T \approx -12$ to $-15°C$, where the difference between the vapor pressures of supercooled water and that of ice reaches a maximum. Drop freezing at lower temperatures, still in mixed phase clouds, often results in nucleation of several crystals per drop. These each form columnar structures assembled in 'bullet rosettes' [13]. Since crystals in clouds move through regions of varying temperature and RH, their shapes reflect the history of the conditions they have encountered, a fact used often to identify origins and trajectories of cloud ice particles.

The rate of growth (or sublimation) of a crystal facet is proportional to the flux of vapor across that facet. The flux of water $F_v(\text{kgH}_2\text{O}(\text{m}^2 - \text{s})^{-1})$ can be written [27]

$$F_v{}^{c,a} = \rho_{v,sat}(T) v_{th} S^i / [1/\alpha^{c,a} + Z_{diff}] \quad (1)$$

where the indices c, a refer to the basal and prism faces, respectively. $\rho_{v,sat}(T)$ is the density of vapor molecules at saturation at temperature T, v_{th} is the thermal velocity of vapor molecules at the particle surface, and S^i is the ambient super (under) saturation with respect to ice. The factors in the denominator on the RHS of eq 1 can be thought of as impedances, as they diminish the flux. The first is $1/\alpha^{c,a}$, where $\alpha^{c,a}$ is the 'condensation coefficient'. This term is a measure of the impedance to vapor passage through the particle surface, and Z_{diff} is the impedance to vapor passage through the air due to diffusive and latent heat effects. Z_{diff} increases with particle size, whereas α is independent of size but depends on the growth mechanism, which is a (usually unknown) function of temperature and humidity at the particle surface and of any crystal defects at the surface. In mixed phase clouds vapor growth of ice particles is rapid (typically $0.1 - 1$ μm s^{-1}) [31, 32]) and the surface impedance term is generally unimportant, but in fully glaciated clouds in the upper troposphere and stratosphere the surface term may be the rate controlling factor [27, 33]. α^c and α^a have been measured in the laboratory at temperatures above $-35°$ [29, 34] and at very low temperatures [35], in experiments carried out in pure vapor in order to diminish the diffusive impedance. (See Fig. 3.) The data reveal temperature dependent oscillations in both condensation coefficients; presumably, these oscillations are related to the temperature dependent ice crystal growth habit oscillations. Typically, α increases quadratically or even faster with supersaturation during growth, but falls off linearly with undersaturation during sublimation. No quantitative theoretical yet exists for the temperature and humidity variations of α for growing or subliming crystals.

Ice particles in clouds grow mostly by vapor deposition until their sedimentation velocities are comparable with the local updraft velocity. Subsequent growth of the precipitating particles occurs by collection of other hydrometeors. Collision products formed from a number of ice crystals are called aggregates, which can be several millimeters in size. Supercooled drops collide with ice particles and can freeze on contact, forming rime on the ice surface that obscures the original crystal shape; prolonged riming creates graupel and even hail particles whose dimensions can reach several centimeters. Most of the precipitation falling to the surface at midlatitudes originates as ice, although it may melt on the way

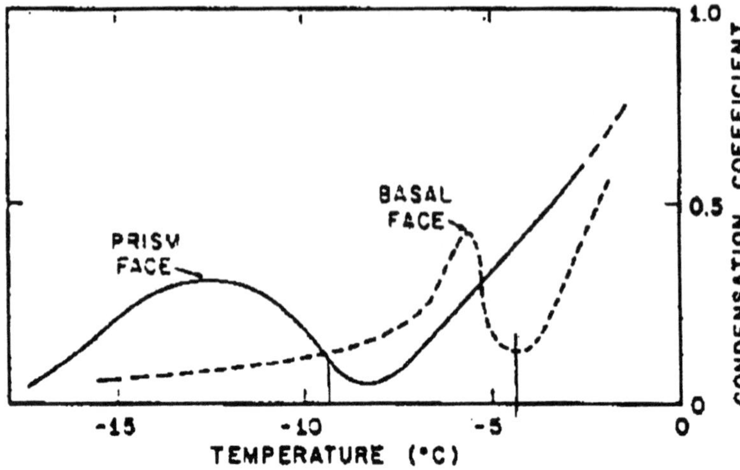

Figure 3: The condensation coefficients α^c, (basal face), α^a (prism face) as functions of temperature measured at constant vapor excess. [29]

down and turn into rain. (See Fig. 2.) Precipitation is the source of surface ice, which is an important factor in global climate, and precipitation at the earth's surface is often the factor that determines the distribution of life forms. Because of its origins in particle collisions in turbulent clouds, precipitation is a highly stochastic and intermittent phenomenon. Remote sensing from satellite is virtually the only method for large scale or global monitoring of precipitation, and this method is still not sufficiently reliable or accurate to meet current needs (see, e.g., [4, 37]).

3 Ice Formation and Evolution in the Upper Troposphere

Both heterogeneous and homogeneous nucleation appear to play roles in ice formation in the upper troposphere. Liquid droplets several microns in radius can be found at temperatures down to about $T \approx -35°C$ [38] (Fig. 2). These are probably concentrated solutions of sulfuric acid, ammonium sulphate, and possibly other nitrogen compounds formed on deliquescent aerosols. Some of the aerosols on which the ice particles nucleate are formed aloft by gas-to-particle conversion in the clear air surrounding clouds ([39]); some originate from volcanos, and some arise from gas and/or particle anthropogenic sources.

The equilibrium RH over a solution droplet of radius r_d and water activity a_w is

$$RH_{eq}(r_d, a_w)(\%) = 100 \times (1 + A/r_d) * a_w \tag{2}$$

The constant A is proportional to σ_{lv}, the surface energy of the solution against vapor. For a dilute solution $a_w \propto B/r_d^3$, where B is proportional to the mass of solute, but the dependence of A and a_w on temperature and solution composition for the solution droplets in the atmosphere are not known. Indeed, even the number of phases is not well characterized for the mixtures of organic and inorganic materials that may be present [40].

Eq 2 is an equilibrium relationship linking a_w in a droplet to the environmental relative humidity. The homogeneous freezing temperature T_{hf} (°C), a nonequilibrium property of the droplets, is also a function of a_w; for many organic and inorganic solutes

$$T_{hf}(a_w) \approx CT_{mp}(a_w)), \tag{3}$$

where $T_{mp}(a_w)$ is the equilibrium freezing point of the bulk solution and C depends on the solute [41]. Thus the relative humidity, temperature and solute concentration at the freezing temperature are linked and at very cold temperatures only highly concentrated, small droplets are still liquid.

Classical nucleation theory has been successfully applied to the study of ice nucleation in pure water using a new equation of state for supercooled water [42]. Application of the theory to atmospheric solution droplets via eq 3 would require that the parameters $C, T_{mp}(a_w)$ be known; however, elucidation of the equilibrium and nonequilibrium physical chemistry of these droplets in the atmospherically relevant thermodynamic regimes represents a formidable challenge in current atmospheric research [43, 44].

Recent field observations [15, 45] suggest that there is an important role for heterogeneous ice nucleation in the upper troposphere, where $1-10$ IN per liter were found at temperatures below -30°C. Examination of the solid residues found in cirrus cloud particles over the North Sea showed the particles were typically several tenths of microns in linear dimension [46]. These were probably involved in the ice nucleation process since their concentrations are sufficiently low at cirrus cloud levels that it is not likely they are scavenged.

The evolution of the ice particle population in upper tropospheric clouds depends not only on the physical chemistry of highly supercooled, concentrated solution droplets and the ambient vapor distribution, but also on dynamic factors. The RH in a rising air parcel is determined by competition between adiabatic cooling, providing a velocity dependent vapor source, and vapor deposition on preexisting particles, which constitutes a vapor sink. Since homogeneous nucleation in general requires higher RH than heterogeneous nucleation, IN can nucleate ice at low updraft velocities, forming particles that deplete the vapor phase, thus possibly preventing homogeneous nucleation [47].

Vapor growth of ice particles in the upper troposphere produces complex and varied crystal shapes [48]. Many of the crystals are hollowed columns and other irregular forms, which implies unstable growth modes that may be associated with high aspect ratios [27]. There have been few laboratory examinations of crystal growth under cirrus conditions with which to compare these field observations [49].

4 Impurities in Tropospheric Ice Particles

Rainwater is typically acid, with a pH of around 5 in 'background' locations, and less in polluted areas. The acidity is primarily a result of the dissolution of CCN and uptake of SO_2, HNO_3 and other gases [50]. It is not known how much of the uptake of impurities by cloud particles occurs in the ice phase, in general. Nonwater substances enter tropospheric ice particles as solid particles, as solutes in liquid that freezes, and directly from the gas phase. The chemical composition of ice particles is determined in part by their growth mechanism [51]. Particles formed by freezing of droplets contain higher concentrations of impurities (and have lower pH) than those grown by vapor deposition, because impurity solubilities are higher in liquid than in ice. However, during freezing the temperature of the ice-water solution rises to 0°C and impurities are rejected from the growing ice. Both of these processes tend to decrease the concentration of impurities in ice relative to that in the supercooled droplets from which they freeze. Furthermore, rimed particles tend to have higher fall velocities than do vapor grown ice particles, which tends to decrease the uptake by the rimed particles. Finally, the composition of atmospheric ice cannot be easily related to that of the environment in which it is found because of long-range transport in the atmosphere [52]. We discuss our current understanding of each of the important uptake mechanisms in turn. See also, for example, [53] and other reviews for more detailed information on this subject.

4.1 Solid Phase Impurities

Mineral particles are often found imbedded in precipitating ice particles. These may be associated with the IN, or with the CCN that were involved in the initial droplet formation; alternatively, they may be have been scavenged or captured by precipitating particles as they fall through clear air (see, for example, [54] and references therein). In-cloud scavenging occurs via thermophoresis and diffusiophoresis; that is, motion of aerosol particles brought about by nonuniformity in temperature or gaseous concentration. In the vicinity of an evaporating hydrometeor thermophoresis transports particles into the hydrometeor, and diffusiophoresis repels them; thermophoresis is the stronger effect for particles less than about 1.5 μm in size, and diffusiophoresis is stronger for larger particles [55]. Thus in a mixed phase cloud evaporating droplets are likely to collect submicron aerosol particles via the dominant thermophoretic process, whereas the ice particles, which are growing via vapor diffusion, collect the large particles via diffusiophoresis. A droplet that evaporates completely scavenges particles from an air volume several orders of magnitude greater than its original volume [56]. Scavenging also occurs as sedimenting ice particles sweep out gases and particles below them. (In this sense precipitation 'washes' the air.) The efficiency of this process depends on the very complex hydrodynamic flow around irregularly shaped ice and snow particles, the nature of the ice particle surface and possible electric charges on the colliding particles.

4.2 Liquid Phase Impurities

Freezing of a solution droplet leaves a small quantity of unfrozen liquid. The composition and pH of this unfrozen solution (and therefore, if it is on the particle surface, the subsequent chemical interactions of the particle with its environment) depend on freezing rate, temperature and initial solution composition. Formation of ice embryos in highly supercooled solution drops may result in exclusion of solute into very concentrated fluid at the ice/vapor interface, in grain boundaries, crystallite boundaries or brine pockets within the ice. At very high freezing rates, a nonnegligible amount of some solutes may be incorporated into the ice. The distribution of water ions and ionic impurities near a freezing front is determined by differential ion incorporation into the ice, differential ion mobility in the ice lattice, and the consequent electrostatic potential gradients [57]. These electrostatic effects may significantly modify the chemistry of the solution that remains when freezing is complete under certain circumstances, ([58, 59]; differential incorporation of ions by the ice lattice may be responsible for the observation that some oxidation-reduction reaction rates are enhanced by several orders of magnitude during freezing ([60]).

4.3 Gas Phase Impurities

Uptake and possible interactions of foreign (nonwater) gases on atmospheric ice particles can modify the chemical composition of the clear air. It has been suggested, but not established, that adsorption may alter the particle growth and sublimation rates [61, 62]. The important gases in this regard are CO_2, SO_2, HNO_3, NH_3 and O_3, plus gases emitted from biomass burning and from natural biological activity. While uptake of certain gases (for example, SO_2)

has long been considered unimportant, recent reanalysis of laboratory studies ([63]) imply that this uptake can be a nonnegligible fraction of the in-cloud sink for the gas.

In principle, the measured temperature and gas concentration dependence of the uptake should provide information on the uptake mechanisms; physisorption onto a dry surface should be linear in gas concentration and increase as temperature is lowered, whereas chemisorption and/or uptake into a reservoir whose volume increases with temperature would yield different trends [63]. However, differences in experimental techniques produce differences in results that can be difficult to interpret [63, 64, 65, 66, 67, 68]. Uptake of SO_2 by ice spheres in packed bed experiments [64, 65] increased with increasing temperatures for $-60 \leq T \leq 0°C$ (see Fig. 4), suggesting that the uptake is modulated by surface or grain boundary melting. The fractional coverage by SO_2 varied less than linearly with the partial pressure of the gas in these experiments, suggesting dissociation of the adsorbed SO_2 either with or without the presence of liquid water. From the laboratory results we can make some simple estimates of the role of ice particles as an atmospheric sink for SO_2. The measured fractional coverage of SO_2 molecules on the ice surface at temperatures above $-30°C$ and a gas partial pressure $p_{so2}/p = 90$ ppb, realistic in heavily polluted regions of the troposphere, was on the order of 0.001 – 0.01 (Fig. 4). If 10% of that were oxidized in the melted particle the concentration of sulfate in a 100 μm drop would be about 5×10^{-2} $\mu g/cm^3$, or about one tenth the concentration found in polluted rainwater. This very rough estimate suggests that the contribution to acid rain by uptake on ice particles may not be negligible, as is usually assumed in numerical models, although it is probably small compared with the uptake into liquid phase particles under the same conditions.

Several sets of laboratory measurements seem to suggest liquid phase chemistry occurs on the surfaces of ice particles at temperatures as low as $-60°C$ [65, 66]. This result may be due to uptake in cracks, or grain boundaries, or it may suggest that (a) surface melting is enhanced by the presence of impurities, and/or (b) the uptake of impurities on the ice surface depresses the melting point and enhances the production of liquid on the crystal surface.

4.4 Cloud Radiative Properties and Ice Microphysics

As shown schematically in Fig. 2, ice clouds interact both with the incoming solar radiation (wavelengths $\lambda \leq 4$ μm) and the outgoing terrestrial radiation, which peaks at around $\lambda \approx 10$ μm. Ice absorbs strongly in the ultraviolet, very little through the visible, in bands in the near-infrared and throughout

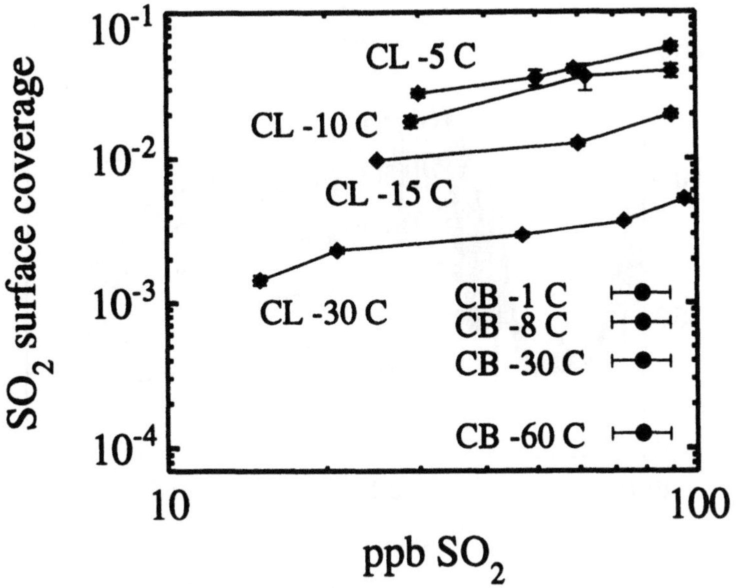

Figure 4: SO_2 surface coverages derived from measurements of Clapsaddle and Lamb (1989) (CL) and Conklin and Bales (1993) (CB). Surface coverage $=1 \equiv 10^{15}$ molecules/cm^2. [63].

the far-infrared, due to lattice vibrations [69]. The interactions of ice particles with radiation are important for two reasons: (i) remote sensing of reflected and emitted radiances from the visible to the microwave are our only means of obtaining global climatological data on the ice particle distributions and parameters, and (ii) the interactions of ice with solar and terrestrial radiation have important climatic implications. The combined shortwave and longwave effects of ice clouds in the upper troposphere heat the earth-atmosphere system; the strongest effect of clouds on the surface radiation budget is warming due to high clouds.

Figure 5 shows the local variation in upwelling LW radiances as measured from aircraft as the plane flew over regions of varying coverage by cirrus clouds. The cold clouds absorb upwelling terrestrial radiation emitted by the relatively warm Earth and atmosphere below the clouds, and they emit radiation at their local temperatures. The resulting decrease in outgoing longwave radiation, which is strongest in the 'atmospheric window' (8–12 μm), is an important component of the so-called atmospheric Greenhouse Effect

The interactions of an ice particle of linear dimension a with radiation of wavelength λ are determined by the complex index of refraction of ice, $m(\lambda)$, and the size parameter $x(a, \lambda) \equiv 2\pi a/\lambda$. For atmospheric ice particles a is usually defined as the radius of the equivalent area (volume) sphere for computations

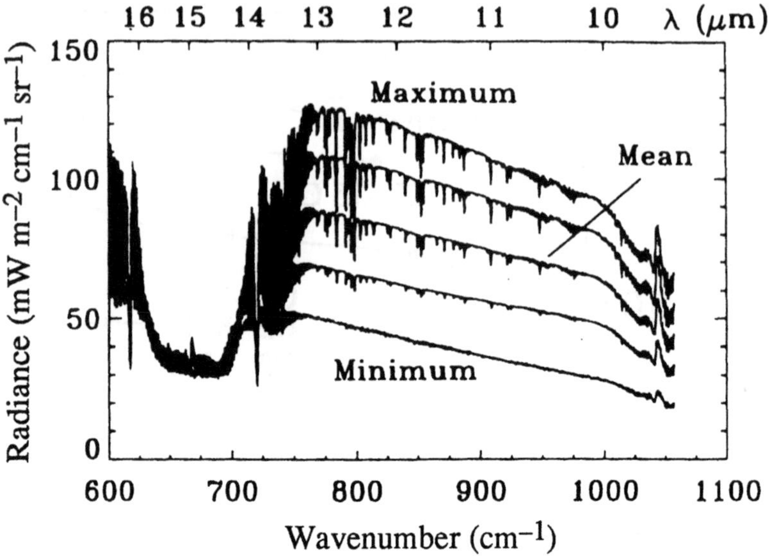

Figure 5: Upwelling LW radiances (mW m^{-2} $cm^{-1^{-1}}$ sr^{-1}) over terrain partially covered by cirrus clouds, from the High Resolution Interferometer Sounder on board an ER-2 aircraft above the clouds. The different curves represent measurements over regions with varying amounts of cloud cover; the cloudiest regions produce the minimum upwelling radiance, whereas the maximum is the clear sky value. The figure shows these curves, the mean of all measurements on a given day, and curves showing ± one standard deviation above and below the mean radiances on that day. [70]

involving visible (longwave) radiation. Typically, 10 μm $\leq a \leq$ 300 μm for upper tropospheric ice particles, so $x(\lambda, a) >> 1$ for visible radiation and for terrestrial (IR) radiation $x(\lambda, a) \leq 1$. In the former case the radiative properties of particles can be calculated in the geometric limit using ray-tracing techniques, and the crystal shapes can be taken into account, whereas in the latter anomalous diffraction theory and other approximations must be used [71, 72].

The effective surface areas and shapes of ice particles in a cloud determine not only the particle light scattering phase functions but also the particle orientations, both of which are important in determining the cloud reflectance at visible wavelengths. The facetted shapes of the particles are responsible for the dramatic halos, sundogs, and other optical phenomena characteristic of some cirrus clouds. For most upper tropospheric ice particles the absorption cross section for IR radiation is approximately proportional to particle volume, so that the emissivity of ice clouds in this wavelength range depends only on the column integrated condensed ice amount. Midlatitude cirrus clouds have emissivities from 0 to nearly 1, and reflectances of 30 – 40 [73]. Most of the IR absorption and much of the reflectivity in the visible are due to small crystals ($a < 20$ μm), which contribute only a few percent of the condensed water mass in upper tropospheric clouds [74, 75]. Numerical models suggest that the small ice crystals remove about 4 W/m^2 from the incoming solar beam—comparable in magnitude with (but opposite in sign from) the predicted global effect of a doubling in CO_2. The numbers, sizes and possibly shapes of these small particles are sensitive to distributions of CCN, IN and/or upper tropospheric gases and aerosols.

The results of multiple scattering calculations based on computed single scattering phase functions of selected types of crystals compare well with observed fluxes in some cases [71, 72, 73]; however, in general the microphysical parameters of ice clouds cannot be measured with the accuracy needed for climate studies [76]. Puzzles remain as to the very nature of certain radiatively important ice clouds; namely, 'subvisual cirrus' (e.g., [77]), clouds of optical depth less than ≈ 0.03 in the upper troposphere, and 'diamond dust'; 10–50 μm facetted crystals and 'clear sky ice precipitation', irregularly shaped frozen particles in the lower Arctic cloudfree sky [78].

5 Atmospheric Ice and Thunderstorm Electrification

Observations reveal that the electrical structure of a developed convective thunderstorm is dipolar or tripolar, consisting of a pocket of negative charge, located on hydrometeors in the 'mixed phase' region between roughly $-10°$C and $-25°$C, a more diffuse positive charge aloft and sometimes a smaller positive charge center lower down. ([79]) In stratiform anvils, where some of the charge is advected in from convective cores, several charge layers, including a robust positive charge layer just at 0°C have been inferred from balloon soundings [80].

The association of in-cloud charge centers with temperature implies that these are microphysical in origin [81]. The large positive charge layer near 0°C in electrified stratiform clouds may be due to charging by melting of the particles; release of entrapped air in bubbles might remove polarization charge on the upper surface of the particles. The negative charge center, on the other hand, appears ([82, 83, 84, 88]) to be the result of rebounding collisions between ice crystals and riming graupel particles, which result in charge transfers of up to several hundred fC/collision. The laboratory results show that under atmospherically relevant thermodynamic conditions, the relatively heavy graupel pellets tend to charge negatively and the lighter, ascending ice crystals charge positively. Gravitational separation of electric charge leads to the buildup of electrical fields that can produce lightning.

The mechanism(s) by which charge is separated during ice-graupel collisions are unknown, despite a long history of attempts to understand this phenomenon (e.g., [79, 88, 86, 90, 87]). For detailed reviews see [85, 89].

The candidate charge transfer mechanisms include;

- The thermoelectric effect, which is due to the higher mobility of hydronium ions than of hydroxyl ions in ice. Excess negative charge builds up near a (relatively warm) surface undergoing vapor growth, and positive charge near an evaporating surface [90]. It is possible that this charge could be transferred during collisions.
- Transfer of charges associated with differences in the surface potentials of the ice surfaces [91]. A riming surface has a contact potential of ≈ -400 mV with respect to a nonriming surface (whether it is evaporating or growing) at temperatures below $\approx -15°$C. This means a particle making contact with the rimed surface might acquire charge.
- Fracture of frosty protruberances on the graupel surface. Laboratory observations show that removal of protruberances is associated with charging of the parent particle [88]; the sign of the charge transferred is consistent with that predicted by the thermoelectric effect but the magnitude is much higher, which may be a geometric effect.
- Workman-Reynolds effect. As ionic solutions freeze, differential incorporation of ions into the ice lattice creates a potential difference between frozen and unfrozen water [57, 92]. Thus the surfaces of droplets freezing on an ice substrate during riming might bear net charge that could be captured by a projectile during collisions.
- Transfer of charge with surface melted fluid [86]. If each ice surface is covered with a film of electrically charged surface melted fluid then transfer of that fluid from one particle to another transfers charge.

A satisfactory explanation for the observations must (a) identify the charge carriers and a charge transfer mechanism consistent with known physical principles and (b) explain the observed dependence of the charge transfer on the growth state of the particles and the environmental conditions. None of these mechanisms is able to meet these criteria quantitatively, but the last agrees qualitatively with the trends observed in the laboratory. The laboratory studies of charge transfer, like those on uptake of impurities by ice particles, stem from a small number of experimental groups and differ to some extent from laboratory to laboratory. Improved control of the experimental parameters is necessary in order to understand the microscopic mechanisms acting in the laboratory, and, eventually, those that act in the atmosphere.

Because the production of lightning stems from the charge transferred between precipitation particles and rising small ice crystals, the distribution of lightning is related to precipitation rate and on the supply of water and ice to the upper troposphere. These relationships have not yet been thoroughly explored.

6 Summary and Conclusions

Recent improvements in atmospheric measuring capabilities and advances in understanding of the underlying physical chemistry have brought about major developments in the field of atmospheric ice microphysics. The contribution of heterogeneous nucleation to ice particle concentrations in mixed phase clouds is fairly well understood. New data from the upper troposphere have revised our understanding of the roles of several ice particle formation mechanisms in that region, and a great deal has been learned about the physical chemistry of some of the relevant concentrated liquid solutions from which the particles form. There has been considerable progress over the past decade both in remote sensing of light scattering by clouds of irregular particles and in its calculation. We now know there is a range of microphysical conditions under which equivalent sphere or spheroid models give satisfactory approximations to observed albedos and infrared radiances of ice clouds.

On the other hand, there are still large gaps in our quantitative understanding of ice clouds and these gaps hinder attempts to forecast the response of the clouds to changing climatic conditions. We do not currently have accurate climatologies of the microphysics of ice clouds. Indeed, as mentioned earlier, we do not know either the total atmospheric ice burden or the geographical distribution of ice mass in the atmosphere. Therefore the most pressing need is for accurate and complete characterization of atmospheric ice particle distributions and lifetimes. In situ measurements are essential for certain parameters such as IN concentrations, but will have to be supplemented by improved remote sensing technologies for global estimates of ice particle distributions and fluxes. We still do not know exactly which molecular species are important in ice cloud evolution in the atmosphere. Therefore a great deal of effort has been expended in the study of molecules (for example, as possible IN), which may in fact be of little atmospheric relevance. To proceed we must collect enough field data to know which molecules and mixtures are important.

Better description of particle distributions in the present atmosphere will not in itself suffice to predict future distributions; the evolution of ice particles in upper tropospheric conditions involves physical and chemical processes in temperature and humidity regimes that have not been quantitatively explored in the laboratory. Particle surface effects are important in most of the phenomena we have discussed, suggesting that traditional laboratory studies of processes occurring on substrates may involve artifacts whose effects are quite large. Detailed laboratory studies are required on ice nucleation and habit evolution under substrate free conditions. Uptake of nonwater substance by atmospheric ice particles has been neglected in most atmospheric chemistry models; to explore the conditions under which uptake on ice may be important, we need laboratory experiments comparing uptake by single crystals with that on polycrystalline ice. Elucidation of the mechanism of charge transfer in ice-ice collisions will require systematic laboratory study of collisions with and without air (to explore the role of dissolved impurities in the electrical phenomena), and detailed observation of collisions between unrimed and rimed ice under highly controlled temperature conditions. These detailed studies of the behavior of single particles in controlled conditions are not only of interest in and of themselves; they provide the basis for appropriate parameterizations of the behavior of populations of particles to be used in large scale numerical models of the atmosphere.

7 Acknowledgments

This paper was written with support from NSF Grant #ATM-9528049 and the Max Planck Institut fur Chemie in Mainz. I am most grateful for helpful comments and unique background material from P. DeMott, A. Heymsfield, D. Lamb, J. Locatelli, A. Rangno and V. Schroeder, and to K. Dewar, M. Krueger and T. Huthwelker for their expert help in processing the manuscript. M.'Michelsen and D. Hartmann kindly provided Fig. 1.

References

[1] Houze, R. Cloud Dynamics. Int'l Geophysics Series 53, eds R. Dmowska and J. Holton. Academic Press, 1993.

[2] Wallace, J. M. and P. V. Hobbs: Atmospheric Sciences; an Introductory Survey. academic Press, London, 1977.

[3] Pruppacher, H. and J. Klett: Microphysics of Clouds and Precipitation. Reidel, 1978.

[4] Baker, M.B., Cloud Microphysics and Climate. Science 276,1072–1078, 1997.

[5] Chahine, M. The Hydrological Cycle and Its Influence on Climate. Nature 359, 373–380, 1992.

[6] Beard, K., Ice Initiation in Warm Base Convective Clouds: An Assessment of Microphysical Mechanisms. Atmos. Res. 28, 125–152, 1992.

[7] Schnell, R. C. and G. Vali. Biogenic Ice Nuclei: Part I. Terrestrial and Marine Sources. J. Atmos. Sci. 33, 1554–1564, 1976.

[8] Vali, G. Atmospheric Ice Nucleation—A Review. J. Rech. Atmos. 19, 105–115, 1985.

[9] Cooper, W. A. Research in Cloud and Precipitation Physics: Review of U. S. Theoretical and Observational Studies, 1987–1990. Revs. Geophys. Suppl. 69–79, 1991.

[10] Vali, G. Ice Nucleation—A Review. 14th Int. Conf. on Nucleation and Atmospheric Aerosols. Helsinki, Elsevier Science Publ. Oxford. 1996.

[11] Szyrmer, W. and I. Zawadski: Biogenic and Anthropogenic Sources of Ice-Forming Nuclei: A Review. Bull. Am. Meteor. Soc. 78, 209–228, 1997.

[12] Bigg, K.: Long term trends in ice nucleus concentrations. Atmos. Res. 25, 397–413, 1990.

[13] Lamb, D. Atmospheric Ice. To be published, Encyclopedia of Applied Physics, VCN Publishers, Inc. 1997.

[14] Bigg, E. K., Ice Freezing Nuclei in the High Arctic. Tellus B 48B, 223–234, 1996.

[15] P. J. DeMott, D. C. Rogers, S. M. Kreidenweis and Y. Chen. The role of heterogeneous freezing nucleation in upper tropospheric clouds: inferences from SUCCESS. Submitted, Geophys. Res. Lett. 1997.

[16] Hallett, J., B. Gardiner, J. Hudson and F. Rogers. Cloud Condensation and Ice Nucleation of a Range of Carbonaceous Aerosols. Preprints, Conf. on Cloud Physics, Snowmass, CO, Amer. Meteor. Soc. 9–12, 1986.

[17] Rosinski, J., P.L. Haagensen, C. T. Nagamoto and F. Parungo. Nature of Ice Forming Nuclei in Marine Air Masses. J. Aerosol Sci. 18, 291–301. 1987.

[18] Charlson, R. J., pers. comm., 1997.

[19] Karcher, B., Th. Peter, U. Biermann and U. Schumann: The initial composition of jet condensation trails. J. Atmos. Sci. 53, 3066–3083, 1996.

[20] Rangno, A. and P. Hobbs: Ice particle concentrations and precipitation development in small polar maritime cumuliform clouds. Q. J. Roy. Met. Soc. 117, 207–242. 1991.

[21] Bower, K., S. Moss, D. Johnson, T. Choularton, J. Latham, P. Brown A. Blyth and J. Cardwell. Q. J. Roy. Met. Soc. 122, 1815–1845, 1996.

[22] Blyth, A. and J. Latham: Development of Ice and Precipitation in New Mexican Summertime Cumulus. Q. J. Roy. Met. Soc. 119,91–121, 1993.

[23] Hallett, J. and S. Mossop. Production of secondary ice particles through the riming process. Nature 249, 26–28. 1974.

[24] Bacon, N., B. Swanson, M. B. Baker and E. J. Davis. Breakup of Sublimating Ice Particles: A Laboratory Study. Submitted, J. Geophys. Res.

[25] Tinsley, B. and K. Beard: Links Between Variations in Solar Activity, Atmospheric Conductivity and Clouds: An Informal Workshop, Bull. Amer. Meteor. Soc. 78, 685–689, 1997.

[26] Nakaya, U, M. Hanajima and J. Mugurama: Physical investigations of the growth of snow crystals. J. Faculty of Sci. Hokkaido Univ., Jpn, Ser. II, vol V, no. 3, 88. 1958.

[27] Nelson, J. and M. Baker, 1996: A New Theoretical Framework for Studies of Vapor Growth and Sublimation of Small Ice Crystals in the Atmosphere. J. Geophys. Res.

[28] Peter, Th. and M. Baker, 1994: Lifetimes of Small Ice Particles in the Atmosphere. Proceedings, NATO Conference on Clouds and Climate, Schloss Ringberg. Eds. Crutzen and Ramanathan.

[29] Lamb, D. and W.Scott, 1972: Linear Growth Rates of Ice Crystals Grown from the Vapor Phase. J. Cryst. Growth 12 21.

[30] Kuroda, T., 1984: Rate determining processes of growth of ice crystals from the vapour phase. J. Cryst. Gr. 12, 21.

[31] Ryan, B.F., E.R.Wishart and D.E.Shaw. The Growth Rates and Densities of Ice Crystals between −3°C and −21°C. J. Atmos. Sci.33, 842–850, 1976.

[32] Takahashi, T., Endoh. G. Wakahama and N. Fukuta. Vapor Diffusional Growth of Free-Falling Snow Crystals between −3 and −23°C. J. Meteor.Soc. Jpn 69,15–30. 1991.

[33] MacKenzie, R. and P. Haynes, 1992: The influence of surface kinetics on the growth of stratospheric ice crystals. J. Geophys. Res. 97, D8, 8057–8064.

[34] Beckmann, W., 1982: Interface Kinetics of the Growth and Evaporation of Ice Single Crystals from the Vapour Phase. J. Cryst. Gr. 58, 425–442.

[35] Haynes, D. R., N. J. Tro and S. M George. Condensation and Evaporation of H_2O on Ice Surfaces. J. Phys. Chem. 96, 8502–8509. 1992.

[36] Chen, J. and D. Lamb, 1994: The theoretical basis for the parameterization of ice crystal habits: growth by vapor deposition. J. Atmos. Sci. 51, 1206–1221.

[37] Kummerow, C. and L. Giglio. J. Appl. Meteorol. 33, 19. 1994.

[38] Heymsfield, A. and L. Miloshevich, 1994: Relative Humidity and Temperature Influences on Cirrus Formation and Evolution: Observations from Wave Clouds and FIRE II. J. Atmos. Sci. 52, 4302–4326.

[39] Perry, K. and P. Hobbs, 1996: Influences of Isolated Cumulus Clouds on the Humidity of their Surroundings. J. Atmos. Sci. 53,159–174.

[40] Kulmala, M., R. J. Charlson, P. Korhonen and A. Laaksonen. Clouds without Supersaturation. To be published, Nature, 1997.

[41] Rasmussen, D. H. Ice Formation in Aqueous Systems. J. Microscopy 128,167–174, 1982.

[42] Jeffrey, C. and P. Austin, 1995: Proceedings, IUGG Conference, Boulder

[43] MacKenzie, R., M. Kulmala, A. Laaksonen and T. Vesala, 1995: J. Geophys. Res. 100, 11,275–11,288.

[44] Luo, B., K. S. Carslaw, Th. Peter and S. L. Clegg. Vapour Pressures of $H_2SO_4/HNO_3/HCl/HBr/H_2O$ solutions to low stratospheric temperatures. Geophys. Res. Lett. 1994.

[45] Kreidenweis, s. M., Y. Chen, D. C. Rogers and P. J. DeMott. Isolating and Identifying Atmospheric Ice-Nucleating Aerosols: A New Technique. To be published, Atmos. Res. 1997.

[46] Strom, J., J. Heintzenberg, K. J. Noone, K. B. Noone, J. A. Ogren, F. Albers and M. Quante. Small Crystals in Cirrus Clouds: Their Residue Size Distribution, Cloud Water Content, and Related Cloud Properties. J. Atmos. Res. 32, 125–141. 1994.

[47] DeMott, P., M. Meyers and W. Cotton: Parameterization and Impact of Ice Initiation Processes Relevant to Numerical Model Simulations of Cirrus Clouds. J. Atmos. Sci. 51, 77–90, 1994.

[48] Heymsfield, A. and L. McFarquhar: High Albedos of Cirrus in the Tropical Pacific WSarm Pool: Microphysical Interpretations from CEPEX and from Kwajalein, Marshall Islands. J. Atmos. Sci. 53, 2424–2451. 1996.

[49] Kobayashi, T. Vapor Growth of Ice Crystals between −40 and −90°C. J. Met. Soc. Jpn. 43, 359–367. 1965

[50] Charlson, R. J. and H. Rodhe. What Determines the Acidity of Rain? Nature 295, 683–685, 1982.

[51] Takahashi, T., T. Endoh, K. Muramoto, C. Nakagawa and I. Noguchi, Influence of the Growth Mechanism of Snow Particles on

their Chemical Composition. Atmospheric Environment 30, 1683–92, 1996.

[52] Maupetit, F. and R. J. Delmas. Snow Chemistry of High Altitude Glaciers in the French Alps. Tellus 46B, 304–24, 1994.

[53] Ravishankara, A. Heterogeneous and Multiphase Chemistry in the Troposphere. Science 276, 1058–1065, 1997.

[54] Precipitation Scavenging, Dry Deposition and Resuspension, I., pp. 505–516. Elsevier Science, N.Y., 1983.

[55] Mitra, S. K., U. Barth and H. R. Pruppacher. A Laboratory Study of the Efficiency with which Aerosol Particles are Scavenged by Snow Flakes. Atmos. Env. 24A, 1247–1254, 1990.

[56] Baker, B. A. On the Role of Phoresis in Cloud Ice Initiation. J. Atmos. Sci. 48, 1545–1548, 1991.

[57] Gross, G., C. Wu, L. Bryant and C. McKee. Concentration dependent solute redistribution at the ice/water phase boundary. II. Experimental investigation. J. Chem. Phys. 62, 3085–3092, 1975.

[58] Bronshteyn, V. and A. Chernov, 1991: Freezing Potentials Arising on Solidification of Dilute Aqueous Solutions of Electrolytes. J. Cryst. Gr. 112, 129–145.

[59] Finnegan, W. , R. Pitter and L. Young, 1991: Preliminary Study of Coupled Oxidation-Reduction Reactions of Included Ions in Growing Ice Crystals. Atmos. Env. 25A, 2351–2534.

[60] Takenaka, N., A. Ueda and Y. Maeda, 1992: Acceleration of the Rate of Nitrite Oxidation by Freezing in Aqueous Solution. Nature 358, 736–738.

[61] Chen, J.-P. and P. Crutzen, 1994: Solute Effects on the Evaporation of Ice Particles. J. Geophys. Res. 99,D9, 18,847–18,859.

[62] Baker, M. B. and J. Nelson. Comment on Solute Effects on the Evaporation of Ice Particles. J. Geophys. Res. 101, 23035, 1996.

[63] Huthwelker, T., Th. Peter, D. Lamb, M. Baker and B. Swanson, 1996: A Theoretical Model for SO2 Uptake into Ice Particles, 1996: Proceedings, ICCP, Zurich.

[64] Clapsaddle, C. and D. Lamb, 1989: The sorption behavior of SO_2 on Ice at temperatures between −30°C and −5°C. Geophys. Res. Lett. 16, 1173–1176.

[65] Conklin, M. H. and R. C. Bales. SO_2 Uptake on Ice Spheres: Liquid Nature of the Ice-Air interface. J. Geophys. Res. 98, 16851–16855. 1993.

[66] Valdez, M. and G. Dawson, 1989: Sulfur dioxide incorporation into Ice depositing from the vapor. J. Geophys. Res. 94, 1095–1103.

[67] Diehl, K., S. Mitra and H. Pruppacher, 1995: A Laboratory Study of the Uptake of HNO3 and HCl Vapor by Snow Crystals and Ice Spheres at Temperatures between 0 and -40°C. Atmospheric Environment 29, 975–981.

[68] Domine, F., E. Thibert, F. van Landeghem, E. Silvente and P. Wagnon, 1994: Diffusion and solubility of HCl in ice: preliminary results. Geophys. Res. Lett. 21, 601–604.

[69] Warren, S. G. Optical Constants of Ice from the Ultraviolet to the Microwave. Appl. Optics 23, 1206, 1984.

[70] S. A. Ackerman, W. L. Smith, A. D. Collard, X. L. Ma, H. E. Revercomb and R. O. Knuteson. Cirrus Cloud Properties Derived from High Spectral Resolution IR Spectrometry during FIRE II. Part II: Aircraft HIS Results. J. Atmos. Sci. 52, 4246–4263. 1995.

[71] Takano, Y. , K.-N. Liou and P. Minnis: The Effects of Small Ice Crystals on Cirrus Infrared Radiative Properties. J. Atmos. Sci. 49, 1487–1493. 1992.

[72] Takano, Y. and K.-N. Liou. Solar radiative Transfer in Cirrus CLouds, Parts I and II. J. Atmos. Sci. 46, 3–36, 1989.

[73] Spinhirne, J., D., W. D. Hart and D. L. Hlavka. Cirrus Infrared Parameters and Shortwave Reflectance Relations from Observations. J. Atmos. Sci. 53, 1438–1458. 1996.

[74] Arnott, W., Y. Dong, J. Hallett, 1994: Role of small ice crystals in radiative properties of cirrus: a case study. FIRE II. J. Geophys. Res. 99, 1371–1381.

[75] Noone, K. B., K. J. Noone, J. Heintzenberg, J. Strom and J. A. Ogren. In situ Observations of Cirrus Cloud Microphysical Properties Using the Counterflow Virtual Impactor. J. Atmos. Oceanic Technol. 10, 294–303, 1993.

[76] Vogelmann, A. and T. Ackerman, 1995: Relating Cirrus Cloud Properties to Observed Fluxes: A Critical Assessment. J. Atmos. Sci. 52, 4285–4301.

[77] Sassen, D. and B.S. Cho (1992): Subvisual-thin cirrus lidar dataset for satellite verification and climatological research. J. Appl. Meteorology 31, 1275–1285.

[78] Curry, J. A., W. B. Rossow, D. Randall and J. L. Schramm. Overview of Arctic Cloud and Radiation Characteristics. J. Clim. 9, 1731–1764. 1996.

[79] Brook, M., M. Nakano, P. Krehbiel and T. Takeuti, 1982: J. Geophys. Res. 87, 1207–1215.

[80] Stolzenberg, M., T. Marshall, W. Rust and B. Smull, 1994: Mon. Wea. Rev. 122, 1777–1797.

[81] Latham, J. The Electrification of Thunderstorms. Quart. J. Roy. Meteor. Soc. 107, 277–298, 1981.

[82] Takahashi, T. Riming electrification as a charge generation mechanism in thunderstorms. J. Atmos. Sci. 35, 1536–1548, 1978.

[83] Jayaratne, E., C. Saunders and J. Hallett, 1983: Q. J. Roy. Met. Soc. 109, 609–630.

[84] Baker, B., M. Baker, E. Jayaratne, J. Latham and C. Saunders, 1987: Q. J. Roy. Met. Soc. 113, 1193–1215.

[85] Illingworth, A. J. Charge Separation in Thunderstorms: Small Scale Processes. J. Geophys. Res. 90, 6026–2032. 1985.

[86] Baker, M. and J. G. Dash, 1994: Mechanism of Charge Transfer between Colliding Ice Particles in Thunderstorms. J. Geophys. Res. 99, D5, 10,621–10,626.

[87] J. Rydock and Williams, E.: Charge separation associated with frost growth. Q. J. Roy. Met. Soc. 117, 409–420. 1991.

[88] Avila, E. and G. Caranti: A Laboratory study of static charging by fracture in ice growing by riming. J. Geophys. Res. 99, 10611–10620, 1994.

[89] Beard, K. V. and H. T. Ochs. Charging Mechanisms in Clouds and Thunderstorms. in The Earth's Electrical Environment. NAtional Academy Press, Washington D.C. 1986.

[90] Latham, J. and B. J. Mason. Electric Charge Transfer Associated with Temperature Gradients in Ice. Q. J. Roy. Met. Soc. XX,523–535. 1961.

[91] Caranti, J. M., A. J. Illingworth and S. J. Marsh. The Charging of Ice by Differences in Contact Potential. J. Geophys. Res. 90, 6041. 1985.

[92] Workman, E. J. and S. E. Reynolds. Electrical phenomena occurring during the freezing of dilute aqueous solutions and their possible relationship to thunderstorm electricity. Phys. Rev. 78, 254–259, 1950.

Physico-Chemistry of Polar Stratospheric Clouds

THOMAS PETER
Max-Planck-Institut für Chemie, Mainz, Germany

Abstract. Liquid and solid particles in polar stratospheric clouds are of central importance for the depletion of stratospheric ozone. Surface-catalyzed and diffusion-controlled bulk reactions on/in the particles, convert halogens, which derive from compounds of mainly anthropogenic origin, from relatively inert reservoir species into forms that efficiently destroy ozone. The microphysics of these particles under cold stratospheric conditions is still uncertain in many respects, in particular concerning phase transitions like freezing nucleation and deposition nucleation. Furthermore, there are indications that the rates of key heterogeneous reactions have not yet been established with sufficient accuracy to enable a reliable diagnosis of observed ozone losses by means of global models. This paper is a shortened version of an Annual Review of Physical Chemistry[1], which surveys this rapidly developing field.

1. Introduction

The stratospheric ozone layer was an early consequence of oxygen production by life on Earth. Through its shielding against harmful ultraviolet radiation it strongly influenced life's development and is still responsible for its continuous protection. Natural and anthropogenic threats to the ozone layer have received increasing attention during the last three decades of this century. Initially homogeneous gas phase processes like NO_x-catalyzed[2] and ClO_x-catalyzed[3] ozone depletion were the focus of scientific interest*. These processes were expected to destroy ozone at high altitudes ($\gtrsim 30$ km) on

*The "odd nitrogen family" is defined as $NO_x=NO+NO_2+2\times N_2O_5+...$ These species are also called "active nitrogen", as they are easily available in ozone-destroying chemical cycles. Deactivation converts NO_x into rather inert "reservoir" gases like HNO_3 and $ClONO_2$. By analogy, $ClO_x=Cl+ClO+2\times Cl_2O_2+HOCl+...$ is called odd chlorine and $HO_x=H+OH+HO_2+...$ odd hydrogen.

NATO ASI Series, Vol. I 56
Ice Physics and the Natural Environment
Edited by John S. Wettlaufer, J. Gregory Dash and Norbert Untersteiner

a global scale. Only after the discovery of the Antarctic ozone hole[4] has the importance of rapid heterogeneous chemistry on the surface and in the bulk of minute particles been recognized, in particular in the polar winter stratospheres[5]. This class of reactions had previously been largely overlooked, and polar stratospheric clouds (PSCs), which provide the necessary heterogeneous surfaces, had been taken for meteorological curiosities. The stratosphere is not prone to cloud formation due to its dryness, and only the extremely cold conditions of the polar winter stratospheres are favorable for clouds. The more surprising it was that these clouds could strongly influence atmospheric chemistry *via* activation of halogen species. Such reactions include

$$ClONO_2 + HCl \xrightarrow{\text{particle}} Cl_2 + HNO_3\,, \tag{1a}$$

$$ClONO_2 + H_2O \xrightarrow{\text{particle}} HOCl + HNO_3\,, \tag{1b}$$

$$HOCl + HCl \xrightarrow{\text{particle}} Cl_2 + H_2O\,, \tag{1c}$$

and deactivation of nitrogen oxides

$$N_2O_5 + H_2O \xrightarrow{\text{particle}} 2\,HNO_3\,. \tag{2}$$

Today we know that these heterogeneous reactions, next to some radical reactions, belong to the fastest stratospheric reactions of all. Similar heterogeneous reactions to those above take place when one or both of the chlorine atoms are exchanged by bromine or iodine. While halogen compounds like $ClONO_2$ and HCl are relatively inert gases, the products (Cl_2, HOCl, BrCl, HOBr, Br_2, etc) photolyze readily and release halogen radicals which can destroy ozone directly in catalytic cycles[6]. Nitrogen oxides, on the other hand, have the potential to deactivate the halogen radicals, e.g. $ClO+NO_2+M \rightarrow ClONO_2+M$, and to decelerate ozone destruction under chemically perturbed, halogen-dominated conditions (M is an inert collision partner). Reactions (1a), (1b), and (2) lead to the formation of rather inert HNO_3, thereby temporarily reducing active nitrogen. This process of "denoxification" is a prerequisite for efficient halogen-induced ozone destruction.

Besides the direct effects of heterogeneous reactions, equally important indirect effects of PSCs on gas phase chemistry were revealed. It was suggested[7,8] that HNO_3 condenses onto the aerosol particles allowing them to grow at temperatures above those at which pure ice crystals can exist*

*This is the temperature at which the H_2O partial pressure of the air is equal to the vapor pressure of ice ($p_{H_2O} = p^{vap}_{H_2O}$), often called the "frost point", T_{ice}. Similar definitions concern the NAT and SAT equilibrium temperatures (T_{NAT}, T_{SAT}). The frost point is often reached in the Antarctic winter, however, in the warmer Arctic synoptic (i.e large scale) temperatures usually remain a few degrees above the frost point.

In contrast to gaseous HNO_3, the condensed HNO_3 does not photolyse easily, so that resupply of nitrogen oxides *via* $HNO_3 + h\nu \rightarrow OH + NO_2$ is cut off. Furthermore, permanent removal of nitrogen species due to sedimentation of HNO_3-containing particles out of the stratosphere (denitrification) enhances the dramatic ozone destruction in Antarctica and might become important for future ozone loss in the northern hemisphere as well. Effectively, the discovery of the ozone hole and its elucidation in terms of such rapid particle-induced processes conveyed the ozone problem from a global scale theoretical prediction into an easily measurable localized phenomenon. Political and economic action has been taken since.

2. PSC Microphysics

Investigations since the mid-1980s strongly indicate that PSC particles consist of merely three species: water, nitric acid and sulfuric acid. Surprisingly, at low stratospheric temperatures (200–180 K) the physico-chemistry even of this distinct ternary system is still poorly known, although substantial progress has been made in understanding its thermodynamics and kinetics during the last ten years. One of the most important recent changes in our understanding of PSC microphysics is the recognition of liquid aerosols in the cold polar stratosphere. Even as recently as 1993 PSCs were thought to consist solely of solid particles, while it is now evident that a frequently observed type of PSC (so-called type-1b) consists of ternary $HNO_3/H_2SO_4/H_2O$ solution droplets. This conclusion was reached by comparing results of thermodynamic models[9,10] of electrolytic solutions with *in situ* particle observations[11].

From lidar* measurements there is clear evidence for both solid and liquid PSCs in the polar stratospheres. The volume depolarization of backscattered light in aerosol-free air is 1.4 % (due to the polarizability of the air molecules), and data below this value indicate liquid type-1b clouds. Ternary solution droplets in such clouds yield substantial backscatter, but only in the parallel polarization plane. In contrast, type-1a clouds have large depolarization ratios but small backscatter. Toon *et al.*[12], referring to lidar measurements of Browell *et al.*[13], were the first to describe these characteristics. They showed that type-1a clouds consist of only a few very large aspherical particles, but the identification of type-1b as ternary liquids was made only later when the composition and size of stratospheric droplets were calculated.

A glaring deficiency in our current knowledge of PSCs is the description of phase transitions like freezing nucleation or deposition nucleation. Results from laboratory freezing studies using solutions with stratospheric

*lidar = LIght Detection And Ranging. Polarized light is emitted by a ground-based or airborne laser and backscattered by air molecules and particles. Spherical objects like droplets do not depolarize the backscattered light. Aspherical objects like air molecules, and in particular crystalline particles, cause depolarization.

equilibrium composition indicate that there is no apparent simple mechanism leading to droplet freezing, at least not above the ice frost point which is rarely reached on a synoptic (i.e. large) scale in the Arctic[14]. Attention must possibly be directed to non-equilibrium processes in order to explain solid PSCs.

The thermodynamics of various solid phases of potential stratospheric importance have been investigated in the past. For the three thermodynamically stable solid PSC phases — water ice, nitric acid trihydrate (NAT $\equiv$ $HNO_3{\cdot}3H_2O$) and sulfuric acid tetrahydrate (SAT $\equiv$ $H_2SO_4{\cdot}4H_2O$) — the thermodynamic properties are well known, whereas the thermodynamics of metastable phases such as nitric acid dihydrate (NAD $\equiv$ $HNO_3{\cdot}2H_2O$) or sulfuric acid hemihexahydrate (SAH $\equiv$ $H_2SO_4{\cdot}6.5H_2O$) are less well determined.

Figure 1 shows the development of an initially liquid sulfuric acid particle exposed to falling temperatures. Two possible evolutions are depicted depending on whether it freezes. The path on the left shows a particle which freezes as sulfuric acid tertahydrate between 195 and 215 K. The frozen particle may then serve as a solid nucleus for NAT at temperatures below the NAT equilibrium temperature[15], T_{NAT} (about 195 K for typical lower stratospheric conditions). Upon further cooling water ice becomes stable below the ice equilibrium temperature, T_{ice}, also called the frost point. This scheme is the conventional 3-stage concept of PSC formation[16] which is based on lidar observations showing, besides the background aerosol particles, two distinct PSC types: the faint type-1, presumably NAT or other HNO_3-rich particles, and the much denser type-2, presumably water ice. This scheme received much attention when PSC measurements in the Antarctic appeared to show the onset of particle growth after the saturation temperature with respect to NAT was reached[17]. However, these Antarctic observations were made in late winter and could not reveal how the PSCs were initially formed.

The path on the right hand side of Fig. 1 shows a particle which does not freeze above the frost point. Instead, it remains liquid and takes up substantial amounts of HNO_3, leading to an $HNO_3/H_2SO_4/H_2O$ ternary solution droplet. In this case freezing is assumed to occur only below the frost point, where water ice precipitates. The two paths yield very different results even below the frost point: a SAT-NAT-ice particle with the acids in the core and the ice on the outside of the particle *versus* a particle with an ice core and the acids as liquid, or eventually frozen solid, on the outside of the particle. Clearly, kinetic properties of such particles would be quite different. Although the 3-stage concept of PSC formation is still used in most of the large scale 2-D and 3-D models, its severe limitations have been recognized.

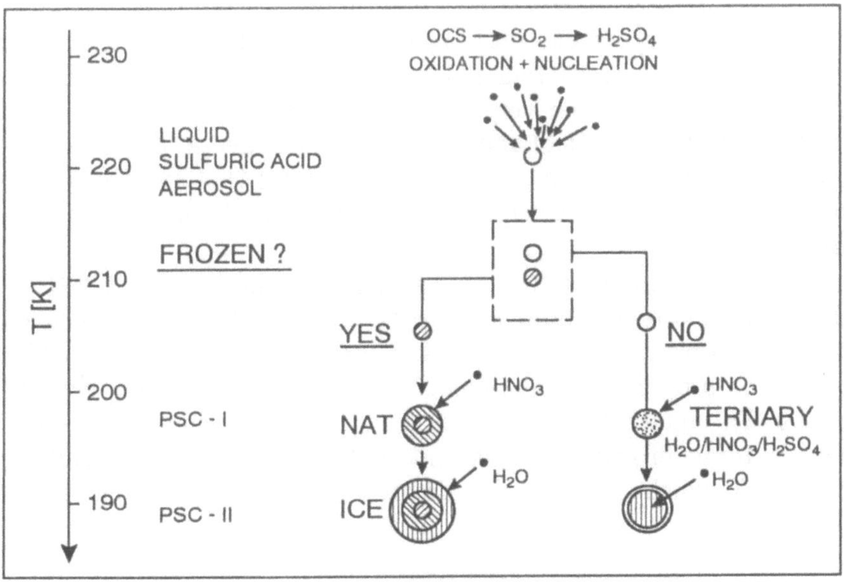

Figure 1. **Two possible pathways for the stratospheric aerosol upon cooling. Left path: conventional 3-stage concept (//// SAT, \\\\ NAT, |||| ice). Right path: aerosol remains liquid, takes up HNO_3 and freezes out water ice below the frost point.**

2.1. THERMODYNAMICS OF LIQUID PSCS

The thermodynamic properties of stratospheric ternary solutions have been described in detail in a recent review by Carslaw *et al.*[23]. Here only some important points are summarized. The thermodynamics of ternary solutions now appear to be reasonably well determined and widely accepted, in contrast to the metastable solid hydrates of potential importance for PSCs. This is due to the employment of thermodynamic models which, based on measurements at higher temperatures, allowed extrapolation to the relevant conditions.

2.1.1. *H_2SO_4/H_2O Binary Solutions*

The existence of a global stratospheric aerosol layer between 15 and 30 km altitude has been known since the early sixties. The aerosol particles consist mainly of H_2SO_4 and H_2O and exist mostly in the liquid state[19]. The thermodynamic properties of the binary H_2SO_4/H_2O system have been well established. The H_2SO_4 vapor pressures of these aqueous solutions are so low (typically $p_{H_2SO_4} < 10^{-14}$ mbar for lower stratospheric conditions) that H_2SO_4 is usually treated as involatile. Only above 30 km altitude, where the temperature in the stratosphere rises above about 245 K, does $p_{H_2SO_4}$ become appreciable, leading eventually to the complete evaporation of the droplets. Under typically lower stratospheric conditions droplet concentrations range from 60 to 80 wt% H_2SO_4 depending on temperature.

2.1.2. $HNO_3/H_2SO_4/H_2O$ *Ternary Solutions*

The onset of HNO_3 uptake into H_2SO_4/H_2O solutions was first inferred from vapor pressure measurements by Molina and coworkers[20]. These authors concluded that under very cold stratospheric conditions the solution droplets would take up H_2O and HNO_3 and then freeze. However, laboratory freezing experiments by Koop *et al.*[14] showed that even bulk samples of such solutions can be brought to extremely low temperatures without freezing.

Our present understanding of the thermodynamics of $HNO_3/H_2SO_4/H_2O$ ternary solutions under stratospheric conditions still depends to a high degree on predictions made by thermodynamic models, which allow to calculate the properties of non-ideal, i.e. highly concentrated, electrolytic solutions. The interactions between the species in such solutions are expressed in terms of activity coefficients (f). For example, the solubility of a species HX which dissolves and dissociates in solution can be calculated according to

$$\mathrm{HX(g)} \overset{k_H}{\rightleftharpoons} \mathrm{HX(aq)} \qquad k_H = \frac{f_{\mathrm{HX}} m_{\mathrm{HX}}}{p_{\mathrm{HX}}^{\mathrm{vap}}} \tag{3a}$$

$$\mathrm{HX(aq)} \overset{K}{\rightleftharpoons} \mathrm{H^+ + X^-} \qquad K = \frac{f_{\mathrm{H}} m_{\mathrm{H}} f_{\mathrm{X}} m_{\mathrm{X}}}{f_{\mathrm{HX}} m_{\mathrm{HX}}} \tag{3b}$$

where k_H is the Henry's law constant relating the vapor pressure $p_{\mathrm{HX}}^{\mathrm{vap}}$ with the molality m_{HX} (in mol/kg) of the undissociated species HX(aq); K is the dissociation constant of HX, and m_{H} and m_{X} are the molalities of the dissociation products. The total solubility of undissociated and dissociated species is given by the effective Henry's law constant $H^\star = (m_{\mathrm{HX}} + m_{\mathrm{X}})/p_{\mathrm{HX}}^{\mathrm{vap}} = k_{\mathrm{H}}/f_{\mathrm{HX}} + Kk_H/f_{\mathrm{X}} f_{\mathrm{H}} m_{\mathrm{H}}$.

The constants k_H and K are known functions of temperature only. In very dilute solutions the activity coefficients f tend to unity (equivalent to no interaction between the species in solution) and the calculation of solubilities is straight-forward: a set of equations (3a,b) for each dissolved species plus an equation for electroneutrality must be solved iteratively. However, for the highly concentrated electrolytic solutions of stratospheric aerosols the interaction between the various constituents (H^+, HSO_4^-, SO_4^{2-}, NO_3^-, H_2O) are strong so that the activity coefficients can assume values of 10^7 and modeling of the f's is required[18]. Ion–Interaction models contain a far-field ion-interaction term and two-body and three-body short-range collision terms which are parameterized by fitting to thermodynamic data. The models can be used to interpolate thermodynamic quantities where measurements exist, and to extrapolate into unknown regimes (e.g. to lower temperature). Such an extrapolation is thermodynamically consistent and therefore much more reliable than straight-forward extrapolation of a single data set.

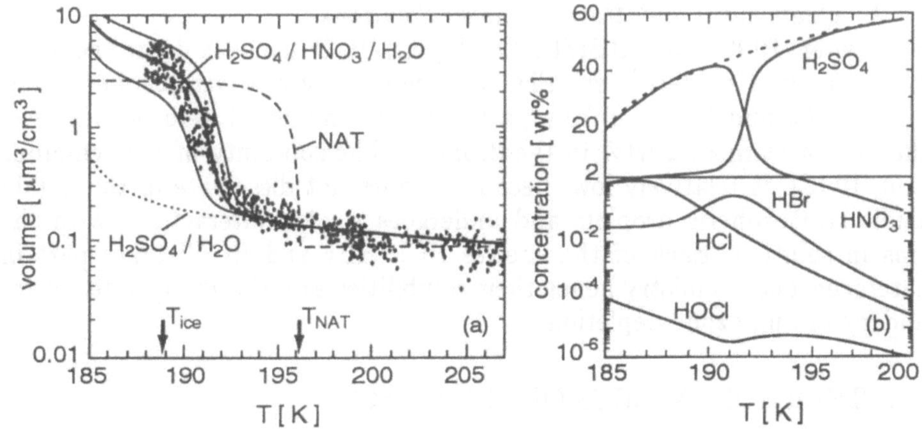

Figure 2. (a) Total particle volume observed by Dye *et al.*[11] and (b) liquid concentrations (wt%) calculated[8] as a function of temperature for a flight of the American high-altitude research aircraft ER-2 on 24 January 1989, northbound from Stavanger, Norway. Flight level was 55 mbar (about 19 km altitude). Amount of H_2SO_4 in the aerosol corresponds to normal background conditions (0.53 ppbv H_2SO_4 total mixing ratio). With decreasing temperatures the particles initially grow due to H_2O uptake by the H_2SO_4/H_2O binary (dotted line for 5 ppmv H_2O). Below 193 K the volume increase is much stronger, and is in agreement with the combined H_2O and HNO_3 uptake by the ternary aerosol (heavy solid line for 10 ppbv HNO_3). NAT particle growth without nucleation barrier is shown by the dashed line. NAT and ice equilibrium temperatures are indicated by vertical arrows. In (b), note the change from linear to logarithmic scales below 2 wt%. Adapted from Carslaw *et al.*[9].

An application of the thermodynamic electrolyte models is shown in Fig. 2. Panel (a) shows measurements of total particle volume (in $\mu m^3 cm^{-3}$) as measured by Dye *et al.*[11] on 24 January 1989 in the Arctic. The total particle volume increases only slowly with falling temperatures above about 193 K, but increases steeply at temperatures below 192 K. Dye *et al.* identified the weak particle growth with falling temperature above 193 K as uptake of H_2O, see dotted curve in Fig. 2a for 5 ppmv H_2O, but invoked freezing to explain the growth at lower temperatures. However, if NAT formed and took up HNO_3 until the condensed and the gas phase were in equilibrium, the particle volume would follow the dashed curve in Fig. 2a. Hence, to explain these observations in terms of NAT formation, some kind of non-equilibrium growth would be required. In contrast, the solid curve shows that the observed PSC volume can be explained in terms of simultaneous uptake of HNO_3 and H_2O by liquid aerosol particles in equilibrium without any phase transition[9]. Figure 2b shows the equilibrium composition of the liquid particles in Fig. 2a. Below 192 K, HNO_3 becomes the dominant condensed acid, and H_2SO_4 drops to below 3 wt%.

2.1.3. *Quaternary and More Complex Solutions*

The system $HOCl/HCl/HBr/HNO_3/H_2SO_4/H_2O$ has been described using ion-interaction models[18], see Fig. 3b. Under very cold stratospheric conditions HCl and HBr can fully partition into the liquid aerosol, but even then reach at most 2 wt% in the droplets. The solubility of hypochlorous acid, HOCl, is relatively low, because it does not dissociate in the highly acidic stratospheric droplets and undergoes weaker interactions with the ions in solution. Each of the species HCl, HBr and HOCl takes part in heterogeneous chemistry hence their solubilities are of primary interest for understanding ozone depletion.

2.2. THERMODYNAMICS OF SOLID PSCS

Crystalline hydrates and water ice are thermodynamically stable phases of the $HNO_3/H_2SO_4/H_2O$ system under stratospheric conditions. Their properties were investigated earlier than liquid PSCs. However, since 1993 the metastable liquid phase has received most attention, and our understanding of both microphysics and heterogeneous chemistry of liquid stratospheric PSCs has outpaced that of the crystalline phases.

2.2.1. *Water Ice*

Mother-of-pearl clouds at 15–25 km altitude have been observed in the Arctic for more than 100 years[21]. They form preferentially over mountains, where rapid vertical perturbations can cause the air to cool adiabatically to very low temperatures, often below the ice frost point (T_{ice}). It is clear that these clouds consist mainly of water ice, because only water is abundant enough in the stratosphere (typically 5 ppmv) so that clouds can become visible (particle radii larger than 1 μm and number densities up to 10 cm^{-3}). The ice particles probably also contain traces of HNO_3, either in the form of NAT cores on which the ice forms by deposition nucleation or as individual NAT clusters due to co-condensation with the water ice. Similarly, traces of HCl and other gases may co-condense. Both sub-types of stratospheric ice clouds, mesoscale mother-of-pearl clouds and synoptic scale ice clouds which frequently occur in the Antarctic, are referred to as type-2 PSCs, in contrast to type-1 PSCs which are rich in HNO_3.

2.2.2. *Nitric Acid Trihydrate (NAT)*

The vapor pressure of NAT has been determined with high accuracy by Hanson and Mauersberger[15]. The measured data behave according to the Gibbs-Duhem relation, which for a two component system reads

$$p_i^n \cdot p_j^m = f(T)\,, \tag{4}$$

where p_i and p_j are the vapor pressures of the two components, n and m are their relative molar abundances in the condensed phase, and f is a function only of temperature. For NAT, $p_{HNO_3} \cdot p_{H_2O}^3 = f(T)$ with $f(T)$ determined

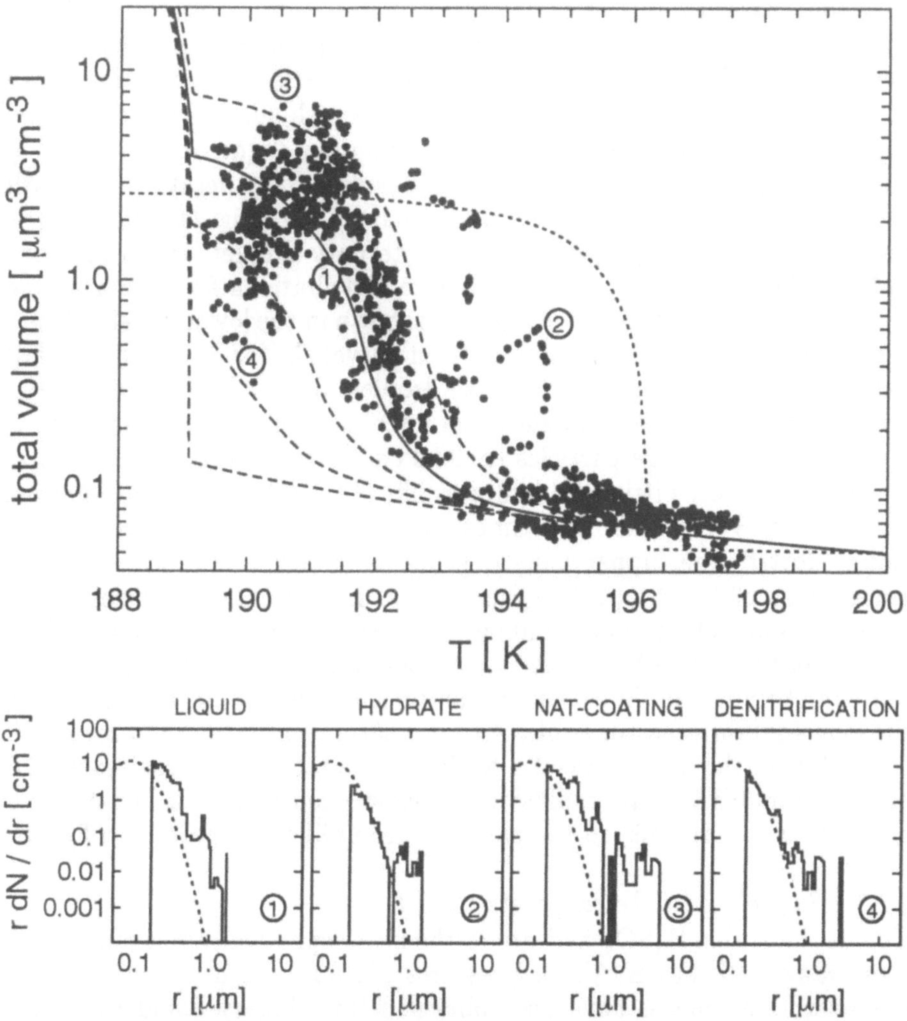

Figure 3. Upper panel: Total particle volume observed by Dye *et al.*[11] under conditions as in Fig. 2 but for 16 January 1989. Calculations for 5 ppmv H_2O. Solid line: ternary solution droplets for 10 ppbv HNO_3. Dashed lines: droplets for 20, 5, 2, 0 ppbv HNO_3. Dotted line: NAT for 10 ppbv HNO_3. Lower panels: examples of the measured particle size distribution (cases 1–4 refer to the corresponding points in the upper panel). (1) Ternary solution droplets at 193.3 K. (2) Type-1a cloud at 194.8 K. (3) Large particles at 190.5 K tentatively identified as NAT-coated ice particles (see text). (4) Particles under strongly denitrified conditions at 190.2 K (Beiping Luo, personal communication).

by Hanson and Mauersberger[15]. For typical HNO_3 and H_2O abundances in the lower polar stratosphere the NAT equilibrium temperature is about 7 K higher than the frost point. It is clear that this is a dramatic difference when taking into account that the Arctic stratosphere on a synoptic scale hardly ever reaches T_{ice}. On the other hand, there is little compelling evidence that solid particles in the polar stratospheres really consist of NAT. One major problem is that chemical *in situ* analyses of PSC particles are not yet available. But even the existing optical particle counter measurements furnish little evidence of fully developed NAT clouds[11]. Figure 3 shows a similar measurement as Fig. 2a but for another flight during AASE (16 January 1989). Although the measured volume/temperature correlation is not nearly as compact as on 24 January 1989 (Fig. 2a), the observed volumes are frequently close to those of ternary droplets (solid line for 10 ppbv HNO_3, 5 ppmv H_2O). Particle size distributions for the data points (1) to (4) in the upper panel of Fig. 3 are shown in the lower panels. Case (1) shows a background distribution (dotted curve) shifted as a whole to larger particle sizes indicating growth of all droplets due to HNO_3 uptake (type-1b characteristics). Case (2) shows an unchanged droplet distribution, but in addition a few large particles ($\simeq 0.02$ cm^{-3}) with radii between 0.5 and 2 μm, presumably nitric acid hydrates like NAT. These particles, however, have a total volume about 5-times smaller than a fully developed NAT cloud (dotted line, upper panel), which is in agreement with the small backscatter of lidar type-1a observations. Either these particles are not NAT but some other compound, as has been suggested by Tabazadeh and Toon[22], or they are NAT particles that had insufficient time to grow to their full equilibrium size. Such kinetic limitation would be in agreement with the long equilibration times of NAT: at a saturation ratio 2 (corresponding to 195.3 K in the present case) a NAT particle requires approximately 15 h to grow from 0.5 to 2 μm radius[23], see below. Cold periods in the Arctic are often shorter than this, so the particles are unlikely to reach equilibrium.

2.2.3. *Sulfuric Acid Tetrahydrate (SAT)*

This is the stable H_2SO_4 hydrate in the stratospherically relevant temperature range 192–215 K. Vapor pressures of SAT have not been measured, but reliable estimates have been obtained from binary-liquid vapor pressures at the SAT melting point using the Gibbs-Duhem relation[24]. SAT has a high melting point under stratospheric conditions of 210–218 K (depending on relative humidity). Less evident, but having at least as important consequences for PSCs, is that SAT deliquesces when temperatures fall below 191.5 K (at 50 mbar), i.e. SAT becomes unstable with respect to the $HNO_3/H_2SO_4/H_2O$ ternary system slightly above the frost point[25]. The presence of NAT can protect SAT against deliquescence.

2.2.4. *Other Acid Hydrates*

A number of other hydrates of potential stratospheric importance have been reported in the literature. All these hydrates are metastable with respect to NAT or SAT, but the formation of metastables preceeding the stable phase is a thermodynamically common feature according to Ostwald's rule[26]. Most prominent metastable hydrates are nitric acid dihydrate (NAD), sulfuric acid hemihexahydrate (SAH), nitric acid pentahydrate (NAP) and a mixed hydrate $H_2SO_4{\cdot}HNO_3{\cdot}5H_2O$. The melting points and vapor pressures of these metastable hydrates have not been determined with satisfactory accuracy to allow reliable comparisons with field data. Estimates of their stability regions are given in Fig. 4.

2.3. GROWTH AND EVAPORATION KINETICS

The thermodynamics of stratospherically relevant hydrates and liquid mixtures have received much attention in past laboratory, field and modeling work. In contrast, the kinetics of particle growth or evaporation under stratospheric conditions has been largely neglected. This is a problem of gas diffusion and surface accommodation, i.e. how rapidly molecules are transported between the free gas phase and a molecular site on the particle surface. Slow diffusion can cause large supersaturations or subsaturations after cooling or warming of air parcels. The driving force is the difference between the partial pressure, p_i^{∞}, and the vapor pressure, p_i^{vap}, of the species i under consideration, e.g. H_2O, HNO_3, HCl, etc. In contrast to the troposphere, where water vapor is present in high abundances so that equilibration times are short and supersaturations low (at most a few percent), in the stratosphere partial pressures and their gradients are much lower, the resulting gas transport much slower and the supersaturations high (factors of ten). While type-2 PSCs need minutes to grow or evaporate due to the still relatively high H_2O partial pressures, type-1 PSCs need hours or days due to the 500-times lower HNO_3 pressures. Thus, the interpretation of field data as thermodynamic measurements is not always warranted, and even the interpretation of laboratory experiments might suffer from the neglect of slow kinetics.

The change in the number of molecules N_i of species i contained in a particle with radius r as a function of time is given by[27]

$$\frac{dN_i}{dt} = 4\pi r \frac{D_i^{\star}}{k_B T}(p_i^{\infty} - p_i^{\mathrm{vap}}(r)) \, , \tag{5}$$

where k_B is the Boltzmann constant and $D_i^{\star}$ the modified gas phase diffusion constant. Equation (5) holds, in principle, for both liquid and solid particles (in the latter case the effective radius should be used for r and an overall factor has to be included if the particle is not spherical, without changing the general picture). According to the Hertz-Knudsen equation[27] the modified gas phase diffusion constant $D_i^{\star}$ takes into account the smooth transition between diffusively controlled gas uptake ($dN_i/dt \propto r$) and kinetically limited uptake in the Knudsen regime ($dN_i/dt \propto \alpha r^2$, α being the mass accommodation coefficient of molecules of species i on the surface). In Eq.(5), $p_i^{\mathrm{vap}}(r) = K(r) p_i^{\mathrm{vap}}(r = \infty)$, where the Kelvin factor K takes account of the change in the vapor pressure due to the surface curvature, which is important for the smallest droplets ($r \lesssim 0.1$ μm). The diffusive mass transport through the gas phase and the accommodation onto the surface as a function of droplet radius are fully described by Eq.(5).*

Besides the strong dependence of equilibration times on composition, the particle radius plays a decisive role. For particles larger than the mean free path (λ) in the gas phase (typically $\lambda = 1\,\mu m$ at 50 mbar), growth/evaporation rates scale linearly with radius (r), for smaller particles even with r^2. Thus, while small particles might not have difficulties establishing rapid equilibrium, larger particles can fall behind and do not take up or evaporate their share. This has important implications for the radial particle distribution and might trigger important processes like droplet freezing (see below).

2.3.1. *Growth/Evaporation of Ternary Droplets*

Size-dependent equilibration of ternary droplets has been described in detail by Meilinger *et al.*[28]. They showed that according to Eq.(5) the e-folding time for composition equilibration after infinitesimal temperature changes is $\tau_i = r^2 \rho_w H_i^* RT/(3D_i^* K)$, where ρ_w is the mass density of water in the solution. For typical lower stratospheric conditions this yields $\tau_{H_2O} \approx 5\,s \times r^2(1 + 1.6\mu m/r)$ for H_2O uptake and $\tau_{HNO_3} \approx 1\,h \times r^2(1 + 1.6\mu m/r)$ for HNO_3 uptake, both scaling linearly with the particle radius (for $r < \lambda$). As argued above, water equilibrates almost instantaneously, while nitric acid equilibration of the larger droplets may take many hours. This can lead to HNO_3 concentrations in the small droplets close to NAT and NAD stoichiometry, so that their freezing is conceivable. Tsias *et al.*[29] concluded that in rapid mesoscale temperature fluctuations with cooling and heating rates of more than 100 K/h the droplets would freeze as NAD.

2.3.2. *Growth/Evaporation of NAT Particles*

Equation (5) also describes the uptake kinetics of solid particles. The time for doubling a particle radius is given by $\tau_2 = 3r^2/2\beta|S-1|$ with a constant saturation ratio $S = p_i^\infty/p_i^{vap} > 1$ (supersaturation), and the time for halving the radius $\tau_{1/2} = \tau_2/4$ with constant saturation ratio $S < 1$ (subsaturation). Here, $\beta = m_i p_i^{vap} D_i^*/\rho k_B T$ is a measure of the lifetime of these particles (m_i is the molecular mass of species i and ρ the solid state density of the particle). Lifetimes for pure ice particles are in the order of minutes, and for NAT particles typically hours or days due to their much lower vapor pressure[30,23].

Unfortunately the mass accommodation coefficient α is unknown for solid surfaces under stratospheric conditions. It contains information on crystal growth habit and virtually nothing is known about this. MacKenzie

---

*Mixing inside droplets is usually much faster than the mass transport in the gas phase under stratospheric conditions[28]. Also, diffusion of latent heat to or from the particles is irrelevant, in contrast to tropospheric particles.

and Haynes[31] investigated theoretically the effect of spiral growth (screw dislocations) on the development of stratospheric ice particles, but it is not clear whether this can also be applied to stratospheric hydrates. Alternatively, it has been argued[32] that the solid surfaces might resemble those of the liquids by developing a quasi-liquid layer similar to the one on ice at tropospheric temperatures. If true, new concepts need to be developed for describing growth and trace gas uptake.

2.3.3. *Growth/Evaporation of NAT-Coated Ice*

Figure 3 also shows observations where the total particle volume is much larger than that of ternary droplets or NAT particles (panel 3), and others which are much smaller than ternaries (panel 4). Tabazadeh and Toon[22] investigated these data and suggested that a total aerosol volume larger than NAT and ternary solutions could be either amorphous solids or hydrates $HNO_3 \cdot nH_2O$ with $n \geq 5$. However, to explain the largest volumes observed (Fig. 4, panel 3) practically all particles would have to be activated as hydrates with $n > 20$. There is no evidence from laboratory data to support the existence of such high hydrates. More importantly, the radial distribution (panel 3) consists of a well-established liquid ternary spectrum plus some large particles with radii enhanced by up to a factor of 5 compared to the liquid particles with the same number density, i.e. with volumes enhanced by two orders of magnitude. This enhancement would correspond to hydrates with $n > 100$, which is even more unlikely. Alternatively, Wofsy *et al.*[33] and Peter *et al.*[23] suggested that such water-rich particles could be ice particles which, by means of a NAT-coating, can exist above the frost point. The NAT-coating leads to protection of the particles, which, despite their large ice-core, behave kinetically like NAT particles. If NAT is supersaturated the particles lifetime can in principle become arbitrarily long.

The occurrence of denitrification might be closely related to the NAT-coating issue, because uncoated ice is not likely to survive long enough to allow for substantial denitrification under the relatively warm Arctic conditions. Panel (4) in Fig. 4 shows such a highly denitrified case: $HNO_3$was measured to be below 5 ppbv in contrast to more than 10 ppbv for panels (1)–(3). This explains the small total volume (compare dashed lines for 5 and 2 ppvb HNO_3 in Fig. 4).

2.4. PHASE TRANSITIONS DUE TO NUCLEATION

Two important kinds of PSC phase transitions require nucleation: freezing nucleation of ice and H_2SO_4 or HNO_3 hydrates from liquid ternary solutions, and deposition nucleation of ice and HNO_3 hydrates on preexisting dry nuclei (without participation of the liquid phase). As mentioned above, an understanding of these phase transitions presents a major obstacle to obtaining a clear picture of solid PSC formation. This situation is complicated not only by the difficulty of performing laboratory nucleation experiments on aerosols under stratospheric conditions, but also by the lack of quasi-Lagrangian measurements revealing the full PSC life-cycle in the field and the insufficiency of the classical nucleation theory, which is the only theoretical approach used so far. Theoretical concepts have been reviewed thoroughly by MacKenzie *et al.*[34].

There is no doubt that water ice forms when temperatures fall 2 K below the frost point due to freezing nucleation in stratospheric ternary solution droplets. At these temperatures the droplets become so dilute that they basically represent water droplets with small amounts of HNO_3, and their freezing rates become higher than $10^{20} cm^{-3} s^{-1}$ which lets even the smallest droplets freeze. Similarly it is expected that at about $T_{ice} - 2\,K$ ice would form on pre-existing solid particles (such as NAT) due to deposition nucleation[16,12,35]

First attempts to explore the freezing rate of SAT from binary H_2SO_4/H_2O solutions based on classical homogeneous nucleation theory showed that SAT is very unlikely to freeze[24]. Later freezing experiments by Koop *et al.*[14] have proven systematically that neither binary nor ternary solutions with compositions corresponding to stratospheric droplets in thermodynamic equilibrium at temperatures from 273 to 188 K freeze homogeneously. This led Koop *et al.*[24] to conclude that the freezing of the hydrates would probably occur heterogeneously on ice below the frost point. However, temperatures in the Arctic stratosphere, in contrast to the Antarctic, are usually not sufficiently low for ice growth, so other mechanisms are possibly responsible for the formation of hydrates above the ice frost point. Potential mechanisms include heterogeneous nucleation on nuclei that are specially suited for hydrate formation[36], the occurrence of amorphous states which freeze only upon warming[22], or the freezing of non-equilibrium droplets in mesoscale temperature fluctuations[29].

Figure 4 shows the "stratospheric particle zoo" with all currently known or speculated transitions between different particles[37]. Phases shown are either marked as liquid (l) or as a sulfuric acid hydrate (SAX, with X = M, T, H, etc., including the mixed hydrate MIX) or nitric acid hydrate (NAX, with X = T, D, P etc.). On the left hand side the diagram shows the stability regions of the solid phases under typical stratospheric conditions at 50 mbar (5 ppmv H_2O, 10 ppbv HNO_3). The right hand side of Fig. 4 shows transitions relating these phases. Thermodynamically allowed transitions that are also kinetically feasible are marked by an arrow ($\rightarrow$), while transitions which are thermodynamically allowed but kinetically so slow that they will not occur in the stratosphere are marked by crossed arrows ($\not\rightarrow$).

The solid arrows in Fig. 4 connect particles in thermodynamic equilibrium, while dashed arrows indicate processes which start from non-equilibrium states. NAT-coating is an example, where the ice is protected by a NAT-layer and not in equilibrium with the gas phase. Three processes

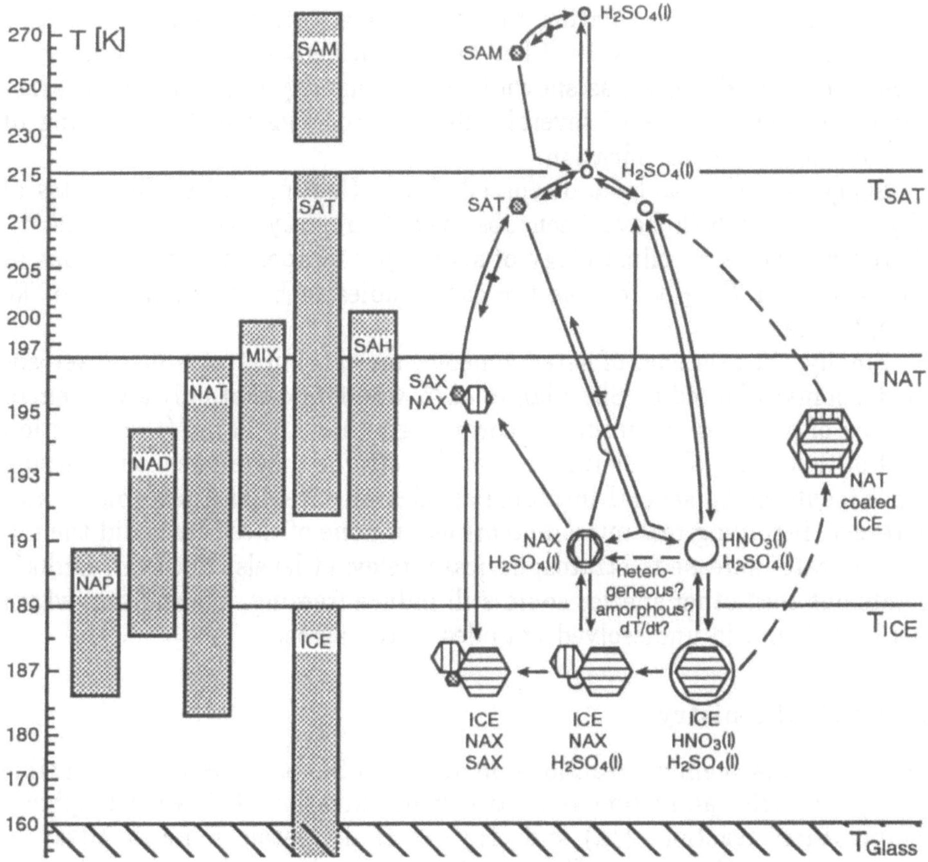

Figure 4. **Stability regions of HNO_3 and H_2SO_4 hydrates and water ice (left) and phase transitions among these solids and with ternary solutions (right). Horizontal lines mark equilibrium temperatures for SAT, NAT and ice. Solid arrows: kinetically favored transitions. Crossed arrows: kinetically hindered. Particles in thermodynamic equilibrium are connected by solid arrows, in non-equilibrium by dashed arrows. Adapted from Koop *et al.*[37].**

have been proposed recently which might lead to freezing above T_{ice} (horizontal dashed arrow in Fig. 6):

(a) Tabazadeh and Toon[22] suggested that HNO_3/H_2O solid phases (so-called type-1c PSCs) form when temperatures stay sufficiently long below T_{NAT}, either as amorphous solids which crystallize upon warming or as water-rich hydrates;

(b) Meilinger *et al.*[28] and Tsias *et al.*[29] suggested that mesoscale temperature fluctuations lead to droplets with very high non-equilibrium concentrations of HNO_3 which possibly freeze as NAT or NAD;

(c) Drdla[36] suggested that there might be heterogeneous nuclei which cause freezing.

Suggestion (a) has been obtained from analyzing ER-2 field measurements by assuming that the observed particles were in thermodynamic equilibrium and by postulating that they constitute an extra phase. The ER-2 flight on 16 January 1989 has been identified as an example for such type-1c clouds. However, as shown in Fig. 3 this day could also be characterized by the presence of several different processes like NAT-coating of ice particles and denitrification.

Suggestion (b) has been discussed above. Heating and cooling rates of $dT/dt = \pm 150$ K/h have been observed frequently (see Tsias *et al.*[29]). However, there is no climatology of such rapid temperature fluctuations in the vicinity of T_{NAT} to enable the case studies to be extrapolated to the synoptic scale.

Finally, the existence of heterogeneous nuclei (c) in stratospheric aerosols has been investigated by Sheridan *et al.*[38], who found that only a very small number of particles had insoluble cores larger than 10–30 nm (the detection limit of their analysis). Biermann *et al.*[39] performed heterogeneous freezing experiments using several insoluble species of potential stratospheric importance, including real micrometeorites. In none of these cases did the nucleation rate increase to stratospherically relevant levels. However, Drdla[36] points out that other species could still induce freezing, like sodium, which has been found in undissolved granules in the aerosol.

3. PSC Chemistry

Heterogeneous chemical reactions on and in* PSC particles are responsible for rapid chlorine activation and nitrogen deactivation. At low stratospheric temperatures reactions (1a) and (1c) can lead to stoichiometric titration of HCl with $ClONO_2$ and HOCl within a few hours. Under conditions, where heterogeneous rates are this fast (i.e. almost at gas/surface collision rates), a precise knowledge of the heterogeneous reaction rates is not required since the chemical activation of chlorine occurs on a short timescale compared to the overall process of ozone depletion. However, changes in the reaction rates of more than about a factor of 10 are important, and can become decisive in situations with incomplete chlorine processing, like at the edge of the polar vortex, or in relatively warm winters when PSC development is weak, or in orographic wave clouds where the processing time is short. Unfortunately, our present knowledge on the most important chemical reaction of HCl with $ClONO_2$ on NAT and SAT (Eq.1a) does indeed suffer from such large uncertainties. The laboratory aspects of atmospheric heterogeneous chemistry have been reviewed in detail by Kolb *et al.*[40].

*Strictly speaking, reactions in the bulk of aerosol particles are homogeneous reactions (e.g. liquid phase reactions), but here the general convention of calling non-gas reactions "heterogeneous" is followed.

Chlorine activation on liquids or solids (Eqs.1) depends on the solubility or adsorptivity of HCl and is therefore strongly dependent on temperature and relative humidity. The interaction of HCl with PSC materials has been a subject of controversy since it was first proposed that PSC-driven chlorine activation could provide a formation mechanism for the Antarctic ozone hole[5]. The solubility of HCl in ice had initially been drastically overestimated, while that in liquid aerosols at low temperatures was underestimated.

In all chlorine activation reactions the availability of water plays an important role either because H_2O is a reaction partner (Eq.1b) or because H_2O facilitates solvation and dissociation of the chlorine species. Within the last five years much progress has been made in understanding the dependences of liquid and solid phase heterogeneous rates on temperature and relative humidity. Based on these improvements together with more refined dynamical transport schemes for the atmosphere, the next few years will show how close we get towards a quantitative simulation of polar ozone destruction.

3.1. CHEMISTRY OF LIQUID PSCS

Hanson *et al.*[41] have applied the theoretical framework for liquid phase reactions in small aerosol particles to stratospheric heterogeneous chemistry. This comprises a treatment of the mass accommodation of the gas species (X) on the droplet surface and its diffusion and reaction with the second species (Y) in the droplet bulk. Heterogeneous reactions are commonly described in terms of the reactive uptake coefficient, $\gamma_{X,Y}$, which is the fraction of X molecules reacting with Y upon impingement on the surface. The fraction $1-\gamma_{X,Y}$ desorbs again without reaction. For a reaction of X with Y in the bulk of a liquid droplet, $\gamma^{bulk}_{X,Y}$, under first order conditions (Y in excess), is given by

$$\frac{1}{\gamma^{bulk}_{X,Y}} = \frac{1}{\alpha_X} + \frac{\bar{v}}{4H^*_X RT\rho_w\sqrt{D_X k_{X+Y} c_Y}} \cdot \frac{1}{f(r/\sqrt{D_X/k_{X+Y}c_Y})}, \quad (6)$$

where α_X is the mass accommodation coefficient (often $\simeq 1$), $\bar{v}$ is the mean thermal speed of X in the gas phase, H^*_X is the effective Henry's law constant of X, D_X its diffusion coefficient in solution, k_{X+Y} is the second order rate coefficient for the reaction in the liquid, c_Y is the (constant) concentration of Y in the liquid, ρ_w is the mass density of water in solution, and R the universal gas constant. The function $f(x) = \coth(x) - 1/x$ takes into account the spherical geometry of the aerosol droplet with radius r. In summary, γ saturates when mass accommodation is limiting ($\gamma = \alpha$), and far below this saturation limit γ is proportional to the solubility of X, to the square root of Y concentration and to the size-function f. This formulation was first tested in the laboratory for a stratospherically relevant system ($ClONO_2$ on H_2SO_4/H_2O) by Hanson and Lovejoy[42].

3.1.1. *Chlorine Activation*

Comprehensive measurements for the reactions of $ClONO_2$ with HCl or H_2O in aqueous H_2SO_4 solutions at stratospheric temperatures have been performed by Hanson and Ravishankara[43]. They succeeded in separating the two competing reactions with HCl and H_2O (Eqs.1a,b). The coefficient γ_{HCl}^{bulk} increases in proportion to $p_{HCl}^{1/2}$ over a wide range of HCl partial pressures and saturates at very high p_{HCl} due to mass accommodation limitation ($\gamma \simeq \alpha \simeq 1$) in accordance with Eq.6, but it is steeper at very low p_{HCl} due to competition with the water reaction. Hanson and Ravishankara[43] realized that the bulk reaction with HCl is not sufficient to interpret their data, and that an additional surface reaction, γ_{HCl}^{surf}, is required to explain the almost linear increase of γ with p_{HCl}. This surface reaction is often the dominant contribution under stratospheric conditions. The reactive uptake coefficient in Eq.(6) takes the generalized form $\gamma^{-1}=\alpha^{-1}+[4H^{\star}RT\rho_w(Dkc_Y)^{1/2}f/\bar{v}+\gamma^{surf}]^{-1}$ with $\gamma^{surf} \propto c_Y$.

Comparison with other measurements[44] of the same reactions reveals differences of almost one order of magnitude between the two data sets. This appreciable discrepancy in the most important liquid PSC reaction should be resolved in future measurements.

The reaction of HCl with HOCl (Eq.1c) has also been measured by Hanson and coworkers. This reaction is of particular importance during the early stage of polar chlorine activation since it adds to the rapid activation of HCl in reaction with $ClONO_2$. Later in the winter, HCl may become locally so depleted that HOCl produced in reaction (1b) may only be activated *via* photolysis. In contrast to the very rapid $ClONO_2$ reactions which are constrained to a thin layer below the gas-liquid interface, the reaction HOCl+HCl often has a diffuso-reactive length $(D_X/k_{X+Y}c_Y)^{1/2} \gg r$ (Eq.6) and can take place deep within the bulk of the droplet.

3.1.2. *Nitrogen Deactivation*

Hydrolysis of N_2O_5 on liquid H_2SO_4/H_2O surfaces (Eq.2) represents a very efficient channel for nitrogen deactivation. This reaction was first proposed by Cadle *et al.*[45], but for a long time discarded because of wrong measurements. Only measurements by Van Doren *et al.*[46] and Hanson and Ravishankara[47] indicated high reactive uptake probabilities ($\gamma \simeq 0.1$) without strong dependence on H_2SO_4 concentration or temperature. Reactions (1a-c) are usually restricted to the polar winter stratospheres, unless temperatures fall to very low values at the mid-latitude tropopause. In contrast, reaction (2) on H_2SO_4/H_2O background aerosols leads to a global redistribution of NO_x and other trace gases that are relevant to ozone destruction, particulary after volcanic episodes. Interestingly, this reaction has also led to an appreciable down-scaling of the risk of high-flying supersonic air traffic, whose NO_x emissions are now thought to be readily converted to HNO_3, as shown by Weisenstein *et al.*[48]. Again, this does not apply to the polar regions where aircraft NO_x is predicted to lead to an increase in PSC formation[35] with potential enhancement of polar ozone losses due to heterogeneous halogen chemistry.

3.1.3. *Other Liquid Reactions*

Bromine plays a significant role in the destruction of ozone despite the fact that the total inorganic Bromine concentration in the stratosphere is typically two orders of magnitude lower than that of inorganic chlorine. The main reason for this disproportionate role is that a higher fraction of the bromine species are present in active, ozone-destroying forms. The species $BrONO_2$ and HBr readily undergo photochemical reactions, so a strong impact of heterogeneous bromine chemistry is not expected during the daytime. However, homogeneous gas phase chemistry models predict $BrONO_2$ to be the dominant bromine reservoir during nighttime. Hanson and coworkers[49] and Abbatt[50] investigated reactions (1a-c) for bromine (i.e. with $BrONO_2$, HOBr and HBr). In particular the $BrONO_2$ hydrolysis was found to be very fast and practically independent of temperature and acidity[49]. When HCl was added, the primary product was BrCl, which was probably formed *via* the sequence $BrONO_2 \rightarrow HOBr \rightarrow BrCl$. Heterogeneous bromine chemistry can potentially enhance [OH], suppress the NO_x to NO_y ratio, and activate HCl. While the heterogeneous reaction $BrONO_2 + HCl \rightarrow BrCl + HNO_3$ followed by $BrCl + h\nu \rightarrow Br + Cl$ requires high solubility of HCl (i.e. $T < 200$ K), the heterogeneous/homogeneous sequence $BrONO_2 + H_2O \rightarrow HOBr + HNO_3$, $HOBr + h\nu \rightarrow Br + OH$ and $OH + HCl \rightarrow H_2O + Cl$ leads to the same result but runs at all temperatures.

While gas phase chemistry leads to much higher levels of active bromine compared to chlorine, this is even more the case for iodine. Solomon *et al.*[51] suggested that iodine might be of importance in ozone depletion. At present there is no information on the amounts of iodine in the stratosphere, but heterogeneous reactions will probably not play a significant role. Conversely, fluorine is almost completely in its deactivated form HF, and also heterogeneous reactions have been found to be immeasurably slow. Hence, fluorine species are not expected to influence stratospheric chemistry.

Burley and Johnston[52] suggested that H_2SO_4 itself might participate in heterogeneous chemistry *via* $NO + NO_2 + 2H_2SO_4 \rightarrow 2NOHSO_4 + H_2O$ leading to nitrosyl sulfuric acid ($NOHSO_4$) in stratospheric aerosol droplets. The reaction $NOHSO_4 + HCl \rightarrow ClNO + H_2SO_4$ might lead to chlorine activation. Nitrosyl chloride readily undergoes photolysis. Temperature variations would promote this mechanism, since $NOHSO_4$ is formed more efficiently at high H_2SO_4 concentrations (high temperature), while HCl dissolves only in more dilute solutions formed at lower temperatures. The true importance of this process is not yet clear.

3.2. CHEMISTRY OF SOLID PSCs

Laboratory studies of gas/solid surface reactions are complicated by the preparation of the laboratory substrate, which influences the morphology of the surface and by the fact that the true morphology of stratospheric particles is not known. The issue of substrate porosity has been the subject of scientific controversy[53,54]. Solid substrates can be produced in the laboratory by freezing a liquid or by deposition from the gas phase. Depending on which method is used and how the substrate is subsequently treated (e.g. annealing), the morphology can be quite different. This affects not only the total surface area available, but also the crystalline structure and the uptake due to diffusion of trace species into grain boundaries. Wolff *et al.*[55] showed that HCl in polycrystalline ice is predominantly located at grain boundaries. Experiments on single crystal samples are currently only available for tropospheric temperatures[56]. In the following these caveats should be kept in mind.

3.2.1. *Chemistry on Ice*

Heterogeneous chemistry on water ice, a proxy for type-2 PSCs, was the first studied under stratospheric conditions. For $ClONO_2$+HCl, HOCl+HCl, N_2O_5+H_2O and reactions with bromine species, high reactive uptake coefficient between 0.02 and 0.3 were found. Many of these measurenments are consistent with $\gamma = 1$ when accounting for experimental uncertainties. Also, there is evidence that a fraction of the reaction product Cl_2 is not formed *via* Eq.(1a), but *via* the sequence (1b)–(1c). The high values of γ for reactions involving HCl are probably related to the dissociation of HCl on ice (i.e. the HCl is chemisorbed, not physisorbed).

3.2.2. *Chemistry on NAT*

The reaction $ClONO_2$+HCl is stratospherically very important because it leads to titration of the two dominant chlorine reservoirs. The reactive uptake is a strong function of H_2O and HCl partial pressures and of temperature. This reaction has been studied on NAT under first order conditions (HCl in excess) by Hanson and Ravishankara[57] using stratospherically realistic HCl partial pressures, and by Abbatt and Molina[58] using HCl pressures typically two orders of magnitude higher. The latter data set covers a wide range of conditions. Also, Abbatt and Molina measured the coverage of HCl on NAT in the absence of $ClONO_2$. Based on this information, Carslaw and Peter[59] used a coadsorption model for the simultaneous HCl and $ClONO_2$ uptake based on these data. This model yields for the reactive uptake coefficient

$$\frac{1}{\gamma_{X,Y}} = \frac{1}{\alpha_X} + \frac{\bar{v}\sigma^2}{4K_X^\star k_B T k_{X+Y}\theta_Y} \tag{7}$$

in close resemblance to Eq.(6) for liquids, where σ is the surface area per adsorption site and $K_X^\star = K_X/(1+K_X p_X)$ is an effective Langmuir constant which tends to K_X for weak absorption and takes self-competition of HCl molecules on the surface into account. In Eq.(7), X is the species with lower gas concentration (here $ClONO_2$) and θ_Y is the fractional coverage of the excess species (here HCl). Carslaw and Peter[59] assumed α_X and $K_X^\star k_{X+Y}$ to be constants and used the HCl coverage data of Abbatt and Molina[58] to confirm Langmuir-like behavior $\theta_Y = K_Y p_Y/(1+K_Y p_Y)$ with K_Y being a function only of relative humidity. Depending on relative humidity, θ_Y can vary by orders of magnitude (Fig. 5b) probably due to the availability of H_2O on the surface allowing the absorptivity of reactants. Understanding this behavior is a major challenge in heterogeneous chemistry. Most importantly, Abbatt and Molina[58] found that the HCl coverage on NAT is linearly proportional to HCl partial pressure ($\theta_Y \propto p_Y$). Figure 5a shows their data (circles) in comparison with the coadsorption model results (lines) as a function of the relative humidity over ice (S_{ice}) and of p_{HCl}. Crosses mark the data of Hanson and Ravishankara[57], which at first sight appear to be in good agreement with the data of Abbatt and Molina and the model. However, Hanson and Ravishankara performed their measurements for p_{HCl} two orders of magnitude lower. Applying the model to the conditions of their experiment yields the dashed line in Fig. 5a which lies as much as two orders of magnitude below the Abbatt and Molina data. According to the model, the reactive uptake coefficients measured by Hanson and Ravishankara are two orders of magnitude higher than those of Abbatt and Molina.

The reason for this contradiction is not clear at present. First, it should be noted that these conclusions are based on a link made by a model. In principle it could be possible that both data sets are correct and the coadsorption picture is wrong. However, the conclusion is based on the linear proportionality of reactive uptake on HCl partial pressure: $\gamma \propto \theta_{HCl} \propto p_{HCl}$ (for not too high S_{ice} and p_{HCl}) which is clearly apparent in Abbatt and Molina's data[58]. A physical reason for a much weaker dependence on p_{HCl} as it is apparent in Hanson and Ravishankara's data[57] is not evident, and further experiments are needed to resolve this issue.

Other chlorine reactions on NAT, like $ClONO_2$+H_2O and HOCl+HCl, have been measured in the same laboratories[57,58]. Data for the HOCl+HCl reaction were also compared using a coadsorption model[59] revealing similar differences by up to a factor of 40 between both measurements. Nonchlorine reactions on NAT, e.g. bromine reactions, have been investigated to some degree and were found to be fast. However, as these reactions are not strongly temperature dependent they are probably more important for mid-latitude *in situ* processes than for polar processes.

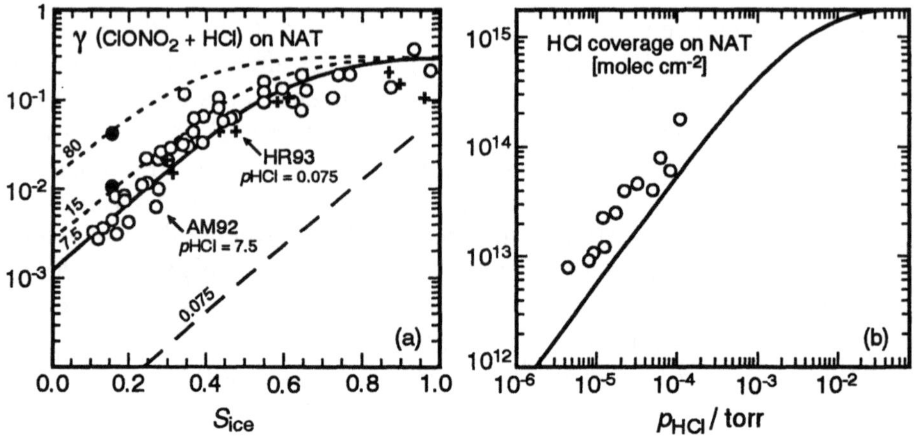

Figure 5. Measurements (a) of the reactive uptake coefficient for $ClONO_2$+HCl on NAT as function of the relative humidity over ice, and (b) of HCl coverage on NAT as function of the HCl partial pressure. Open circles: Abbatt and Molina[160] for $p_{HCl}=7.5\times10^{-6}$torr. Solid circles: same for $p_{HCl}=15\times10^{-6}$torr and 80×10^{-6}torr. Crosses: Hanson and Ravishankara[159] for $p_{HCl}=0.075\times10^{-6}$torr. Lines were calculated using the coadsorption model of Carslaw and Peter[59].

3.2.3. *Chemistry on SAT and Other Solids*

Here the case is similar to NAT. Again, model results[59] suggest for the dominant reaction $ClONO_2$+HCl on SAT a significant discrepancy (up to a factor of 20) between the various measurements. Work by Hanson and Ravishankara[57] indicates that the reaction $ClONO_2$+HCl on NAD is similarly fast as that on NAT. There are no studies of chemical reactions on any of the other metastable hydrates.

4. Conclusions

One of the burning questions of current ozone research is whether an ozone hole of Antarctic dimensions might form in the northern hemisphere within the next 5–10 years. Chemically induced ozone depletion during the recent northern hemisphere winters 1993/4–1995/6 reached alarming values[60,61], with local losses up to 30–40 %. However, given the recent stagnation and expected slow decrease of chlorofluorocarbon concentrations[62] in the troposphere, which will lead to falling chlorine levels also in the stratosphere at the beginning of the next decade, it might seem unlikely that a northern ozone hole similar to the one in the Antarctic could still develop. On the other hand, this conclusion depends on the assumption that other parameters, like the global temperature or the chemical composition of the atmosphere, remain constant. The increase of greenhouse gases in the atmosphere leads to stratospheric cooling and preliminary model simulations suggest that this could lead to an increase in denitrification, and possibly to enhanced polar ozone destruction[63]. In addition, significant increases of NO_x and H_2O in the stratosphere caused by the future operation of high-flying supersonic aircraft would be expected to influence ozone[64,65]. Recent analyses have shown that the rapid heterogeneous conversion of the aircraft NO_x to HNO_3 (Eq.2) may mean that global ozone changes are probably not as significant as previously thought[48]. However, in the polar regions aircraft emissions could lead to more than a doubling in PSC occurrence[35]. Three-dimensional high-resolution models are necessary to fully judge the consequences of stratospheric air traffic on the polar routes.

This review has shown that proper diagnosis and prognosis of the depletion of polar ozone depends crucially on our understanding of stratospheric particles. In particular we need to:

(1) understand thermodynamics of solid metastables,
(2) understand phase transitions like freezing nucleation and deposition nucleation and their consequences for denitrification,
(3) reconcile the existing measurements of chemical reactivity of HCl and $ClONO_2$ on NAT and SAT surfaces.

The urgency of point (3) is emphasized by box model studies which show that the current uncertainties in heterogeneous chemistry might amount to a difference of 0.5 ppmv in ozone mixing ratios during a typical Arctic winter[66]. Obtaining better constrained kinetic data for chlorine chemistry might be even more important than measuring new, but weaker reactions involving halogens other than chlorine. Depending on the result the reactive uptake coefficients of heterogeneous liquid phase reactions are either similar[67] or even much higher than the corresponding solid surface reaction. Furthermore, the results ought to be implemented into atmospheric models at all scales. Up to now this has been delayed because of the remaining uncertainties in particle microphysics.

While points (1)–(3) should be addressed in the laboratory supported by process-related modeling, there are three major goals for future measurements in the field:

(4) perform *in situ* chemical analysis of PSC particles,
(5) perform measurement flights along quasi-Lagrangian trajectories,
(6) understand the role of mesoscale temperature fluctuations in the stratosphere.

It is indeed remarkable that even one decade after the suggestion that reactions of chlorine compounds take place on PSCs, *in situ* analysis of these particles is still pending. Techniques capable of performing such analysis are underway and could lead to a quantum-leap in understanding PSC physico-chemistry. Furthermore, flight campaigns up to now were usually designed to achieve the best chemical tracer measurements, rather than particle measurements. The latter would benefit greatly from performing measurements along the streamlines of the wind (quasi-Lagrangian motion). Finally, orographically induced PSC pattern in the northern hemisphere are often practically stationary so that quasi-Lagrangian flight pattern allow the particles' life-cycle to be followed from the original H_2SO_4/H_2O droplets to the melting and evaporation of the PSCs.

Acknowledgements: I thank all members of the Nachwuchsgruppe for their help in preparing this review, in particular Athanasios Tsias for his editorial support. This paper is a shortened version of Ref. 1, with friendly permission from the Annual Review of Physical Chemistry, Volume 48, copyright 1997, by Annual Reviews Inc.

References

1. Peter Th. 1997. *Annu. Rev. Phys. Chem.* 48: 785-822
2. Crutzen PJ. 1970. *Quart. J. Roy. Met. Soc.* 96: 320-325
3. Molina MJ, Rowland FS. 1974. *Nature* 249: 810-812
4. Farman JC, Gardiner BG, Shanklin JD. 1985. *Nature* 315: 207-210
5. Solomon S, Garcia RR, Rowland FS, Wuebbles DJ. 1986. *Nature* 321: 755-758
6. Solomon S. 1990. *Nature* 347. 347-354
7. Toon OB, Hamill P, Turco RP, Pinto, J. 1986. *Geophys. Res. Lett.* 13: 1284-1287
8. Crutzen PJ, Arnold F. 1986. *Nature* 324: 651-655
9. Carslaw KS, Luo BP, Clegg SL, Peter Th, Brimblecombe P, Crutzen PJ. 1994. *Geophys. Res. Lett.* 21: 2479-2482
10. Tabazadeh A, Turco RP, Drdla K, Jacobson MZ, Toon OB. 1994. *Geophys. Res. Lett.* 21: 1619-1622
11. Dye JE, Baumgardner D, Gandrud BW, Kawa SR, Kelly KK, Loewenstein M, Ferry GV, Chan KR, Gary BL. 1992. *J. Geophys. Res.* 97: 8015-8034
12. Toon OB, Browell EV, Kinne S, Jordan J. 1990. *Geophys. Res. Lett.* 17: 393-396
13. Browell EV, Butler CF, Ismail S, Robinette PA, Carter AF, Higdon NS, Toon OB, Schoeberl MR, Tuck AF. 1990. *Geophys. Res. Lett.* 17: 385-388
14. Koop T, Biermann UM, Raber W, Luo BP, Crutzen PJ, Peter Th. 1995. *Geophys. Res. Lett.* 22: 917-920
15. Hanson D, Mauersberger K. 1988. *Geophys. Res. Lett.* 15: 855-858
16. Poole LR, McCormick MP. 1988. *J. Geophys. Res.* 93: 8423-8430
17. Fahey DW, Kelly KK, Ferry GV, Poole LR, Wilson JC, Murphy DM, Chan KR. 1989. *J. Geophys. Res.* 94: 11299-11316
18. Carslaw KS, Peter Th, Clegg SL. 1997. *Revs. Geophys. Res.* in press
19. Junge CE, Chagnon CW, Manson JE. 1961. *J. Meteorol.* 18: 81-108
20. Molina MJ, Zhang R, Wooldridge PJ, McMahon JR, Kim JE, Chang HY, Beyer KD. 1993. *Science* 261: 1418-1423
21. Stanford JL, Davis JS. 1974. *Bull. Americ. Meteorol. Soc.* 55: 213-219
22. Tabazadeh A, Toon OB. 1996. *J. Geophys. Res.* 101: 9071-9078
23. Peter Th, Müller R, Crutzen PJ, Deshler T. 1994. *Geophys. Res. Lett.* 21: 1331-1334
24. Luo BP, Peter Th, Crutzen PJ. 1994. *Geophys. Res. Lett.* 21: 1447-1450
25. Koop T, Carslaw KS. 1996. *Science* 272: 1638-1641
26. Worsnop DR, Fox LE, Zahniser MS, Wofsy SC. 1993. *Science* 259: 71-74
27. Pruppacher HR, Klett JD. 1978. *Microphysics of clouds and precipitation*, D. Reidel Publ. Co., Dordrecht
28. Meilinger SK, Koop T, Luo BP, Huthwelker T, Carslaw KS, Crutzen PJ, Peter Th. 1995. *Geophys. Res. Lett.* 22: 3031-3034
29. Tsias A, Prenni AJ, Carslaw KS, Onasch TP, Luo BP, Tolbert MA, Peter Th. 1997. *Geophys. Res. Lett.* 24: 2303-2306
30. Toon OB, Turco RP, Jordan J, Goodman J, Ferry G. 1989. *J. Geophys. Res.* 94: 11359-11380
31. MacKenzie AR, Haynes PH. 1992. *J. Geophys. Res.* 97: 8057-8064

32. Molina MJ. 1994. In *Chemistry of the Atmosphere: Its Impact on Global Change*, ed.: JG Calvert, pp. 27-38, Oxford: Blackwell Sci.
33. Wofsy SC, Salawitch RJ, Yatteau JH, McElroy MB, Gandrud BW, Dye JE, Baumgardner D. 1990. *Geophys. Res. Lett.* 17: 449-452
34. MacKenzie AR, Kulmala M, Laaksonen A, Vesala T. 1994. *Geophys. Res. Lett.* 21: 1423-1426
35. Peter Th, Brühl C, Crutzen, P.J. 1991. *Geophys. Res. Lett.* 18: 1465-1468
36. Drdla K. 1996. *Applications of a model of polar stratospheric clouds and heterogeneous chemistry* Ph.D. thesis, Univ. of California, Los Angeles, pp. 396
37. Koop T, Carslaw KS, Peter Th. 1997. *Geophys. Res. Lett.* 24: 2199-2202
38. Sheridan PJ, Brock CA Wilson J.C. 1994. *Geophys. Res. Lett.* 21: 2587-2590
39. Biermann UM, Presper T, Koop T, Mößinger J, Crutzen PJ, Peter Th. 1996. *Geophys. Res. Lett.* 23: 1693-1696
40. Kolb CE, Worsnop DR, Zahniser MS, Davidovits P, Keyser LF, Leu MT, Molina MJ, Hanson DR, Ravishankara AR, Williams LR, Tolbert MA. 1995. *Laboratory studies of atmospheric heterogeneous chemistry* in: Progress and Problems in Atmospheric Chemistry, Ed. Barker JR, Word Scientific, Singapore
41. Hanson DR, Ravishankara AR, Solomon S. 1994. *J. Geophys. Res.* 99: 3615-3629
42. Hanson DR, Lovejoy ER. 1995. *Science* 267: 1326-1328
43. Hanson DR, Ravishankara AR. 1994. *J. Phys. Chem.* 98: 5728-5735
44. Zhang R, Leu MT, Keyser LF. 1995. *Geophys. Res. Lett.* 22: 1493-1496
45. Cadle RD, Crutzen PJ, Ehhalt D. *J. Geopys. Res.* 80: 3381-3385
46. Van Doren JM; Watson LR, Davidovits P, Worsnop DR, Zahniser MS Kolb, CE. 1991. *J. Phys. Chem.* 95: 1684-1689
47. Hanson DR, Ravishankara AR. 1991. *J. Geophys. Res.* 96: 17307-17314
48. Weisenstein DK, Ko MKW, Rodriguez JM, Sze N-D. 1991. *Geophys. Res. Lett.* 18: 1991-1994
49. Hanson DR, Ravishankara AR, Lovejoy ER. 1996. *J. Geophys. Res.* 101: 9063-9069
50. Abbatt JPD. 1995. *J. Geophys. Res.* 100: 14009-14017
51. Solomon S, Garcia RR, Ravishankara AR. 1994. *J. Geophys. Res.* 99: 20491-20499
52. Burley JD, Johnston HS. 1992. *Geophys. Res. Lett.* 19: 1363-1366
53. Keyser LK, Leu MT, Moore SB. 1993. *J. Phys. Chem.* 97: 2800-2801
54. Hanson DR, Ravishankara AR. 1993. *J. Phys. Chem.* 97: 2802-2803
55. Wolff EW, Mulvaney R, Oates K. 1989. *Geophys. Res. Lett.* 16: 487-490
56. Dominé F, Thibert E, Van Landeghem F, Silvente E, Wagnon P. 1994. *Geophys. Res. Lett.* 21: 601-604
57. Hanson DR, Ravishankara AR. 1993. *J. Geophys. Res.* 98: 22931-22936
58. Abbatt JPD, Molina MJ. 1992. *J. Phys. Chem.* 96: 7674-7679
59. Carslaw KS, Peter Th. 1997. *Geophys. Res. Lett.* 24: 1743-1746
60. von der Gathen P, Rex M, Harris NRP, Lucic D, Knudsen BM, Braathen GO, de Backer H, Fabian R, Fast H, Gil M, Kyrö E, Mikkelsen IS, Rummukainen M, Stähelin J, Varotsos C. 1995. *Nature* 375: 131-134
61. Müller R, Crutzen PJ, Grooß JU, Brühl C, Russell JM III, Gernandt H, McKenna DS, Tuck AF. 1997. *Nature* 389: 709-712
62. Montzka SA, Butler JH, Myers RC, Thompson TM, Swanson TH, Clarke AD, Lock LT, Elkins JW. 1996. *Science* 272: 1318-1322
63. Pitari G, Palermi S, Visconti G, Prinn RG. 1992. *J. Geophys. Res.* 97: 5953-5962
64. Crutzen PJ. 1971. *J. Geophys. Res.* 76: 7311-7327
65. Johnston H. 1971. *Science* 173: 517-522
66. Carslaw KS, Peter Th, Müller R. 1997. *Geophys. Res. Lett.* 24: 1747-1750
67. Ravishankara AR, Hanson DR. 1996. *J. Geophys. Res.* 101: 3885-3890

A Minimal Model of Sea Ice and Climate

Alan Thorndike

Department of Physics, University of Puget Sound, Tacoma, WA 98416, USA

Abstract. A simple model of the Northern Hemisphere climate is presented that emphasizes the role of sea ice. The model is used to investigate the response of the climate to increased concentration of greenhouse gases, the predicted amplification of climatic warming in the arctic, the possibility of multiple equilibria, and the waiting times between transitions from one equilibrium to another.

Keywords. Arctic climate, heat balance, radiative equilibrium

1 Introduction

Sea ice and its variations have always been of importance in the economic life of northern peoples. Well into the twentieth century, fishing, sealing, and whaling have been the primary sources of income for countries such as Iceland and Norway. The occurrence of ice in the fjords of Iceland has been recorded for more than a thousand years. The former Soviet Union maintained an agency of several thousand people solely dedicated to the study of northern meteorology and oceanography and methods of sea ice forecasting in support of shipping along the Northern Sea Route.

More recently, the extraction of fossil fuels from the continental shelves surrounding the Arctic Basin, and the technology of building ice-worthy tankers and LNG ships have become an issues of interest.

The fecundity of polar and sub-polar waters is, in part, explained by the variable sea ice cover. The freezing process enriches the surface water with brine, which renders the stratification of upper ocean layers unstable. This leads to convection and the upwelling of nutrient-rich water to the surface where, in the presence of sunlight, the food chain of plankton-fish-marine mammals is initiated.

The convection induced by ice formation has another far-reaching consequence: it was shown that it is in small ocean areas of the Greenland, Iceland, and Labrador Seas that enormous quantities of Atlantic Deep Water are formed, which can be traced along the global "conveyor belt" to the South Atlantic and to the Indian and Pacific Oceans. Unlike the atmosphere, the World Ocean below a layer affected by wind stress, circulates in large gyres that cross the equatorial region. Theoretical arguments

NATO ASI Series, Vol. I 56
Ice Physics and the Natural Environment
Edited by John S. Wettlaufer, J. Gregory Dash and Norbert Untersteiner

suggest that a greatly diminished ice cover in the Arctic may change this and create circulation cells on either side of the equator.

The incorporation of sea ice processes into interactive models of the global climate have proven difficult because the "On-Off" character of surface freezing leads to large changes of the ice cover resulting from small changes of the surface heat balance. Interactive models are complex and require large computing resources, and detailed studies of the output are time-consuming and expensive. Another way of studying the relative importance of the participating processes in the atmosphere, ice, and ocean are simple models, one of which is described here.

2 The Model

The first law of climate is that the earth is in radiative equilibrium. The second is that the atmosphere and ocean transport heat from warmer to cooler regions. Combined with the fact that the albedo of the Arctic Ocean depends on whether its surface is frozen or free of ice, these laws form a minimal model of climate. As Budyko, North, and many others have shown, these three ideas have interesting consequences. The purpose of this paper is to explore a simple form of the three idea model—one that highlights the interaction between the arctic and the rest of the hemisphere—and to add one embellishment: stochastic forcing.

Figure 1 illustrates the structure of the model. Refer to the list of symbols for definitions. The low latitude cell (latitude below 70N, subscript 1) and the polar cell (subscript 2) have surface temperatures T_i, relative surface areas g_i, surface albedos α_i, incoming solar radiation at the surface F_i, atmospheric longwave emissivity parameter β, explained below, and poleward heat advection D_i. The two surface temperatures are the dependent variables of the model. Denote the area weighted average of any quantity Q by $\langle Q \rangle = g_1 Q_1 + g_2 Q_2$, and let $\Delta Q = Q_1 - Q_2$ denote the meridional contrast.

The most important physical process in the model is blackbody radiation. The T^4 dependence is approximated here by the linear function A + BT where T is measured in degrees Celsius. Therefore A=320 Wm^{-2} is the rate of energy radiated from a black surface at the freezing point. B=4.6 $Wm^{-2}K^{-1}$ is the slope of the blackbody curve at the freezing point.

The heat balances for both cells determine the equilibrium state.

$$(1-\alpha_i)F_i + \beta(A + BT_i) + D_i / 2 = A + BT_i \qquad i = 1,2 \tag{1}$$

These two equations state that, in each cell, the surface radiates away as much heat (right hand side) as it receives (left hand side). The term involving F on the left hand side is the solar radiation absorbed at the surface. The downwelling longwave radiation from the atmosphere is a fraction β of the energy radiated upward by the surface plus half the meridional transport D. This result is taken from a toy model of the radiative equilibrium between the atmosphere and the surface, Thorndike 1992. β is a measure of the longwave emissivity of the atmosphere. In today's atmosphere β is about 2/3. It is determined largely by the amount of water vapor and carbon dioxide in the atmosphere. It will increase by about 1% as the carbon dioxide level doubles.

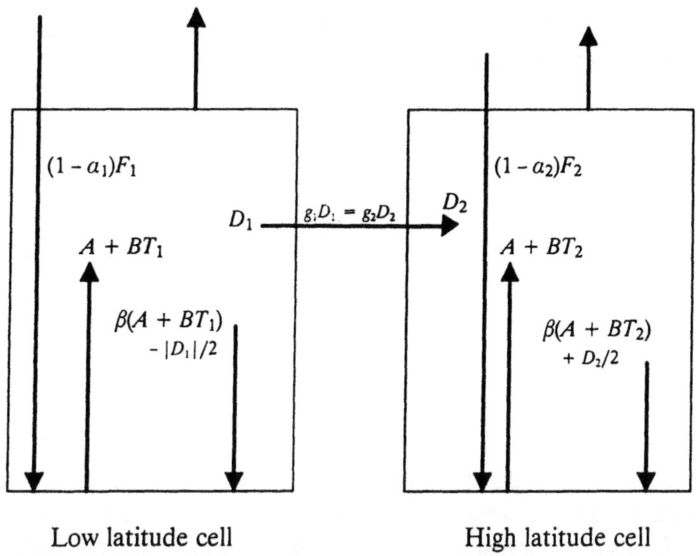

Fig. 1. The two cell model includes solar radiation (1–α)F, blackbody radiation from the surface A+BT, longwave atmospheric radiation β(A+BT), and horizontal advection D. The longwave radiation from the atmosphere to space does not figure into the model.

The meridional transport is assumed to be proportional to the temperature contrast

$$D_1 g_1 = -k\Delta T \qquad D_2 g_2 = k\Delta T \tag{2}$$

The constant k parameterizes all atmospheric dynamics. It can be estimated by taking approximate values for the other quantities in (2). Taking D_2=100 Wm^{-2}, Nakamura and Oort, 1988, $g_2 = 0.05$, and $\Delta T = 30$ K, leads to k=1/6 $Wm^{-2}K^{-1}$. Notice that D_1 is only about 5 Wm^{-2}. The poleward heat transport is a large term in the energy balance for the polar cell, but a small term for the low latitude cell. It is positive in the pole cell, thus warming the atmosphere, which radiates half of that heat down to the surface, and half to space. In the low latitude cell, D is negative, so the downward radiation at the surface, and the radiation at the top of the atmosphere are both reduced by $|D_1|/2$.

It is convenient to define

$$f \equiv (1-\alpha)F - (1-\beta)A \qquad b \equiv (1-\beta)B \qquad \kappa \equiv \frac{k}{2g_1 g_2} \tag{3}$$

f is the difference between the net shortwave radiation and the net longwave radiation at a surface at the freezing point.

Equations (1) are linear in the two temperatures T_1 and T_2. They are coupled by the transport terms D that depend on the temperature contrast. The solutions assume the simplest form after the coordinate change $(T_1,T_2) \rightarrow (<T>,\Delta T)$. Then

$$<T> = \frac{<f>}{b}$$

$$\Delta T = \frac{\Delta f}{b+\kappa}$$

$$T_1 = <T> + g_2 \Delta T \qquad (4)$$

$$T_2 = <T> - g_1 \Delta T$$

The $<T>$ equation shows that the average temperature is determined by the average radiation balance $<f>$. The absence of k implies that the average temperature is unaffected by poleward heat transport. The ΔT equation shows that the temperature contrast between the two cells is determined by the contrast in the radiation balance, softened by the poleward heat transport. Using numerical values for the parameters shows that the radiative cooling ($b = 1.53\ Wm^{-2}K^{-1}$) and the poleward advection ($\kappa=1.75\ Wm^{-2}K^{-1}$) are equally important. In other words, the meridional temperature contrast in the coupled system is about half what it would be if it were determined solely by radiative equilibrium and no advection. Figure 2 shows the solutions for T_1 and T_2 as functions of k. See the Table 1 for numerical values of T_1 and T_2, using a value of k for today's climate.

There are two choices for the polar cell albedo. If the coupling is weak, small k, the arctic is cold, and an icy albedo is appropriate. But if k is somewhat larger, the more vigorous poleward heat transport may warm the arctic above the freezing point and then an ocean albedo is appropriate. One way to proceed is simply to solve the problem twice, first using the cold polar albedo $a_{2c} = 0.65$, and again using the warm polar albedo $a_{2w} = 0.1$. If the first solution gives a polar cell temperature, T_{2c}, above freezing, it is rejected for being inconsistent with the assumed cold albedo. Similarly, reject the second solution if T_{2w} is below freezing.

Table 1. Model solutions for the three equilibria

state	$\alpha_{1,2}$	T_1	T_2	<T>	ΔT	f_1	f_2	<f>	Δf
cold	0.3, 0.65	20	−12	18	32	33	−72	28	105
warm	0.3, 0.1	21	5.7	20	15	33	−17	31	50
saddle	0.3, 0.27	21	0	20	21	33	−34	30	67

Temperatures in [C], heat fluxes in [Wm^{-2}]. For the unstable saddle, T_2 must be zero, allowing equations (1) to be solved for T_1 and α_2.

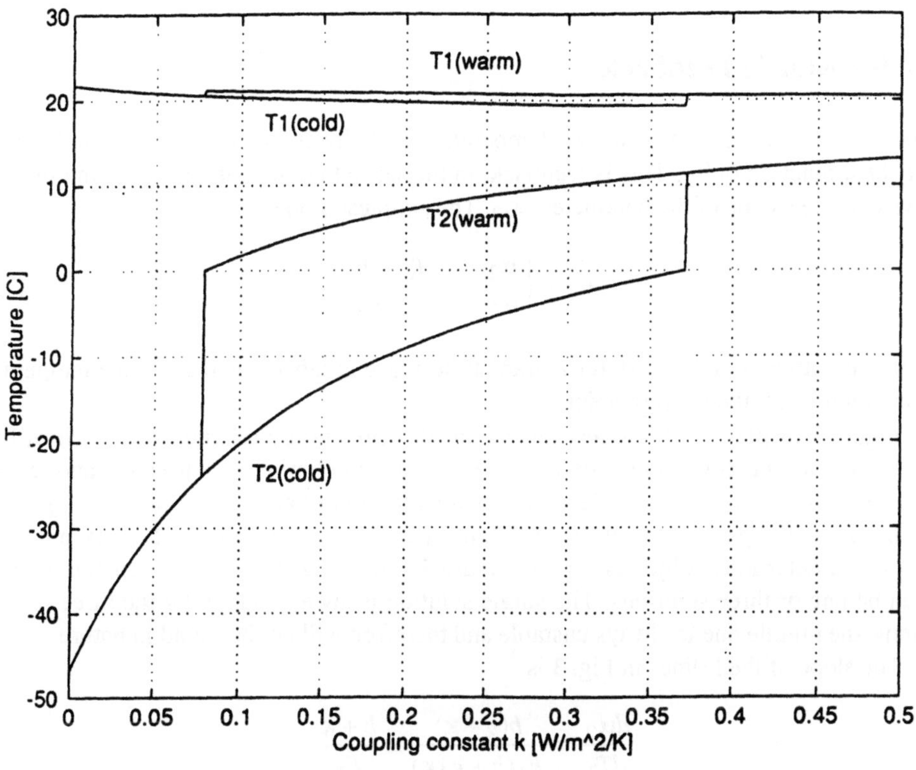

Fig. 2. Solutions for the model temperatures T_1 and T_2 [C], as functions of the coupling parameter k [$Wm^{-2}K^{-1}$], which provides for transport of heat from the low latitude cell to the polar cell. For a range of k values, two solutions are possible.

As Fig. 2 shows, there is a range of k values for which both the warm and cold solutions are acceptable. The range includes the estimate for $k=1/6\ Wm^{-2}K^{-1}$ in the present climate. So, despite its simple structure, this heat balance argument is rich enough to account for two climate states consistent with the present solar forcing F_i and atmospheric composition β.

To restate the mathematical structure of the minimal model at this point: we have two coupled, linear, algebraic equations (1) in two unknowns, T_i, and several parameters, α_i, β, k, and forcing F_i. The equations can be uncoupled by changing to $\langle T \rangle$, ΔT coordinates. The equations have a unique solution for each choice of parameters and forcing. The two choices for the polar cell albedo therefore produce two solutions. To be acceptable, the solutions must give polar cell temperatures that are consistent with the assumed polar cell albedo. In the next section, this test for consistency between the polar cell albedo and temperature is built into the heat balance equations, and the linearity is lost.

3 Ice-Albedo Feedback

The constraint between polar cell temperature and albedo can be enforced by letting α_2 be a function of T_2. This is feedback, in the sense that the output, T_2, of the linear system affects one of the parameters, α_2. The simplest model is

$$\alpha_2 = \begin{cases} \alpha_{2c} & T_2 < 0, \\ \alpha_{2w} & T_2 > 0. \end{cases} \tag{5}$$

An alternative model allows for a gradual decrease in albedo as the surface temperature approaches the melting point.

Figure 3 displays both models for the albedo-temperature relation. Any such $\alpha(T)$ destroys the linearity in (1). One way to proceed is to eliminate T_1 from (1) and rearrange in the form $\alpha_2(T_2) = L(T_2, \ldots)$, in which the right hand side is linear in T_2 and depends on the parameters in the problem. Intersections between the L lines and the α_2 curve determine solutions for T_2. Figure 3 shows that for fixed parameters there can be one or three solutions. The single solution is always stable. Of the three solutions, the middle one is always unstable and therefore will not be found in nature.

The slope of the L lines in Fig. 3 is

$$\frac{d\alpha_2}{dT_2} = \frac{b(b+\kappa)}{F_2(b+g_2\kappa)} \cong \frac{b+\kappa}{F_2} \tag{6}$$

which is about 3%/K.

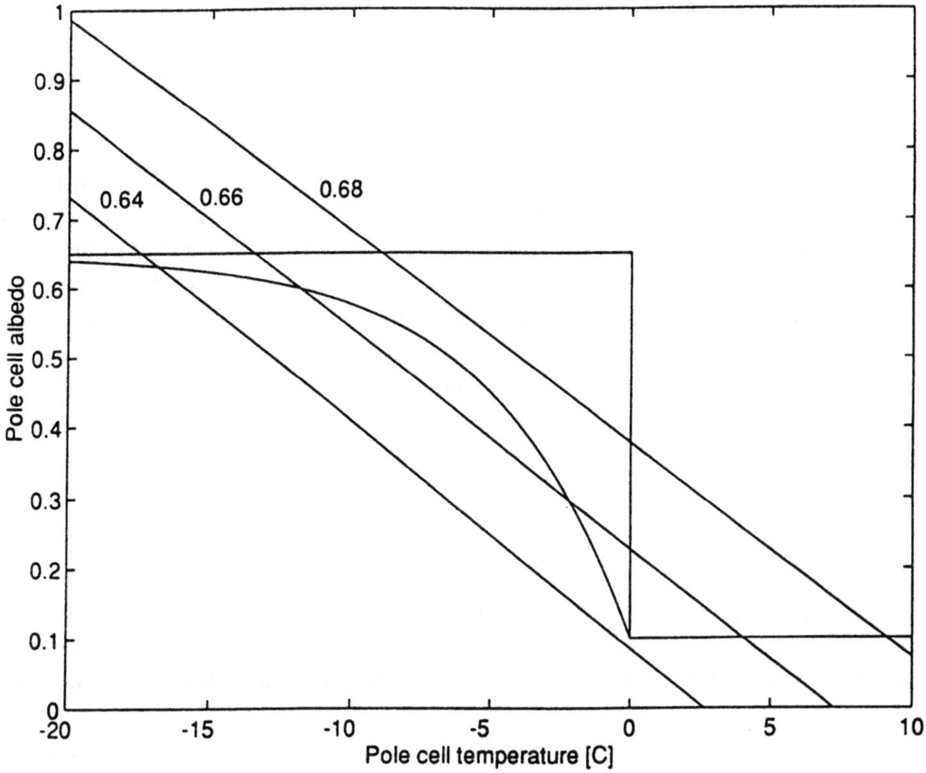

Fig. 3. The two horizontal lines show the pole cell albedo varying discontinuously with surface temperature T_2. In the curved line, the discontinuity has been spread over about 10 K. The three straight lines are called L lines in the text. The intersections of an L line and the albedo curve are solutions to the model equations.

The response of the system to increased concentration of greenhouse gases is found in Fig. 3. The increased emissivity of the atmosphere is modeled as an increase in β, which displaces the state to a higher L line. So long as the polar cell remains ice covered, its albedo holds fast at α_{2c} and the cold solutions to the linear equations apply. With increasing β, the temperatures T_1, T_2, $\langle T \rangle$, and ΔT all increase—the low latitude cell warms more rapidly than the pole cell (polar attenuation of global warming!). But eventually, as the ice concentration begins to decrease, the albedo of the polar cell starts to drop. When the state reaches the L line that has a point of tangency to the albedo curve, the cold solution disappears, and the climate jumps to the warm, ice free state.

The figure suggests that changes in β, and therefore in the downwelling longwave radiation, of a few percent are enough to trigger a transition from a definitely ice covered to a definitely ice free climate. The downwelling longwave is roughly βA, about 200 Wm^{-2} in the present climate. Radiative transfer models imply that the change in the longwave radiation due to doubling the atmospheric CO_2 is 2–4 Wm^{-2}, which corresponds to increasing β from 0.67 to 0.68.

The qualitative behavior of the system arises from the way the polar albedo depends on the polar temperature. If the transition from an ice albedo to an ocean albedo is much more gradual than is portrayed by the two α(T) functions in Fig. 3, the system will not experience sudden changes. The critical condition to avoid the catastrophic behavior is that the α(T) curve must be less steep than the L lines. This would require that the change in albedo of 55% must occur gradually over a temperature range of greater than about 18K. It seems likely that the change to an ice free arctic would occur over a smaller temperature change. One reason for thinking so is that in today's climate the change from the perennial ice pack to ice free ocean occurs over a change in latitude of about 5 degrees and a change in annual mean temperature of about 5K.

If the polar albedo is constant, increasing emissivity increases the meridional temperature contrast: the low latitude cell warms faster than the high latitude cell. In fact,

$$\frac{\partial \Delta T}{\partial \beta} = \frac{\Delta T}{1+\kappa / B} > 0. \qquad \text{(constant albedo)} \tag{7}$$

But as the polar albedo begins to fall, the polar temperature rises more rapidly. The sensitivity of the polar temperature to a change in β depends on the angle of intersection between an L line and the α(T) curve, the sensitivity increasing as the angle becomes more acute. So as the state shifts toward the L line tangent to the albedo curve, it is inevitable that the polar temperature eventually rises faster than the low latitude temperature, and it makes sense to speak of polar amplification of the greenhouse effect.

4 Time Dependence

Time dependence can be added to the model equations (1). This will allow us to justify the earlier statements regarding stability of the equilibrium states and to investigate waiting times for transitions between equilibria.

Sediment cores from the Arctic Ocean suggest that the ocean has been continuously ice covered for several hundred thousand years. This forces us to ask why perturbations to the system or secular changes in system parameters have not triggered transitions between the equilibria. The time dependent equations are

$$c\dot{T}_i = (1-\alpha_i)F_i - (1-\beta)(A+BT_i) + D_i/2 \qquad i=1,2 \tag{8}$$

Here c $[Jm^{-2}K^{-1}]$ is the energy required to raise the temperature of a unit area of either cell one degree. With the simplest model (5) for the polar cell albedo, the three equilibria are

$$\langle T\rangle_{c,w} = \frac{\langle f\rangle_{c,w}}{b}$$

$$\Delta T_{c,w} = \frac{\Delta f_{c,w}}{b+\kappa}$$

$$T_{1s} = \frac{f_1}{b+\kappa g_2} \tag{9}$$

$$T_{2s} = 0$$

As before, the subscripts c and w refer to the cold and warm solutions and s refers to the unstable saddle that separates them.

In the neighborhood of each stable solution, define a potential U(<T>, ΔT) such that d<T>/dt = –dU/d<T> and dΔT/dt = –dU/dΔT. As a consequence of the linear nature of the problem, these potential have parabolic minima at the stable equilibria (Fig. 4). They can be patched together at the saddle, but they will still inherit, from α_2, a discontinuity along the $T_2 = 0$ line.

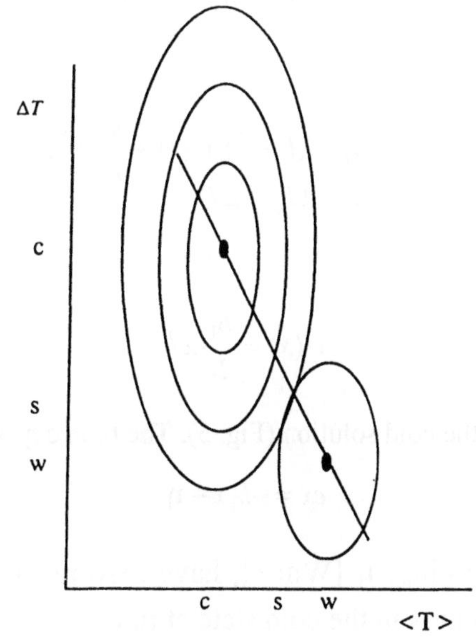

Fig. 4. The potential U as a function of <T> and ΔT.

$$
\begin{aligned}
2cU_c &= b(<T> - <T>_c)^2 + (b+\kappa)(\Delta T - \Delta T_c)^2 \\
2cU_w &= b(<T> - <T>_w)^2 + (b+\kappa)(\Delta T - \Delta T_w)^2 + \\
&\quad b(<T>_s - <T>_c)^2 + (b+\kappa)(\Delta T_s - \Delta T_c)^2 - \\
&\quad b(<T>_s - <T>_w)^2 + (b+\kappa)(\Delta T_s - \Delta T_w)^2
\end{aligned}
\tag{10}
$$

The potential differences $U_c(s) - U_c(c)$ and $U_w(s) - U_w(w)$ are measures of the stability of the cold and warm states. They present barriers to transitions from one stable equilibrium to the other. If the barriers are low, random climate fluctuations could trigger a transition. If the barriers are high, the climate will remain stuck in one stable state until parameters change.

To investigate the effect of random climate fluctuation, postulate an additional random forcing in the time dependent equations, thereby introducing a fourth idea into the model. To simplify matters somewhat, let us restrict attention to a line in $<T>$, ΔT space passing through the three equilibria, (Fig. 4). On this line

$$
<T> = -\left(1+\frac{\kappa}{b}\right)\Delta T + \frac{f_1}{b} \tag{11}
$$

With the definitions

$$
\begin{aligned}
b_1 &\equiv (b+\kappa)[1+(1+\frac{\kappa}{b})g_2^2] \\
x &\equiv \Delta T_c - \Delta T
\end{aligned}
\tag{12}
$$

the potential becomes

$$
U(x) = \frac{b_1}{2c}x^2 \tag{13}
$$

in the neighborhood of the cold solution (Fig. 5). The time dependent equation is now

$$
c\dot{x} = -b_1 x + \eta \tag{14}
$$

Let the random forcing, η [Wm^{-2}], have zero mean and autocovariance $R(t)$. Then, if the system is in the cold state at time $t = 0$, the state at time t is

$$
x(t) = \frac{1}{c}\int_0^t e^{-b_1(t-s)/c}\eta(s)ds \tag{15}
$$

and the variance of x is

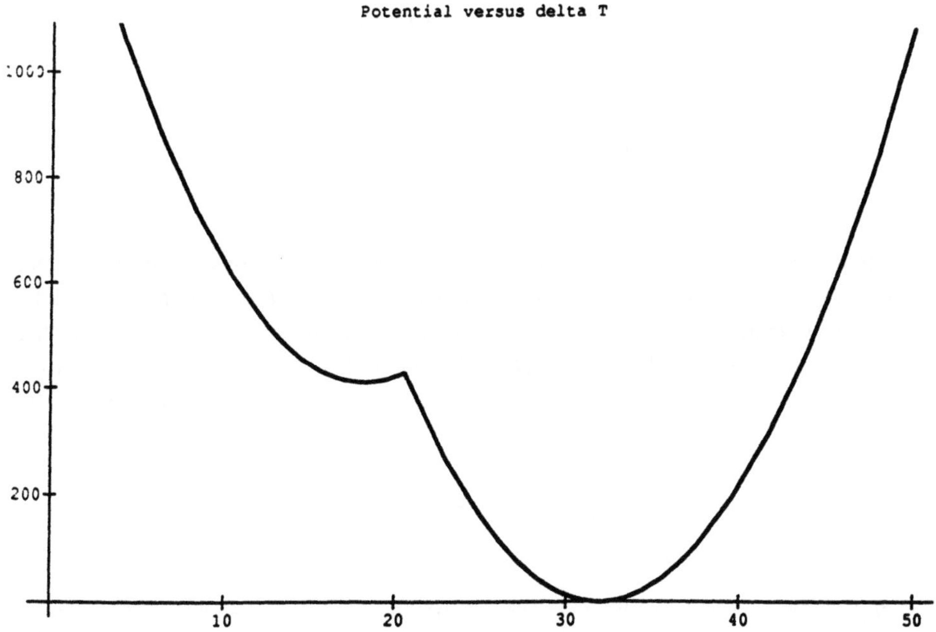

Fig. 5. The potential U as a function of ΔT, along the line through the equilibrium states in Fig. 4.

$$\sigma^2(t) = \frac{1}{b_1 c}\int_0^t R(s)\left[e^{-b_1 s/c} - e^{-b_1(2t-s)/c}\right] ds \tag{16}$$

For example, if the random noise is Gaussian with autocovariance $R(t)=Q^2\exp(-t/\tau)$, we obtain

$$\sigma^2(t) = \frac{2Q^2}{c^2}\frac{1}{\frac{b_1}{c}-\frac{1}{\tau}}\left[\frac{1-e^{-\left(\frac{1}{\tau}+\frac{b_1}{c}\right)t}}{\frac{1}{\tau}+\frac{b_1}{c}} - \frac{1-e^{-2b_1 t/c}}{\frac{2b_1}{c}}\right] \tag{17}$$

which is approximately

$$\sigma^2(t) \approx \begin{cases} \left(\frac{Qt}{c}\right)^2 & \text{for small t, and} \\ \left(\frac{Q}{b_1}\right)^2 \frac{1}{1+\frac{c/b_1}{\tau}} & \text{for large t.} \end{cases} \tag{18}$$

The state is described by a Gaussian probability density with zero mean and a variance, $\sigma^2(t)$, that increases with time, eventually approaching the large time asymptote.

We are interested in the probability of a transition from the cold to the warm state, which will be denoted by $P_Y(t) = \text{Prob}(x > Y)$, where $Y = \Delta t_c - \Delta T_s$. We can evaluate this probability using an approximation to the integral of one tail of a Gaussian distribution

$$P_Y(t) = \int_Y^\infty G(x;0,\sigma)dx \approx \frac{\sigma}{Y\sqrt{2\pi}} \exp -\frac{1}{2}\left(\frac{Y}{\sigma}\right)^2 \tag{19}$$

P_Y approaches a steady value as σ rises to its large t limit. The rate of approach to this limit is set by whichever of the two rates, $(b_1/c + 1/\tau)$ or $2b_1/c$, in the expression for σ, is slower. Let t^* denote the larger of the characteristic times $c\tau/(c+b_1\tau)$ and $c/2b_1$. The rate of flow of probability from the cold state, past the x=Y barrier, is roughly $P_Y(\infty)/t^*$. The time that must elapse for there to be significant probability of such a transition is the reciprocal

$$\text{waiting time} \approx \frac{t^*}{P_Y(\infty)} \tag{20}$$

Therefore, we have the following estimate

$$\begin{aligned} &\text{waiting time} \approx \max(c/b_1, \tau)\chi \exp\frac{\chi^2}{2} \\ &\chi \equiv \frac{\Delta T_c - \Delta T_s}{\sigma(\infty)} \end{aligned} \tag{21}$$

The time c/b_1 is the response time of the physical system, and τ is the time scale for random climate fluctuations. Both are of order 1 to 10 years. The dimensionless factor involving χ, however, can be quite large: it increases from 1.6 to about 10^6, as χ increases from 1 to 5. In other words, if the separation between the cold state and the potential barrier is five standard deviations of the stochastic response, the waiting time is millions of years. Without placing much confidence in numerical values, we can estimate χ as (32–21)/2 = 5.

5 Secular Variations in Forcing

Another way a transition from the cold to the warm equilibrium can occur is for the model parameters to change in such a way that the potential minimum corresponding to the cold state disappears. It is interesting to ask, for example, if the changes in shortwave forcing F_1 and F_2, which must have accompanied cyclic changes in the earth's orbital parameters, are large enough to cause such a transition.

In a warming climate, the temperatures corresponding to the cold solution will move closer to those corresponding to the saddle solution. The stable solution disappears when the two solutions are equal. This condition is satisfied when

$$\frac{\Delta f}{< f >} = -\frac{b+\kappa}{g_1 b} \tag{22}$$

With standard parameters, the shortwave fluxes must each increase about 30%, say from $(F_1, F_2) = (200, 100)$ to $(260, 130)$, to cause the cold solution to vanish. For the arctic cell, the maximum increase in annual average insolation from orbital variations is about 10 Wm^{-2}, somewhat shy of the 30 Wm^{-2} this analysis suggests is needed to make a cold icy arctic impossible.

6 Discussion

The point of the minimal model is to show that a short list of processes (solar radiation, blackbody radiation, and horizontal advection), constrained by the single physical assumption (no heat accumulates at the surface), with scanty resolution in space (two cells) and no resolution in time can account for many observed or suspected properties of the system. The bare bones character of the model emphasizes which causes are responsible for which effects. It is not a revelation to find that the average temperature is proportional to the average solar heating (4), but it is interesting to know which physics appear in the constant of proportionality $(1-\beta)B$, and which do not. Similarly, our intuition is strengthened by seeing how the infrared optical properties of the atmosphere (β) and its dynamical properties (k) collaborate in $b+\kappa$ in (4) to regulate the meridional temperature contrast. The appearance of the geometrical factors g_i helps us understand why it makes a difference that the high latitude cell is much smaller than the low latitude cell.

The minimalist sees only one property of the arctic ice pack: its albedo. To first order, the climate is not interested in growing or melting ice or moving it around.

It is useful to write the time dependent equations (8) in matrix form

$$X \equiv \begin{bmatrix} T_1 \\ T_2 \end{bmatrix}, \quad M \equiv \begin{bmatrix} -b - g_2\kappa & g_2\kappa \\ g_1\kappa & -b - g_1\kappa \end{bmatrix}$$

$$Z \equiv \begin{bmatrix} f_1 \\ f_2 \end{bmatrix} \tag{23}$$

Then

$$c\dot{X} = MX + Z \tag{24}$$

The eigenvalues of M determine the stability of the equilibrium solutions. They are $-b$ and $-b-\kappa$, both real and negative. They reveal the stabilizing effect of the long-wave radiation (in b) and the poleward advection (in κ). This allows us to evaluate the effects of other feedback mechanisms. For example, suppose we think some model parameter really depends on one of the temperatures, as in $\alpha_2(T_2)$. To examine the effect of this feedback, take a linear approximation to the function that is valid near an equilibrium state. This will introduce new terms in M, and this may alter the eigenvalues. The feedback mechanism being considered may sound dangerous, warmer temperature leads to lower albedo which leads to even warmer temperature ..., but to assess the damage we need to inspect the eigenvalues, finding for example

$$\begin{aligned} -b &\rightarrow -b+\alpha' \\ -b-\kappa &\rightarrow -b-\kappa+\alpha' \end{aligned} \tag{24}$$

The point is that including a positive feedback mechanism in the model may or may not alter the stability of the equilibrium solutions. It is necessary to evaluate the eigenvalues to see if the new terms are enough to change the sign of the eigenvalues. Including the new process does not add a new positive eigenvalue, it alters the existing negative eigenvalues. In the case of the feedback between albedo and temperature, the α' is small near the cold and warm solutions, so these retain their stability. But α' is large at the equilibrium near the freezing point and this equilibrium is unstable.

The potential function U governs the time dependent stochastic system. The potential barrier against transitions from the cold to the warm state appears to be several times the standard deviation of the interannual climate noise. This makes the waiting time for a transition to occur very long.

The same processes included in the minimal model are treated in more complex simulations of climate. The minimal model suggests ways to examine results from other climate models that may help develop intuition about the physical system and about both classes of model. For example: do conventional climate models produce a poleward advection of heat proportional to the temperature contrast? Is the downwelling longwave radiation proportional to the upwelling radiation at the surface? Are there multiple equilibria? How stable are equilibria against perturbations to parameters? The minimal model gives answers to these questions in terms of physical parameters. Similar parameters can be defined and varied in conventional numerical models to test the analytical relationships presented here. It must be an easy matter to use results from large simulations to define quantities analogous to <T>, ΔT, <f>, Δf and so on. Could it be that they behave *exactly* as predicted by the minimal model?

List of Symbols

g	surface area, dimensionless, $g_1 = 0.95$, $g_2 = 0.05$.
a	albedo, $a_1=0.3$, $a_{2c} = 0.65$, $a_{2w} = 0.1$.
F	solar radiation at the surface, $F_1 = 200$, $F_2 = 100\ Wm^{-2}$
A+BT	approximation to blackbody curve near T = 0 C $A = 320\ Wm^{-2}$, $B = 4.6\ Wm^{-2}K^{-1}$.
β	atmospheric long wave emissivity parameter: 2/3.
b	$(1-\beta)B$
k	advection parameter: $0.17\ Wm^{-2}K^{-1}$.
κ	$k/2g_1g_2$
D	advective heat flux, $g_2D_2 = k(T_1-T_2)$: Wm^{-2}.
c	thermal inertia of each cell: $Jm^{-2}K^{-1}$.
Q	standard deviation of interannual variability: K
U	potential function: $dT/dt = -dU/dT$
τ	time scale for random climate perturbations
σ	standard deviation of x

References

Budyko, M. I., The effect of solar radiation variations on the climate of the earth, *Tellus* XXI, 5, 611–619, 1969.

Nakamura, N. and A. H. Oort, Atmospheric heat budgets of the polar regions, *Journal of Geophysical Research*, 93(D8), 9510–9524, 1988.

North, G. R., The small ice cap instability in diffusive climate models, *Journal of Atmospheric Science,* 41, 3390–3395, 1984.

Thorndike, A. S., A toy model linking atmospheric thermal radiation and sea ice growth, *Journal of Geophysical Research*, 97(C6), 9401–9410, 1992.

Forecasting Ice on Lakes, Estuaries and Shelf Seas

Anders Omstedt

Swedish Meteorological and Hydrological Institute
SMHI, S-601 76, Norrköping, Sweden
E-mail: aomstedt@smhi.se

1 Introduction

The presence of ice changes the environmental conditions in fresh and salt water bodies in a number of dramatic fashions. Therefore, forecasting ice is of great importance for the inhabitants of cold climate regions. This is the case in, for example, safe navigation, winter roads, off shore exploration, fishing and weather forecasting. The purpose of the chapter is to outline the modelling of cold water bodies and discuss how the models can be used in the development of forecasting systems. Instead of beginning with the fully three-dimensional coupled atmosphere-ice-ocean/lake equations, we start from simple physically based models and develop more advanced models as the complexity of the problems demands.

2 Modelling

2.1 Shallow Lakes

In many cold regions, the number of lakes is very large. For example, in Sweden the number of lakes, larger than 100 x 100 m^2, is more than 90 000. Except for some larger lakes, they are quite shallow and almost vertically well mixed, Fig. 1. The modelling approach then is to develop slab models, which we can classify as zero-dimensional models, as we do not need to resolve any spatial dimension. On the basis of the heat conservation principle, the lake temperature can be modelled according to:

$$\frac{dT}{dt} = -\frac{1}{(\rho C_p D)}\left[F_h + F_e + F_{lu} + F_{ld} + F_{sun} - F_w\right] \tag{1}$$

where T is the water temperature, ρ the water density, C_p the specific heat of water (at constant pressure), D the mean depth. The heat fluxes are: F_h the sensible heat flux, F_e the latent heat flux, F_{lu} + F_{ld} the net long-wave radiation, F_{sun} the short wave radiation and F_w the heat flux from deeper water layers or from the sediments. All fluxes are defined to be positive from the surface towards the atmosphere (positive z direction).

NATO ASI Series, Vol. I 56
Ice Physics and the Natural Environment
Edited by John S. Wettlaufer, J. Gregory Dash and Norbert Untersteiner

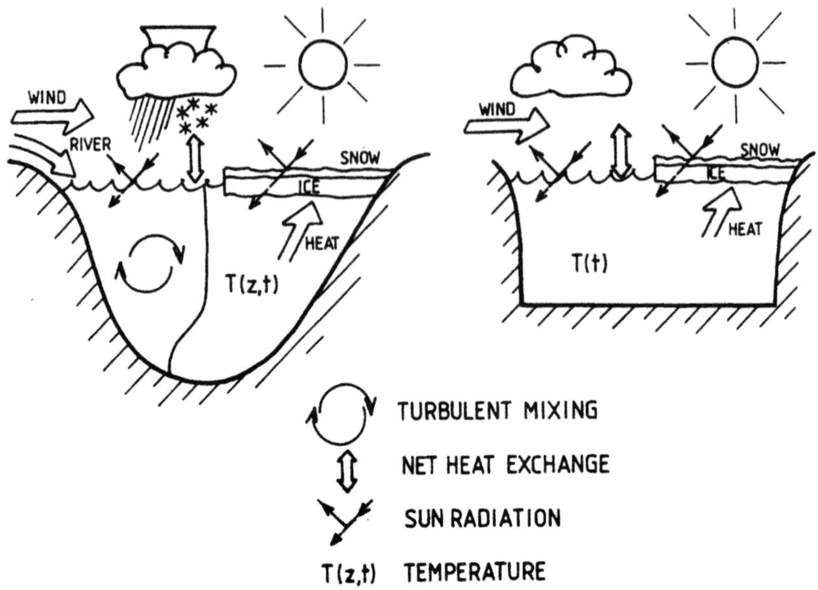

Fig. 1. Illustration of different physical processes acting in a lake and corresponding model approach.

In the ice modelling, we again rely on the heat conservation principle. For the initial ice formation, one can assume a linear (with respect to time) model according to:

$$\frac{dh_i}{dt} = \frac{F_{np}}{(L_i \rho_i)} \; with \; F_{np} = \begin{cases} F_n & F_n > 0 \\ 0 & F_n \leq 0 \end{cases} \tag{2}$$

where h_i is the ice thickness, F_n the net heat flux from the water surface, L_i the latent heat of ice and ρ_i the density of ice. The equation for the growth of columnar ice, neglecting heat flow from deeper layers, reads:

$$\frac{dh_i}{dt} = \frac{k_s k_i}{(k_i h_s + k_s h_i) L_i \rho_i}\left(T_f - T_a\right) \tag{3}$$

where h_s is the snow thickness, T_f the freezing temperature of the water, T_a the air temperature, k_s and k_i are the thermal conductivity of snow and ice. From the equation one can notice that the equation does not hold for the case of zero ice and snow thicknesses. This is one reason why we need to treat the initial ice growth separately. From the equation we can also notice that the ice growth increases as square root of time.

For shallow and small lakes we do not need in general to consider ice ridging, but ice melting and break-up must be realistically simulated. The melting is modelled by considering the net radiation balance between the short-wave radiation, that reaches the ice surface (F_{st}) and the short wave radiation that penetrates the ice and goes into the water (F_{sb}). The equation then reads:

$$\frac{dh_i}{dt} = \frac{F_{st} - F_{sb}}{L_i \rho_i} \tag{4}$$

Thus, the ice thickness decreases linearly with time (the fluxes are defined positive towards the atmosphere).

The present model has been applied to the parameterization of lake thermodynamics in a high resolution weather forecasting model, Ljungemyr et al.(1996). In the study it was demonstrated that the slab model could simulate temperatures in shallow lakes as accurately as a more advanced vertically resolved lake model. By representing all lakes in each meteorological grid cell by lake surface area and four lakes with different depths, the large amount of lakes in Sweden were parameterized and coupled to the meteorological model.

2.2 Deep Lakes

In deep lakes the vertical stratification becomes important and thus a one-dimensional modelling approach is needed. Based upon heat conservation the lake temperature satisfies:

$$\frac{\partial T}{\partial t} = \frac{\partial}{\partial z}\left[\frac{\nu_T \partial T}{\Pr_T \partial z}\right] + \Gamma_{sun} \tag{5}$$

where z is the vertical space coordinate positive upwards, ν_T the kinematic eddy viscosity, Pr_T the turbulent Prandtl number, Γ_{sun} the source term associated with solar radiation.

The source term associated with solar radiation, Γ_{sun}, is:

$$\Gamma_{sun} = \frac{1}{\rho C_p}\left\{F_{sun}(1-\delta)(1-A_i) + F^i_{sun} A_i\right\}\beta e^{-\beta(D-z)} \tag{6}$$

where F_{sun} and F^i_{sun} are the short wave energy flux to the water and through the ice respectively, A_i is the ice concentration, (1-δ) the fraction of sun radiation that penetrates into deeper layers, β the absorption coefficient and D the water depth.

The boundary condition for open water conditions reads:

$$\frac{\nu_T}{\Pr_T}\frac{\partial T}{\partial z} = \frac{F_n}{\rho C_p} \quad at\ z = 0 \tag{7}$$

where F_n is the net heat flux to the air/water interface. The net heat flux to the upper layer of the lake is:

$$F_n = F_h + F_e + F_{lu} + F_{ld} + \delta F_{sun} \tag{8}$$

where δ is the fraction of the sun radiation that is absorbed in the surface top layer.

The boundary conditions for ice-covered conditions are:

$$T = T_f \qquad at\ z = 0 \tag{9}$$

$$\frac{\nu_T}{\mathrm{Pr}_T}\frac{\partial T}{\partial z} = \frac{F_w}{\rho C_p} \qquad at\ z = 0 \tag{10}$$

where T_f is the freezing temperature and F_w the heat flux at the ice/water interface.

The stability of lake water is due to a rather complicated equation of state. Lake water is by no means pure water. Instead, dissolved substances may effect the thermodynamic properties. Also pressure effects need to be considered in deep lakes. The approach taken is to apply the equation of state for sea water to lake water. The equation of state and the equation for the temperature of maximum density in deep lakes have been examined by Farmer and Carmack (1981). In the study re-stratification near the temperature of maximum density was examined on the basis of data from Babine Lake, Canada. The estimated salinity in Babine Lake was about 0.065 (‰). Of more importance to the temperature of maximum density was, however, the pressure effect which decreased with pressure at a rate of about 0.002 (°C/m). Following Farmer and Carmack (1981) and neglecting salinity, the equation of state for lake water reads:

$$\rho = \rho_0\left\{1 + P\left(C_0 - \alpha_1(T - T_0)\right) - \alpha_2(T - T_0)^2\right\} \tag{11}$$

where P is the pressure, T the temperature and C_o, α_1, α_2, T_o , ρ_o are constants, see Farmer and Carmack (1981). Vertical mixing in a deep lakes, Lake Baikal and Lake Ontario, have been further analyzed by Carmack and Weiss (1991) and Omstedt and Murthy (1994), respectively.

Several models for the calculation of the eddy viscosity are available, here we apply a two-equation turbulence model, the so called k-ε model. This model has been used in several lake studies (e.g. Svensson 1978 and Sahlberg, 1988, Omstedt, 1984, Elo, 1994) and provides us with a model that, in a realistical way, simulates the seasonal variation of lake temperatures.

Measured and calculated temperature profiles under melting lake ice, illustrates that large amounts of heat can be stored in the water beneath the ice before break-up, Fig. 2. This indicates that the molecular sub-layers below the ice interface are important, a fact that implies that wall functions need to be introduced in the turbulence modelling.

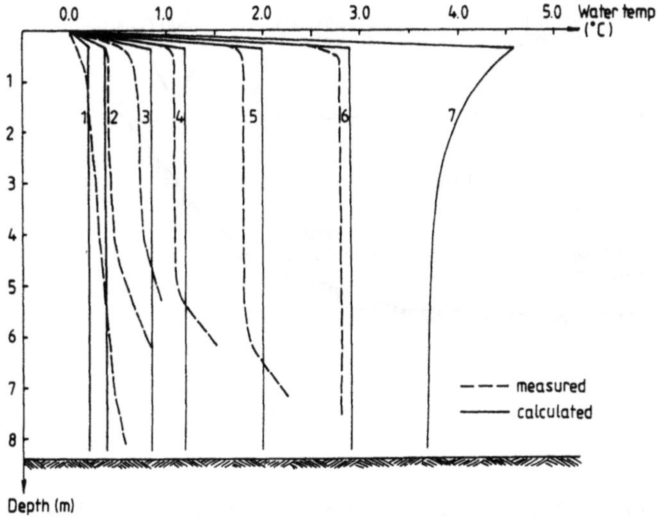

Fig. 2. Measured and calculated temperature profiles at seven different occasions (from Sahlberg, 1988).

In large lakes we also need to consider ice ridging. This can be simulated by adding an ice front model to the vertically resolved lake model. The change of the horizontal extent of the ice cover in a lake is then assumed to be due to ice drift and thermodynamics (Omstedt, 1990a). The equation reads:

$$\frac{dx_f}{dt} = U_i - \frac{x_f}{h_i}\frac{F_n - F_w}{\rho_i L_i} \tag{12}$$

where xf is the position of the ice edge (Fig. 3).

The ice drift equation reads:

$$U_i = \begin{cases} U_i^{free} - \dfrac{P_i}{x_{\text{dim}} - x_f} & \textit{on shore winds } U_i \geq 0 \\ U_i^{free} & \textit{off shore winds } U_i < 0 \end{cases} \tag{13}$$

where U_i^{free} is put equal to 2 % of the wind speed, x_{dim} is the horizontal dimension of the lake and P_i is the ice strength. In the parametrization of ice strength we follow Hibler (1979) and write:

$$P_i = P_* h_i e^{-C_i(1-A_i)} \tag{14}$$

where P_* and C_i are constants.

The basic assumptions in the equation are that the ice is deforming by wind blowing towards the shore and drifting freely during offshore winds. Note that $A_i = (x_{dim} - x_f) / x_{dim}$ where x_{dim} is the horizontal dimension of the lake.

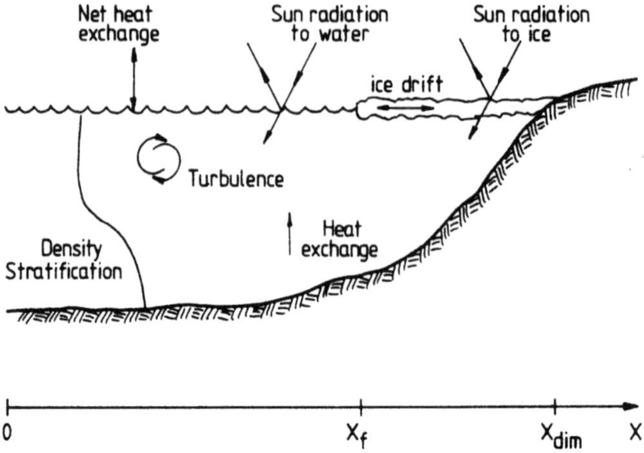

Fig. 3. A sketch of an ice edge model (from Omstedt, 1990a).

In Fig. 4 (Ljungemyr, pers. comm.), the one dimensional lake model presented above is compared to the zero dimensional model discussed in Section 2.1. The lake considered is the eastern part of Lake Vänern, a large and deep lake in the central part of Sweden. From the figure we can notice that the slab model does not simulate summer temperatures correctly, but comes close during autumn and winter periods when the whole lake is vertically mixed.

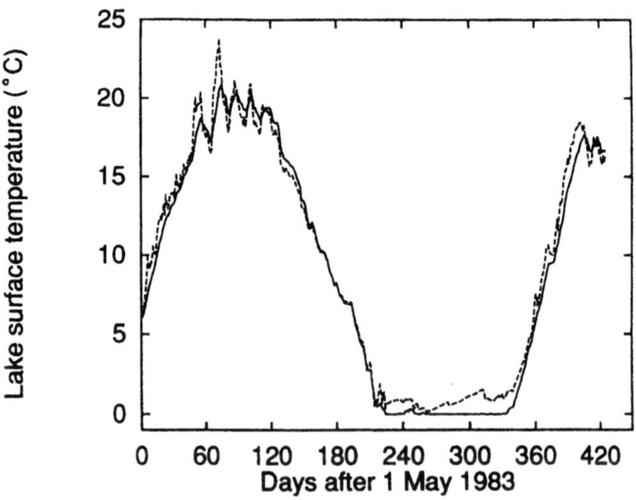

Fig. 4. Temperature comparison between siPmulated data using a well-mixed model (full lines) and a vertically resolved model (dashed line) (from Ljungemyr et al., 1996).

2.3 Estuaries

Freezing estuaries can be found all around the Arctic Ocean and also in sub-arctic areas in the Atlantic and the Pacific Oceans. The ice conditions in estuaries can be quite dramatic with rapid freezing and melting. During out-flow events, cold and brackish surface water may form a thin layer which may accelerate the freezing. On the other hand, during stormy inflow events with warm ocean water, even a thick ice-cover may melt fast. For a review of ice covered estuaries see Omstedt (1997).

In estuaries, the vertical stratification as well as the in- and out-flows therefore become important, Fig. 5. We thus need to consider a vertically resolved model that takes horizontal advection into account. A useful approach is to introduce one- and one half dimensional models, that divides the water body into sub-basins, resolving the vertical dimension and considering geometrical and dynamic restrictions, such as sounds, sills and fronts in the parameterization of the horizontal exchange. The conservation of temperature for each sub-basin then reads:

$$\frac{\partial T}{\partial t} + W\frac{\partial T}{\partial z} = \frac{\partial}{\partial z}\left[\frac{\nu_T}{\mathrm{Pr}_T}\frac{\partial T}{\partial z}\right] + \Gamma_H + \Gamma_{sun} \tag{15}$$

where z is the vertical space coordinate positive upwards, T the water temperature, W the vertical mean velocity calculated from the differences between the in- and out-flows, ν_T the kinematic eddy viscosity, Pr_T the turbulent Prandtl number, Γ_H the

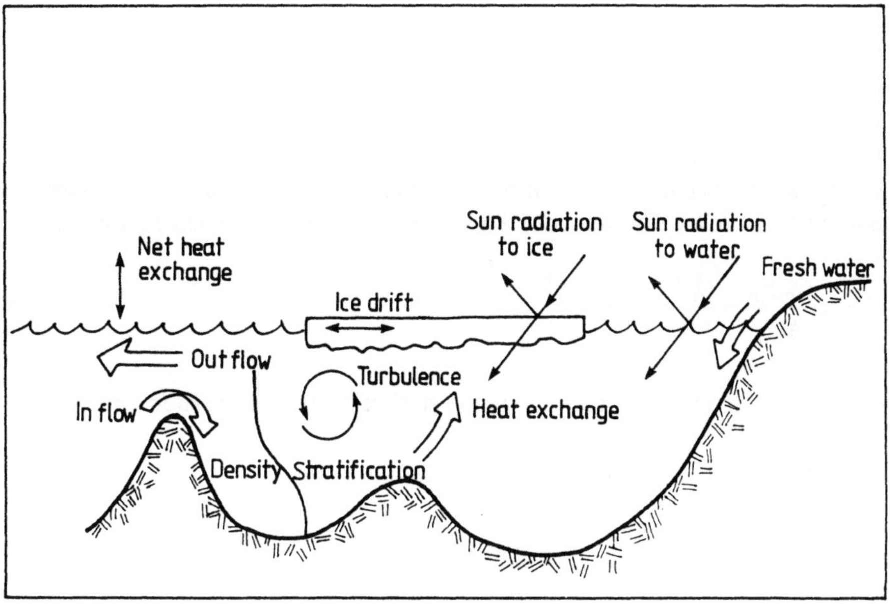

Fig. 5. Schematic representation of an estuary.

sink/source term associated with in- and out-flows and Γ_{sun} the source term associated with solar radiation (see equation 6). In the conservation equations we only deal with friction associated with vertical boundaries, thus the horizontal diffusion terms are not dealt with.

The sink/source term Γ_H, associated with the in- and out-flows to each sub-basin in its finite-difference form, reads:

$$\Gamma_H = \frac{Q_{in}T_{in}}{\Delta V_{in}} - \frac{Q_0 T}{\Delta V_0} \tag{16}$$

where Q_{in} and Q_o are the in and out-flows, T_{in} is the in-flow temperature and ΔV_{in} and ΔV_0 are the volumes associated with the in- and out-flows respectively.

The boundary conditions for ice-covered conditions are:

$$T = T_f(S_0) \qquad at\ z = 0 \tag{17}$$

$$\frac{\nu_T}{\Pr_T}\frac{\partial T}{\partial z} = \frac{F_w}{\rho C_p} \qquad at\ z = 0 \tag{18}$$

where T_f is the freezing temperature, S_0 the interfacial salinity and F_w the heat flux at the ice/water interface. At the ice/water interface we thus assume that both phases are in local equilibrium and utilize the empirical freezing temperature equation (Millero, 1978).

The conservation of salinity in each sub-basin reads:

$$\frac{\partial S}{\partial t} + W\frac{\partial S}{\partial z} = \frac{\partial}{\partial z}\left[\frac{\nu_T}{Sc_T}\frac{\partial S}{\partial z}\right] + \Gamma_S \tag{19}$$

where Sc_T is the turbulent Schmidt number and Γ_S the sink/source term associated with in- and out-flows. The source/sink term, Γ_S, in its finite-difference form reads:

$$\Gamma_S = \frac{Q_{in}S_{in}}{\Delta V_{in}} - \frac{Q_0 S}{\Delta V_0} - \frac{Q_f S}{\Delta V_S} \tag{20}$$

where Q_f is the river runoff and ΔV_s is the surface volume associated with the river runoff.

The boundary condition for open surface conditions are:

$$\frac{\nu_T}{Sc_T}\frac{\partial S}{\partial z} = S_S(E - P_{re}) \qquad at\ z = o \tag{21}$$

where S_S is the surface salinity and P_{re} and E are the precipitation and evaporation rates, respectively.

The boundary conditions for ice-covered conditions read:

$$\frac{\nu_T}{Sc_T}\frac{\partial S}{\partial z} = \frac{\rho_i}{\rho_0}\frac{dh_i}{dt}(S_0 - S_i) \qquad at\ z = o \tag{22}$$

where ρ_i is the ice density, ρ_0 the reference water density, h_i the ice thickness and S_i the ice salinity.

The conservation equations for momentum, assuming horizontally homogeneous flow, need to be solved for each sub-basin. In straits or channels the boundary layer considered could often be represented by a channel flow model with wind stress, while in larger sub-basins an Ekman flow model can be applied.

The equation of state for saline water is a complex function of temperature, salinity and pressure and the reader is referred to Gill (1982, Appendix 3). It should also be noticed that in estuaries the equation of state may need to be modified, as the dissolved river salt may influence the density (Millero and Kremling, 1976).

In Fig. 6, a temperature time history from the entrance area of the Baltic Sea is illustrated. From the salinity data, we can notice large variations, corresponding to inflow of saline water and outflow of brackish water. The corresponding time-temperature curves are also illustrated in the figure.

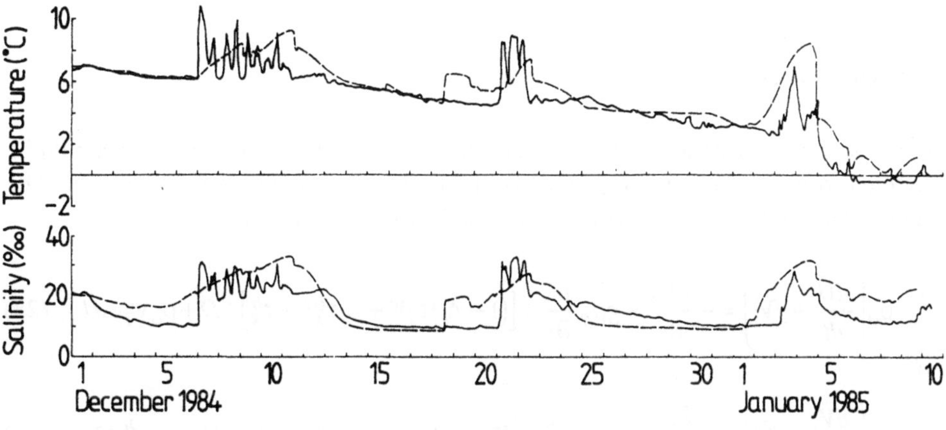

Fig. 6. Measured (solid lines) and calculated (dashed lines) data from the entrance area of the Baltic Sea (from Omstedt, 1987).

The one- and one half-modelling approach has been applied to the Baltic Sea by Omstedt and Nyberg (1996). By dividing the water body into 13 sub-basins, and modelling each sub-basin as a vertical resolved model with in- and out-flows, they were able to model the seasonal and interannual variation realistically. In the model ice advection out or into a sub-basin were added. The effect of ice in reducing the flow through straits were also considered. A verification example from the model is illustrated in Fig. 7.

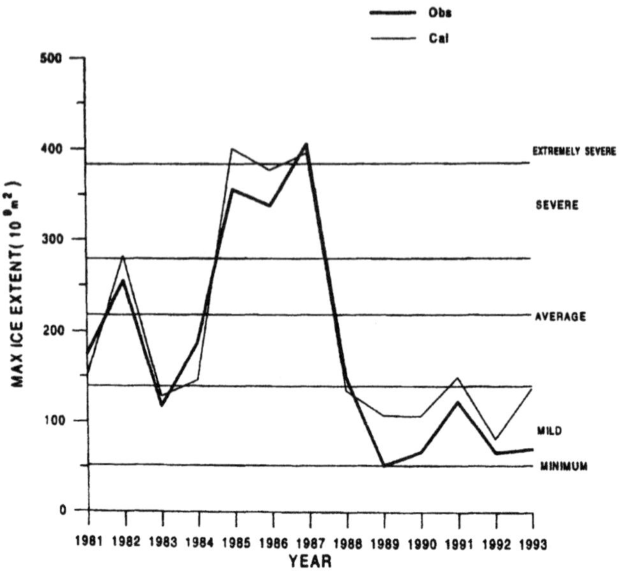

Fig. 7. Observed (thick line) and calculated (thin line) seasonal maximum ice extent for the Baltic Sea (from Omstedt and Nyberg, 1996).

2.4 Shelf Seas

In freezing shelf seas with weak vertical stratification, storm surge models should be coupled to ice models, giving us a two-dimensional coupled ice-ocean model. The depth integrated shallow water equations read (Kowalik and Murty, 1993):

$$\rho_o\left(\frac{d\tilde{U}}{dt}-f\tilde{V}\right)=-\frac{\partial Pa}{\partial x}-g\rho_o\frac{\partial\zeta}{\partial x}+\left[(1-A_i)\tau_x^{aw}+A_i\tau_x^{iw}-\tau_x^b\right]/D+\rho_o\nu_T^h\Delta\tilde{U} \quad (23)$$

$$\rho_o\left(\frac{d\tilde{V}}{dt}+f\tilde{U}\right)=-\frac{\partial Pa}{\partial y}-g\rho_o\frac{\partial\zeta}{\partial y}+\left[(1-A_i)\tau_y^{aw}+A_i\tau_y^{iw}-\tau_y^b\right]/D+\rho_o\nu_T^h\Delta\tilde{V} \quad (24)$$

where Pa is the air pressure, ζ is the water level, D the water depth, ν_t^h the horizontal eddy viscosity and τ_x^{aw}, τ_y^{aw}, τ_x^{iw}, τ_y^{iw}, τ_x^{b}, τ_y^{b} are the x and y components of the air stress over water, ice-water stress and bottom stress, respectively. The vertically averaged velocities read:

$$\tilde{U}=\frac{1}{D}\int_{-H}^{\zeta}Udz \quad (25)$$

$$\tilde{V} = \frac{1}{D} \int_{-H}^{\zeta} V dz \tag{26}$$

The continuity equation then reads:

$$\frac{\partial}{\partial x} \tilde{U} D + \frac{\partial}{\partial y} \tilde{V} D + \frac{\partial \zeta}{\partial t} = 0 \tag{27}$$

The two-dimensional ice drift model equations read:

$$\rho_i \left(\frac{dU_i}{dt} - fV_i \right) = -\frac{\partial P_a}{\partial x} - \rho_i g \frac{\partial \zeta}{\partial x} + \left(\tau_x^{ai} - \tau_x^{iw} \right) / h_i + F_x \tag{28}$$

$$\rho_i \left(\frac{dV_i}{dt} + fU_i \right) = -\frac{\partial P_a}{\partial y} - \rho_i g \frac{\partial \zeta}{\partial y} + \left(\tau_y^{ai} - \tau_y^{iw} \right) / h_i + F_y \tag{29}$$

where U_i, V_i are the x and y components of ice drift, τ_x^{ai}, τ_y^{ai} are the x and y components of the air stress over ice and F_x, F_y represents the internal stress propagated through the ice. The coupling between atmosphere and ice is by the wind stress, and between ice and water by the ice-water stress.

The equation of state for ice connects ice concentration A_i and ice thickness h_i by the following relation:

$$m_i = \rho_i h_i A_i = \rho_i \left(h_l + h_r \right) A_i \tag{30}$$

where h_l and h_r are the level and the equivalent thickness of ridged ice, respectively.

In shelf seas, with only first year ice present, a useful approach to represent the ice thickness is by a three-level approach. The levels are: open water ($1-A_i$), level ice thickness (h_l) and ridged ice thickness (h_r). They are calculated according to:

$$\frac{d}{dt} \left(A_i, h_l, h_r \right) = \left(\psi_A, \psi_l, \psi_r \right) \tag{31}$$

where ψ_A, ψ_l and ψ_r are functions describing open water changes, ice growth/decay and rafting and ridging. For the calculations of these functions, see Leppäranta (1981).

Two-dimensional coupled dynamic ice-ocean models have been applied to the Bohai Sea and the Baltic Sea. An interesting study by Zhang and Leppäranta (1995), illustrated that this type of model could well predict the sea level variations in the Baltic Sea during ice covered conditions, Fig. 8. The reducing influence of sea ice on the piling-up of water due to wind has been well documented in the Baltic Sea and indicates that for freezing shelf seas one need to consider the ice-water coupling when forecasting water level variations.

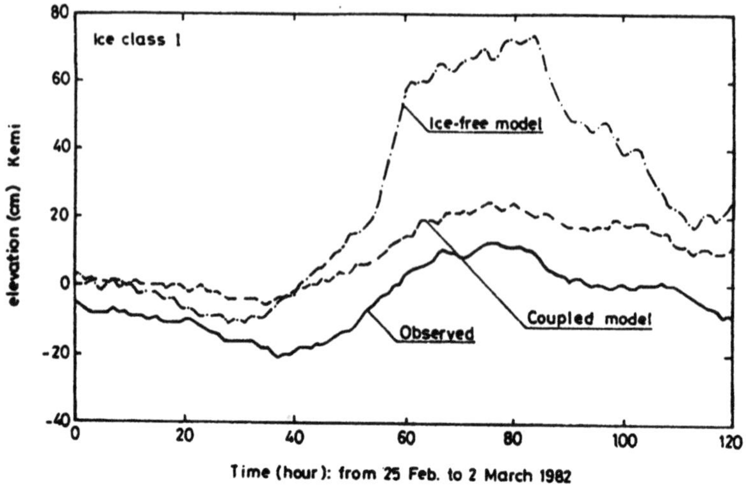

Fig. 8. Modelled and observed sea level differences between two sea level stations in the northern part of Baltic Sea (from Zhang and Leppäranta, 1995).

For dynamic studies, horizontally resolved two-dimensional ice-ocean models can be used, but when modelling the thermodynamic response (cooling, ice formation, growth, decay and warming), we need in general also to resolve the vertical structure. Instead of resolving all three dimensions, a two- and one half-dimension model approach can often be applied. We could then add a vertically resolved boundary layer model to the two-dimensional coupled ice-ocean model. This approximation was developed by Haapala and Leppäranta (1996) and Omstedt and Nyberg (1995) when modelling the Baltic Sea. In the former model, the dynamics were due to a storm surge model, but the thermodynamics were resolved with four vertical layers. In the latter model the dynamics were calculated from a storm surge model, but the thermodynamics were resolved with a vertically resolved boundary layer model applied to different regions of the Baltic Sea. In general, one should apply vertically resolved boundary layers models to all grid cells. For many applications it is, however, only necessary to apply vertical resolved models to different thermodynamic regions, dividing the model area into archipelagos, coastal zones and open sea areas as indicated by sea surface maps. In the forecasting system in Sweden we now only consider 31 thermodynamic regions, Fig. 9, representing thermodynamics in about 1000 grid cells. In Fig. 10, observed and modelled maximum ice extent for the winter 1993/1994 is illustrated. The vertical resolve boundary layer model applied in these studies is the same as was discussed in Section 2.2.

Three-dimensional coupled ice-ocean models, are under development and have been applied in ocean forecasting systems, Preller and Posey (1996). For shelf seas they are now in an active research stage and will become important tools in future forecasting systems, particularly when coupled atmosphere and ocean models are introduced into the different national forecasting institutes.

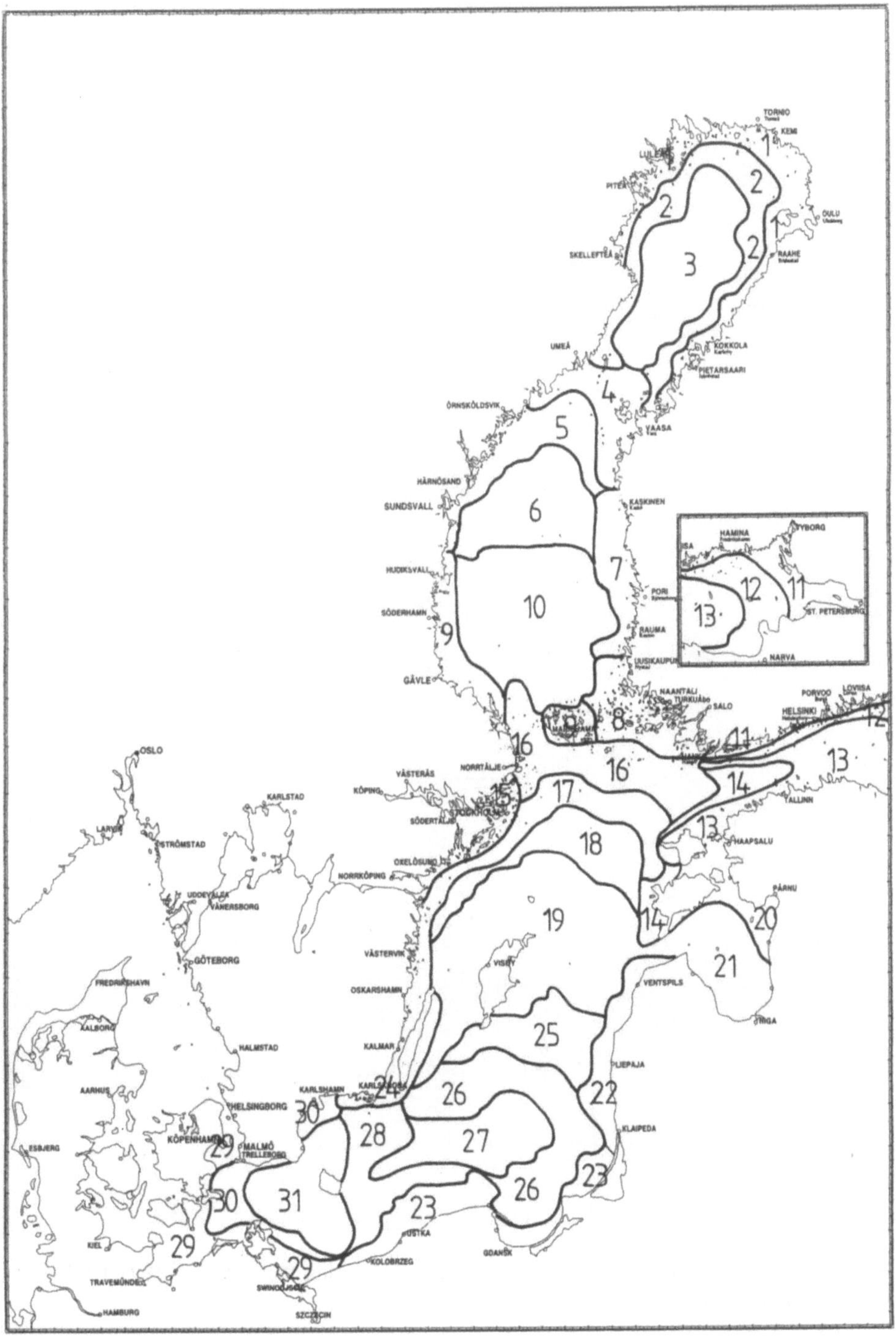

Fig. 9. Division of the Baltic Sea into 31 thermodynamic regions (from Gustafsson et al., 1998).

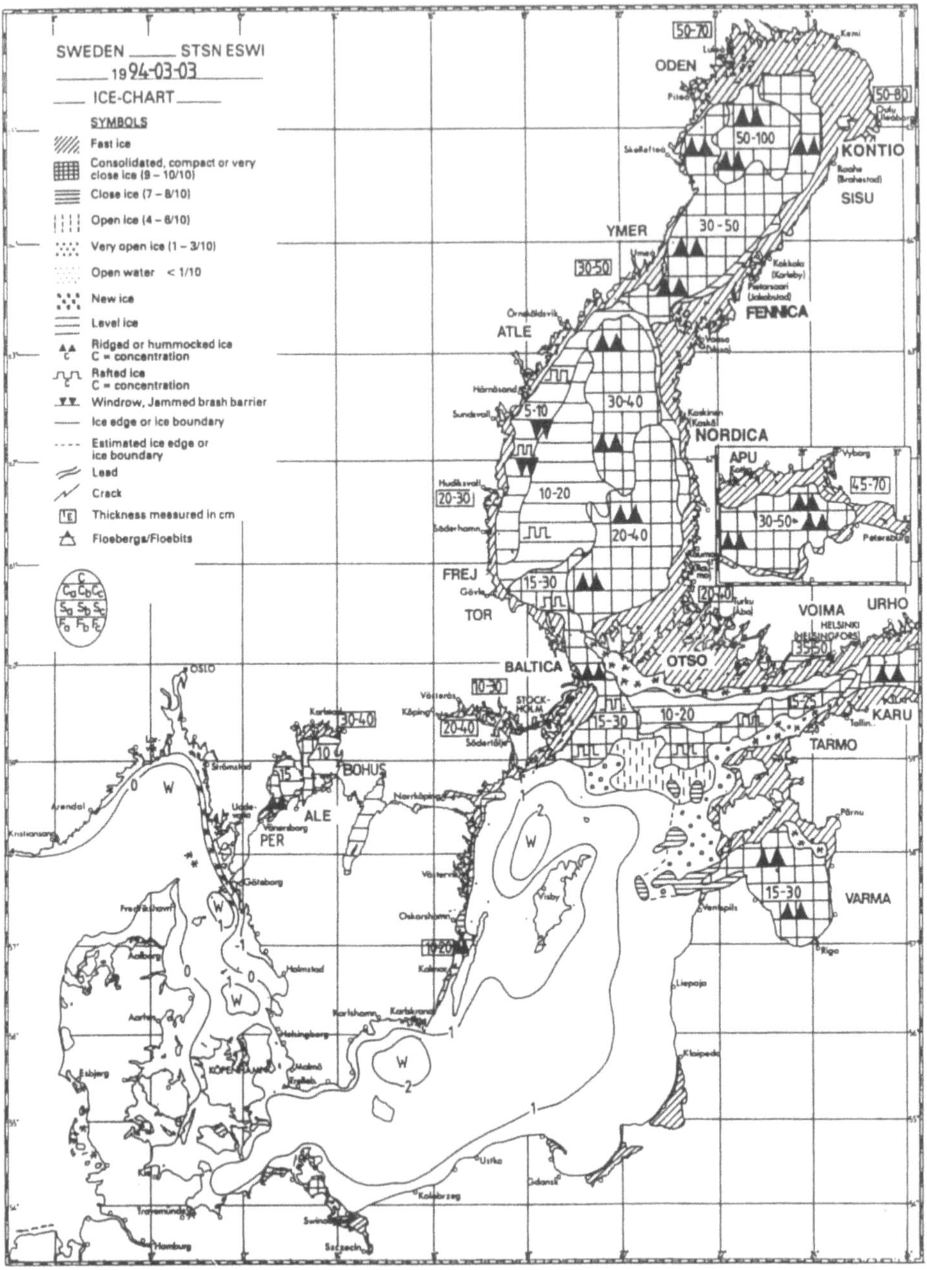

Fig. 10. Observed and modelled maximum ice extent during the winter of 1993/1994 (from Omstedt and Nyberg, 1995).

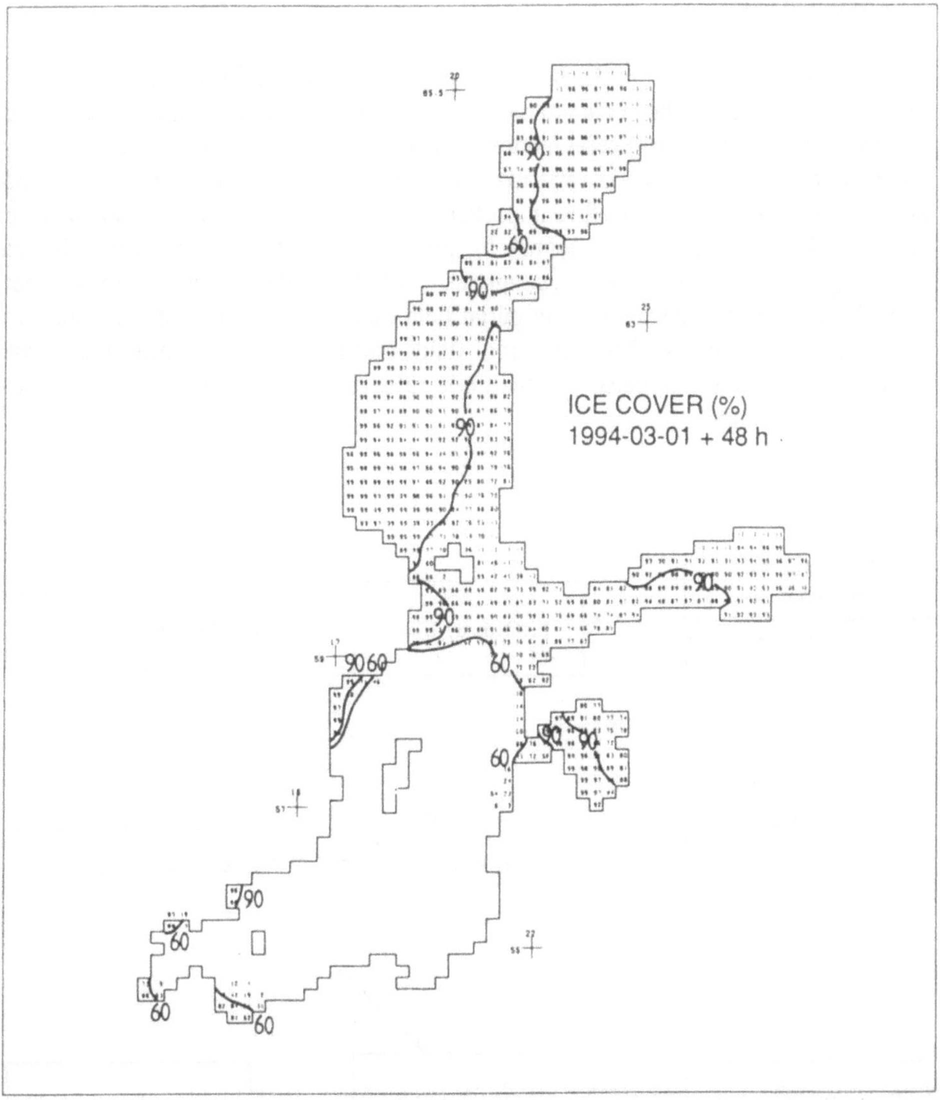

Fig. 10. Continued.

3 Forecasting

Real time observations from the lakes and oceans are rare, and it is therefore important to integrate all available data to come up with the best estimate of the state of the studied water body. From several modelling studies we know that we can realistically simulate ice drift and seasonal variations as water cooling, ice formation, growth and decay. These models are therefore important complements to the observation system and could in a systematic way integrate the present available knowledge about the sea state. Figure 11 illustrates how ocean models could be used together with data. From ship observations in late autumn the initial field of salinity and temperature needs to be measured. These data form the initial fields and from now on during the whole winter season, the ocean density structure is calculated by the model forced by observed forcing fields.

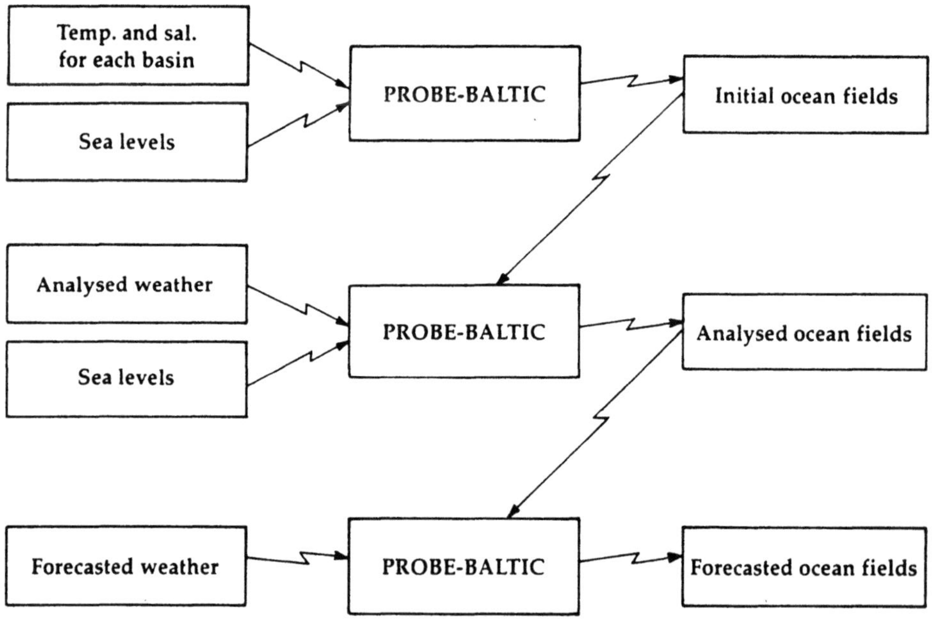

Fig. 11. Schematic presentation on up dating and forecasting system (from Omstedt, 1990b).

The atmosphere, the ice and the ocean constitute a physical system with strong coupling. An example is (illustrated in Omstedt and Nyberg, 1995) from the early winter with a situation when rapid ice formation/ break-up occurred in the northern Baltic Sea. In late December, 1993, the Bothnian Bay was almost covered with thin ice. Due to strong winds, the ice broke up and two days later the ice concentration was reduced. The reason was high winds that mixed warmer water from deeper parts up into the surface layer. Sea surface properties in shelf seas may thus rapidly change.

This is also easy to study during summer when upwelling may drastically change the sea surface temperatures.

Another example on the atmosphere-ice-ocean coupling is given in Fig. 12, from January 1987, when cold air was blowing out over the Baltic Sea with strong cloud formation and snow bands. In a study by Andersson and Gustafsson (1994), it was demonstrated that the snow bands over the Baltic Sea were dependent on the geometry of the coastlines, which may drastically change due to the sea ice.

Data assimilation has been used over many years in meteorological forecasting and is now also introduced in ocean models, particularly with respect to sea level forecasts. Before we discuss this, it is important to give some warnings about data assimilation. For many applications it is probably better not to introduce data assimilation in the models. The accuracy of the models is then easy to check, at the same time as improvements of the physics can be introduced. Data assimilation

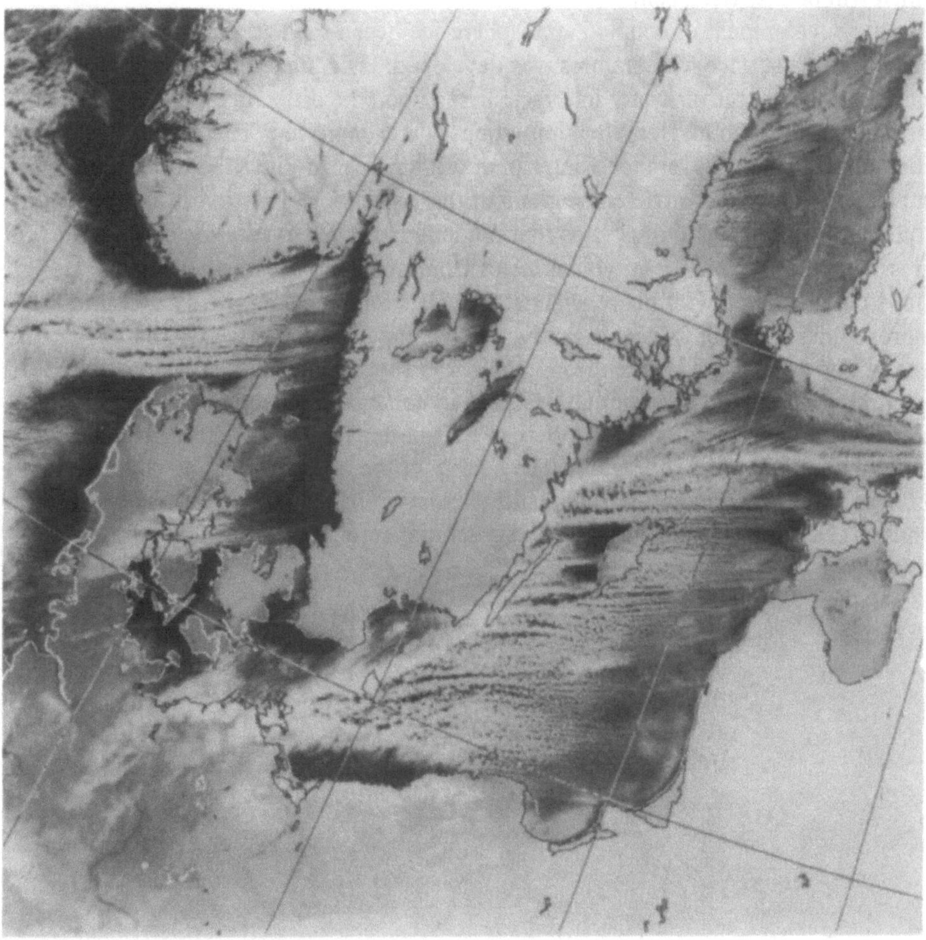

Fig. 12. Infrared satellite image from NOAA-9 at 1235 UTC 11 January 1987.

should therefore only be introduced in a later phase of the development and only if it is necessary. It is wiser to have a discrepancy between calculated and observed data and try to understand why, instead of hiding bad models through data assimilation.

In meteorological models many different types of data are assimilated using different methods. A recently review of the problems related to data assimilation for mesoscale limited area atmosphere models is given by Gustafsson (1997). In the ocean, where the observations are of quite a different quality great care must be taken. The only data, from my point of view, that can today be used in operational ocean modelling are sea levels and sea surface temperatures. We are probably far from introducing CTD data or ice data into our data assimilation schemes. Different methods for assimilating water level data into ocean models are available, where for example Kalman filter and adjoint methods have been used, Heemink et al.(1997) respectively Meier and Krauss (1997). For sea surface temperature a simple method of assimilating data into an ice forcasting system has been developed by Gustafsson et al (1998), which will be discussed below.

In the development of coupled atmosphere-ice-ocean models a simple assimilation method of sea surface temperatures was developed. By adding a flux correction to the energy balance at the air/water interface, based upon the difference between observed and modelled cooling rates, the simulations were improved. The system was used when re-analysing a severe ice winter in the Baltic Sea. In Figs. 13a to 13c, three examples are given from different regions of the Baltic Sea. The first from the central Bothnian Bay, Fig. 13a, illustrates that modelled and observed sea surface temperatures were close, which is a typical result from open sea subareas. In the next, Fig. 13b, from a subarea with many islands, the Åland archipelago, the data assimilation improve the simulation. A significantly positive impact of the data assimilation can be seen in Fig. 13c, which is taken from region 25 (Fig, 9) in the eastern part of the Baltic proper. Here the data assimilation helps to enhance the cooling of the sea water, that is too slow in the model run. We can thus notice an improvement using data assimilation in the model simulation. The quality of the re-analysis using coupled models, was assessed through the successful simulation of the convective snowbands in January 1987, Fig. 14, that generated a substantial amount of snow along the Swedish east coast.

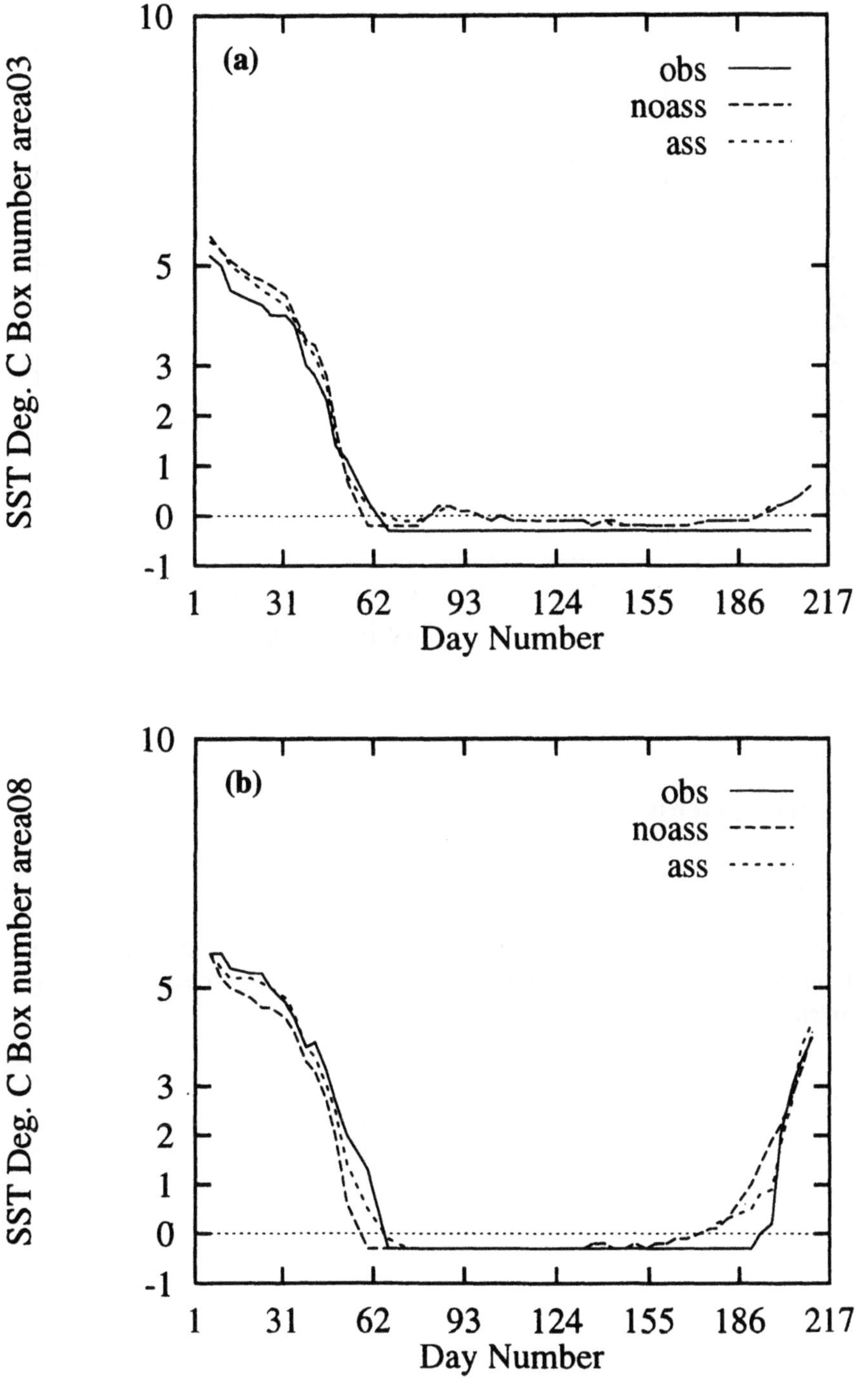

Fig. 13. Observed and calculated temperatures using no data assimilation and data assimilation from (a) Central part of Bothnian Bay, (b) Arhipelago Sea and (c) Eastern Baltic proper (from Gustafsson et al., 1997).

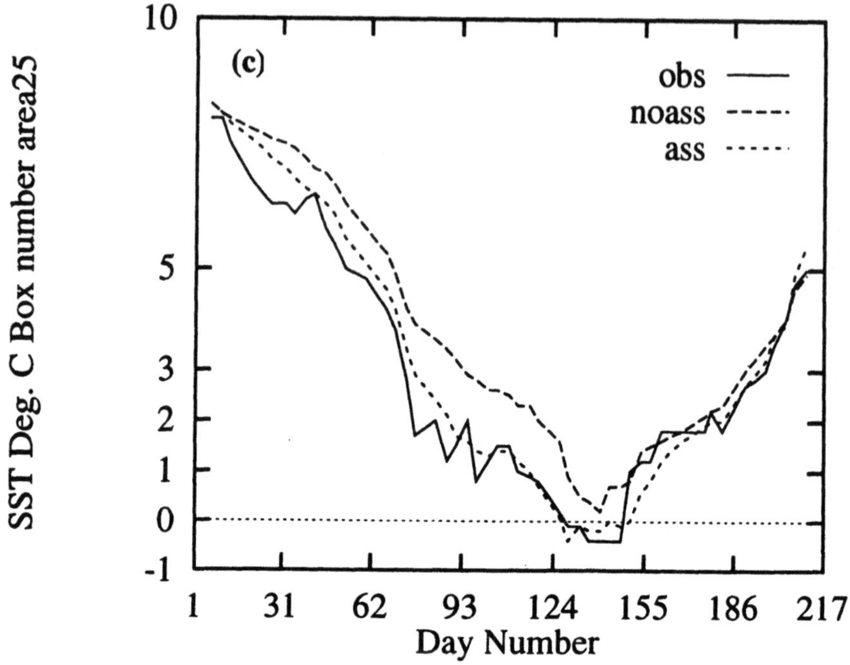

Fig. 13. Cont.

Acknowledgments

This chapter was much due to John Wettlaufer, who invited me to the summer school, and also due to Norbert Untersteiner who arranged the Maratea summer school 1981. I would also like to thank Alan Thorndike for some valuable comments on an earlier draft, and Gunilla Mild and Vera Kuylenstierna for word-processing and linguistic improvements.

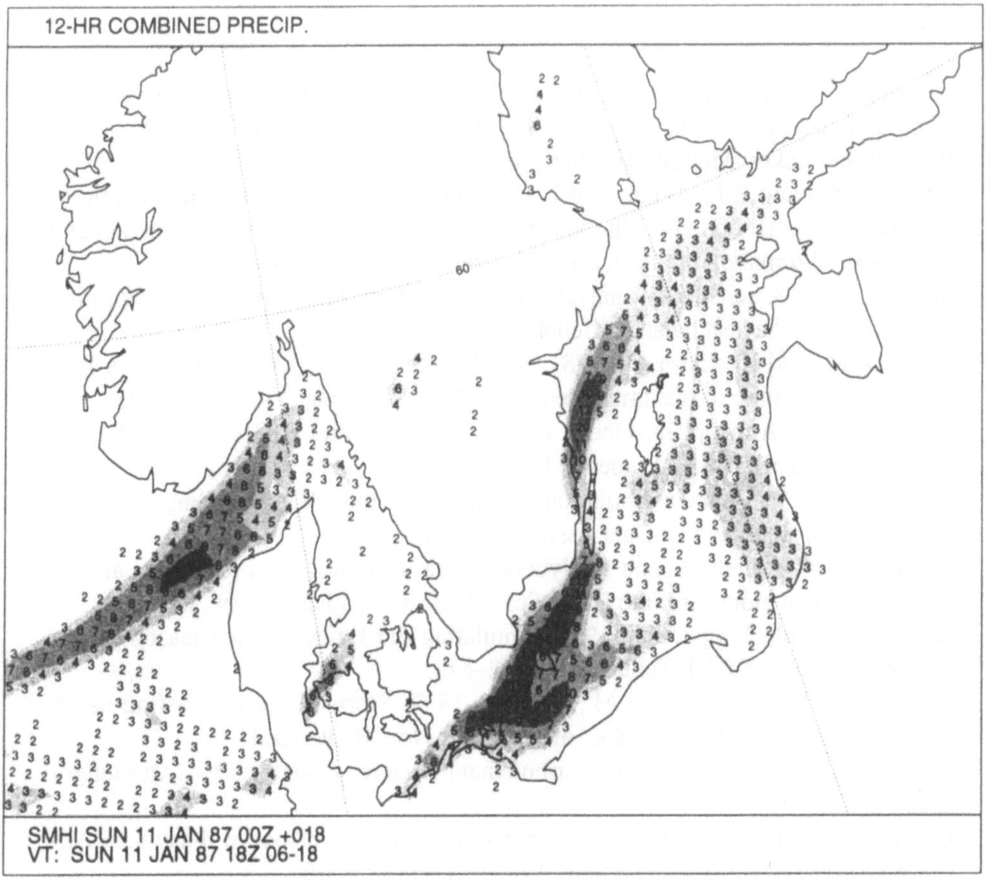

Fig. 14. Simulated snow bands from the reanalysis study (from Gustafsson et al., 1998).

References

Andersson, T. and N., Gustafsson, 1994. Coast of departure and coast of arrival. Two important concepts for the formation and structure of convective snow bands over the Seas and Lakes. Monthly Weather Review, 122, 1036–1049.

Carmack, E.C., and R.F., Weiss, 1991. Convection in Lake Baikal: An example of thermobaric instability. P.C. Chu and J.C. Gascard (Editors), Deep Convection and Deep Water Formation in the Oceans. Elsevier Science Publishers B.V.

Elo, A-R., 1994. A sensitivity analysis of a temperature model of a lake examining components of the heat balance. Geophysica, 30, 79–92

Farmer, D. M. and E., Carmack, 1981. Wind mixing and re-stratification in a lake near the temperature of maximum density. J. Phys. Oceanogr., 11, 1516–1533.

Gill, A.E., 1982. Atmosphere-Ocean Dynamics. International Geophysical Series, Vol. 30, New York, USA, Academic Press.

Gustafsson. N., 1997. Data assimilation for mesoscale models. Proceedings from the 1996 ECMWF seminars on Data Assimilation, in press.

Gustafsson, N., Omstedt, A. and L. Nyberg , 1998. Coupling high resolution atmosphere and ocean models for the Baltic Sea. Monthly Weather Review, in press.

Haapala, J. and M., Leppäranta, 1996. Simulating the Baltic Sea ice season with a coupled ice-ocean model. Tellus, 48 A, 622–643.

Heeink, A.W., Bolding, K. and M., Verlaan, 1997. Storm surge forecasting using Kalman filtering. J. Meteorol. Soc. of Japan, 75, No. 1B, 305–318.

Hibler III, W.D., 1979. A dynamic thermodynamic sea ice model. J. Phys. Oceanogr., 9, 815–846.

Kowalik, Z. and T. S., Murty, 1993. Numerical modeling of ocean dynamics. Advanced Series on Ocean Engineering, Vol. 5, World Scientific.

Leppäranta, M., 1981. An ice drift model for the Baltic Sea. Tellus, 33, 583–596.

Ljungemyr, P., Gustafsson, N. and A., Omstedt, 1996. Parameterization of lake thermodynamics in a high-resolution weather forecasting model. Tellus, 48 A, 608–621.

Millero, F.J., 1978. Freezing point of sea water. 8th report of the joint Panel on Oceanographic Tables and Standards. UNESCO Tech. Paper Mar. Sci., 28(annex 6), UNESCO, Paris, France.

Millero, F.J. and K., Kremling, 1976. The densities of Baltic Sea waters. Deep Sea Res., 23, 1129–1138.

Omstedt, A., 1984. A forecast model for water cooling in the Gulf of Bothnia and Lake Vänern. SMHI, Report RHO 36.

Omstedt, A., 1987. Water cooling in the entrance of the Baltic Sea. Tellus, 39A, 254–265.

Omstedt, A., 1990a. A coupled one-dimensional sea ice-ocean model applied to semi-enclosed basin. Tellus, 42A, 568–582.

Omstedt, A., 1990b. Real-time modelling and forecasting of temperatures in the Baltic Sea. SMHI Reports, RO 12, 1–28.

Omstedt, A., 1997. Freezing estuaries and semi-enclosed basins. In Physics of Ice-covered Seas, in press.
Omstedt, A. and C.R. Murthy, 1994. On currents and vertical mixing in Lake Ontario during summer stratification. Nordic Hydrology, 25, 213–232.
Omstedt, A. and L., Nyberg, 1995. A coupled ice-ocean model supporting winter navigation in the Baltic Sea. Part 2: Thermodynamics and meteorological coupling. SMHI Reports, RO 21. SMHI, S-601 76 Norrköping, Sweden.
Omstedt, A. and L., Nyberg, 1996. Response of Baltic Sea ice to seasonal, interannual forcing and climate change. Tellus, 48A, 644–662.
Preller, R.H. and P.G., Posey, 1996. Validation test report for a navy sea ice forecasting system: The polar ice prediction system 2.0. USA Naval Research Laboratory, Stennis Space Center, MS 39529-5004, USA.
Sahlberg, J., 1988. Modelling the thermal regime of a lake during winter season. Cold Region Science and Technology, 15, 151–159.
Svensson, U., 1978. A mathematical model of the seasonal thermocline. Lund Inst. of Technology, Dept. of Water Res. Eng., Report No. 1002, Lund, Sweden.
Zhang, Zh.-H. and M., Leppäranta, 1995. Modeling the influence of sea ice on sea level variations in the Baltic Sea. Geophysica, 31(2), 31–45.

On Productivity in Ice-Covered Polar Oceans

Ole Jørgen Lønne

The University Courses on Svalbard, P.O. Box 156,
N-9170 Longyearbyen, Norway

Keywords: productivity, sea ice, polar, marine, sympagic, biota

1 Introduction

At maximum extent the sea ice covers some 8% of the Southern Hemisphere and 5% of the Northern Hemisphere and thus has a significant influence on the productivity of large ocean areas. The structure, extent and distribution of sea ice not only influences biogenic production within the ice system, but also affects pelagic and benthic systems under the ice and near ice edges. The sea ice in Polar Oceans provides a unique environment utilised by a diverse group of organisms ranging in size from single-celled algae within the ice to meter-long algal assemblages, dominated by diatoms, attached to the lower surface of the ice, and from protozoans to marine mammals. Sea ice (sympagic) biota play an important role in high Arctic food webs. Ice algae may serve as an inocculum for marginal ice zone blooms and form the base of food webs culminating in polar cod (Lønne and Gulliksen 1989), seals (Nilssen et al. 1995) and seabirds (Lønne and Gabrielsen 1992). Matter entering this lipid-rich food web may reach higher trophic levels in areas geographically distant from its place of origin, due to passive transport of sea ice biota in Arctic ice drift (Falk-Petersen et al. 1990).

2 The Sea Ice Habitat

Sea ice is in almost continual motion, unless anchored to the bottom or obstructed by land. Winds and currents are principal driving forces. These forces may cause the ice to break up or be pushed together depending on the direction of the force relative to the ice field. Where the ice diverges, the underlying water is exposed to light and space is created for the production of new ice. New ice areas, between older floes, are thus made available for sea ice organisms. Converging and shearing motions cause the ice to fracture and pile up. Deep keels or layers of sandwiched floes, separated by water, may be formed below the upper surface. The area of the sea/ice interface is enlarged and a wider range of microhabitats created for sea ice organisms.

Sea ice fields undergo radical changes in structure and composition in response to seasonal melting and freezing processes. In some areas the ice melts completely, releasing sea ice organisms to the underlying water column, while in other areas there may be a partial melting resulting in loss of organisms or a concentration of organisms on remaining floes. During freezing, salt and other impurities is partly rejected

NATO ASI Series, Vol. I 56
Ice Physics and the Natural Environment
Edited by John S. Wettlaufer, J. Gregory Dash and Norbert Untersteiner

and partly trapped in the ice as it thickens. Through different mechanisms these pockets of salty water become interconnected, producing 50–300 vertical brine channels per square meter of the lower ice surface. These drainage channels connect the interior of the ice to the water column below, thus enlarge the ice/water interface substrate which is available to biological production.

In the Arctic, ice-melt occurs at the ice/atmosphere interface, resulting in melt ponds on the upper ice surface. These ponds may occupy more than 60% of the ice surface and range in salinity from fresh to fully saline seawater (Maykut 1985). They are inhabited by a variety of organisms e.g., bacteria, diatoms, flagellates and ciliates (Spindler 1994).

In the Antarctic, the sea ice is often depressed due to the weight of the overlying snow cover. This allows seawater to penetrate laterally and create suitable conditions for colonisation by sea ice organisms (Horner et al. 1992).

Together these processes create a complex space in three dimensions with its own set of characteristics attracting unique assemblages of organisms. Three categories of habitats and communities are defined, namely the surface, interior and bottom. Often a sub-ice habitat/community immediately beneath the ice is recognised (Horner et al. 1992). A schematic representation of biological communities found within sea ice is given in Fig. 1 and also discussed in e.g. Horner et al. (1992), Spindler (1990) and Syvertsen (1991).

Most of the Antarctic pack ice is seasonal, varying between 20×10^6 km^2 at its maximum extent to 4×10^6 km^2 at the end of the summer. In the Arctic about half of the ice covered area during maximum extent (14×10^6 km^2) remains through the summer (7×10^6 km^2) (Maykut 1985). This implies that most of the ice in the Southern Ocean has to be inhabited anew every year, while in the north lateral dispersion between floes is possible and the sea ice can serve as a habitat for several years. Longer living organisms may complete their whole life cycle within the ice.

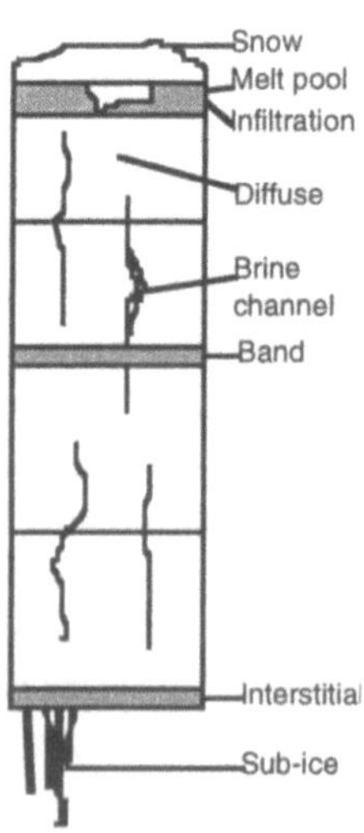

Fig. 1. Schematic representation of algal communities found in sea ice (redrawn from Horner et al. 1988).

3 Sea Ice Associated Production

Several estimates of global annual phytoplankton primary production are now at hand, and the use of satellite radiometers such as the CZSC has increased the resolution of such data. Production estimates rely basically on regional and seasonal distribution of biomass, which again is related to ocean colour. Present figures indicate a total annual production in the ocean of 45–50 Gt C year^{-1} (Longhurst et al. 1995) (Table 1). But see also Behrenfeld and Falkowski (1997) for a comparison of recent estimates using different production models. Similar estimates for land plants (45–68 Gt C year^{-1}) (Lurin et al. 1994) suggest that the annual production on land and that in the sea may be in the same order of magnitude.

Primary production in polar regions (and elsewhere within the marine environment) can basically take place in three different environments; on the bottom, in the water column, and in association with floating sea ice. In the same way as water quality and water depth put limitations on the production taking place on the bottom, so is production in the water column affected both under and in the near vicinity of sea ice.

Early measurements indicated the central Arctic Ocean (English 1961) to be a barren area with an annual production of about 1 g C m^{-2} y^{-1}. More recent studies estimates the total annual production to be at least an order of magnitude higher (Legendre et al. 1992) and that a significant portion of the production takes place in the Central Arctic Ocean (Wheeler et al. 1996).

Table 1. Total primary production (adapted from Longhurst et al 1995).

	Area	Production rate	
	10^6 km^2	gCm^{-2}y^{-1}	10^{14} gCy^{-1}
Poleward of the polar fronts	20.8	385	64
Southern Ocean	1.93	398	7.7
Arctic Ocean	1.66	645	10.7
Total all oceans	328	282	450–500

Production estimates based on ocean colour only take into account production that is reflected in changes in surface colour. Longhurst et al. (1995) argue that production in the ice and under the ice is negligible and can be set to zero. Their estimates of the production in the polar oceans (Table 1) ranges from 385 g C m^{-2} y^{-1} as a total average to 645 g C m^{-2} y^{-1} in the Arctic Ocean, but since those estimates only take into consideration areas not covered with ice both the area and the total production in ice covered waters is not comparable to more detailed estimates. Production in open high latitude water associated with ice is, however, comparable to world ocean estimates, and open water within the Arctic ice covered region is comparable to high production coastal regions.

The areas with high primary production in these estimates are however small compared to the areas covered with ice. Key questions are therefore whether it can be justified to ignore ice-associated production or not, and to what extent the presence of ice enhances production relative to areas without ice.

A recent review on primary production estimates in ice-covered regions, based on standing stock and ^{14}C productivity data indicate that 25% of the total Arctic primary production takes place in sea ice (Table 2). In the Antarctic, the proportion of primary production taking place in the ice is even higher (33%) (Legendre et al. 1992). It is also worth noting that they estimated the production under ice in the Antarctic to account for less than 1% of the total primary production.

Table 2. Estimated primary production associated with Antarctic and Arctic sea ice (adapted from Legendre et al 1992).

		Area 10^6 km^2	Annual production 10^{14} gCy^{-1}
Antarctic			
Water column in ice edge blooms		1.5–2	1.41
Water column under ice cover		5–10	0.03
Within ice			0.7
Surface	0.4	10	
Freeboard	0.2		
Interior	0.02		
bottom	0.08		
Total Antarctic			2.14
Arctic			
Water column – shelf areas		4.9	1.32
Water column – offshore		8.2	0.73
Within sea ice		12	0.73
Total Arctic			2.78

By partitioning the production within the sea ice cover according to those habitats/ communities shown in Fig. 1, it is indicated that in the Antarctic more than 50% of the production in ice takes place in the surface layer of the ice and about 10% is associated with bottom communities. Bottom-ice communities are, however, some of the most productive found in the Antarctic (Arrigo et al. 1990), but considered by Legendre et al. (1992) to be restricted to fast ice areas. Observations of their existence in the pack ice as seen in Fig. 2 indicate that the contribution from these communities are underestimated.

Measurements of biomass in ice are scarce and the methods are at best rough. It is difficult to get access to ice in polar regions on a regular basis and to estimate the area of ice/water surfaces available for production. This must be taken into consideration when the presented data are assessed.

Low or even missing values are not necessarily indicating no production. Visual observations from the Arctic of long and dense algal strands mostly consisting of diatoms (Melnikov 1997, own obs.) indicate that production can be substantial even in thick multiyear ice. Wheeler et al. (1996) found, during a trans-Arctic crossing, that although water-column production decreased in the poleward direction, production by ice algae increased. In the central Arctic, the mean ice algal production was similar to that found in Arctic first-year ice at lower latitudes. They estimated the total annual production to be ~ 10 g C m^{-2}. Estimates of ice associated primary production are thus increased with an order of magnitude since English (1961) first figures and production in multiyear ice doubled since Legendre et al.'s (1992) compilation of available measurements in 1992.

Another aspect of ice-associated production and that of the lower surface of the ice in particular, is the availability of the algae to consumers. Food of the ice/water interface (either internally or at the lower surface) is more accessible than food suspended in the water column. The high densities and patchy distribution of sympagic fauna in Arctic multiyear ice (Fig. 3) and dense swarms of juvenile krill under the ice in the Antarctic are examples of organisms that utilise these food sources.

The biomass of the invertebrate sympagic fauna in the Arctic is closely related to the age and history of the ice habitat, more as a consequence of the life strategies of the organisms than the size of the primary production. Permanent residents (autochthonous sympagic fauna) who comprise the bulk of the biomass are more common in the permanent sea ice zone, and the resulting biomass are significantly higher. The average biomass value in the seasonally ice-covered Barents Sea is on average less than 0.1 g C m^{-2}, while the corresponding value in the permanent ice pack further north is about 0.5 g C m^{-2} (Lønne and Gulliksen 1991a,b).

Fig. 2. Underwater photograph of the underside of thick first-year ice in the Weddell Sea (74°39'S, 32°40') (Jan. 21, 1990). The picture is about 50 cm across. Conspicuous species present in the thick mat of ice algae, which is loosely attached to the lower surface of the ice, are *Entomoneis* sp., *Nitzschia stellata* and *Synedropsis recta.* (Photo: B. Gulliksen)

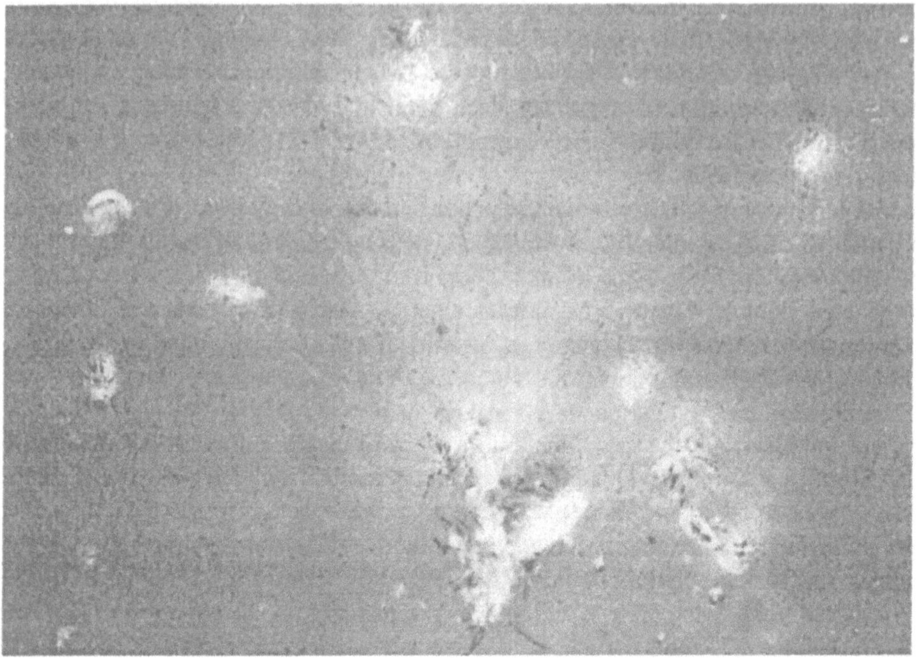

Fig. 3. Underwater photograph of the lower surface of multiyear ice taken north of Svalbard in August 1986. The picture is about 40 cm across. The sympagic amphipod *Gammarus wilkitzkii* is the most conspicuous species, but the amphipod *Onisimus sp.* is also present. (Photo: B. Gulliksen)

Despite our limited knowledge of the sea ice system and its associated organisms, the sea ice is a key polar ocean marine habitat that also affects benthic and pelagic production.

During ice melt ice algae are released into the water column. At the same time, fresh melt water creates a low-density layer upon the high-salinity seawater. The two layers are separated by a discontinuity layer where mixing is negligible. This ensures that ice algae and phytoplankton are not transported to greater depths where light conditions are inadequate. Under such conditions the presence of ice and the seeding by algae released from the ice are important factors for the start and early development of ice edge blooms (Sakshaug and Skjoldal 1989). Depending on local conditions, polynias and other openings in the ice may have similar productivity to that of ice edge zones.

In the Antarctic, there are no identified permanent residents in the ice and species such as the krill, amphipods and fish are at times also found in the water column or on the bottom.

Zooplankton able to perform vertical migrations, and for larger organisms also horizontal migrations, concentrate on the lower surface of the ice to feed on the developing blooms. This again attracts larger organisms such as fish e.g., the capelin in the Arctic that aggregate to feed on zooplankton along the ice edge.

For seals and birds these blooms represents areas with maximum concentrations of food. In addition the ice provides a place where they can rest and breed safe from marine predators

As a result of ice melting a strong pulse of organic material also sinks ungrazed to the bottom. In general, pelagic-benthic coupling appear to be particularly tight in the Arctic (Carey 1991, Grebmeier and Barry 1991) where a greater portion of biogenic material may reach the bottom than at lower latitudes. As benthic production in general depends on material sinking down from above, one may expect a richer benthic fauna where such release of material from the ice is high and a recurring phenomenon such as under polynias and in melting areas. Studies in the Northeast Water Polynya off Northeast Greenland (Hobson et al. 1995) have shown that a major component of the benthic community is supported by ungrazed particulate organic material reaching the benthos.

4 Ice Movement

Sea ice motion not only affects the physical properties of the ice, but also the large-scale movement and distribution of ice fields and its associated biologically produced material. Ice motion in the Arctic Basin has been well documented from the trajectories of manned drifting stations and automatic data buoys (Untersteiner and Thorndike 1982). Based upon analysis of such data, Colony and Thorndike (1984) have shown a general pattern of anticyclonic motions of sea ice in the Amerasian Basin and a southward drift toward Fram Strait in the Eurasian Basin. The loss of sea ice to other marginal seas is small compared to the export through the Fram Strait. On an annual basis, roughly one tenth of the ice in the basin drift into the Greenland Sea where it eventually melts (Maykut 1985). Presumably 80–90% of the exported ice cover originates within the perennial ice zone (Rigor and Colony 1997). The balancing formation of ice takes place in the marginal seas, either in the fall by the freezing of the seasonally ice free waters or by wintertime advection of ice away from coasts of the marginal seas across the minimum ice edge boundary and into the perennial zone (Rigor and Colony 1997). The ice, having a possible lifetime of 3–4 years and may travel thousands of kilometres in net terms thus transport the associated sea ice biota across great distances.

These large-scale patterns have at least two potentially important implications: The anticyclonic drift in the Amerasian Basin serve to maintain organisms in an area with a permanent ice and thus allowing a community of permanent residents to evolve. The transpolar drift transport and aggregate biogenic material produced over a large area as the ice melts in the North Atlantic, offering a significant input to these biologically rich surface waters (Lønne 1992). Similar enrichment of organic matter can also be found in other marginal seas, such as the Barents Sea, where the position of ice melt is confined to a large extent to the oceanic front system (Loeng 1991).

Different samples of pack ice exiting into the North Atlantic may have quite different histories, characterised by thickness, age, place of origin, crystal structure, chemical composition, inclusion of pollutants and associated sea ice biota. Changes in the extent, physical properties or drift pattern of the polar ice pack brought about by e.g., climatic change may significantly alter the structure of the sympagic community and its role in high arctic food webs.

Ice movements in the Antarctic are simpler, being generally clockwise around the continent. Ice concentration in the Antarctic is fairly low, averaging 60–80% throughout the year (Zwally et al. 1983). The large area of open water is a result of divergence with a mean northward direction. During the light season, leads act as moving windows which admit substantial amounts of light to the upper ocean and, hence, the lower surface of neighbouring ice. Multiyear ice in the Antarctic is restricted to the Weddell Sea, parts of the Ross Sea and the Bellingshausen and Amundsen Seas. The age of the ice is on average much younger than in the Arctic, with a residence time in the Weddell Sea is of < 2–3 years.

The differences in sea ice characteristics in the Antarctic and the Arctic have important implications for organisms inhabiting the ice, determining the availability of space, of light and nutrients to the habitats created, and the predictability of the presence of sea ice within and among years. In the Arctic, we find organisms that are associated with the ice through all stages of their life cycles (autochthonous sympagic organisms) such as the amphipods *Gammarus wilkitzkii* and *Apherusa glacialis*. In the Antarctic, however, organisms are only temporarily associated with the ice (allocthonous sympagic organisms) (Gulliksen and Lønne 1991).

References

Arrigo, K. R., G. S. Dieckmann, M. Gosselin and C. W. Sullivan. 1990. Studies on the nutrient status in sea ice an underlying platelet layer. *Antarct. J. U.S.* 25: 185–88.

Behrenfeld, M. J. and P. G. Falkowski. 1997. Photosynthetic rates derived from satellite-based chlorophyll concentration. *Limnol. Oceanogr.* 42, no. 1: 1–20.

Carey, A. G. 1991. Ecology of North American Arctic shelf benthos; a review. *Cont. Shelf Res.* 11: 865–83.

Colony, R. and A. S. Thorndike. 1984. An estimate of the mean field of Arctic sea ice motion. *J. Geophys. Res.* 89, no. C6: 10,623–10,629.

English, T. S. 1961. Some biological oceanographic observations in the central North Polar Sea, drift station Alpha, 1957–1958. *Sci. Rep. No. 15.* Washington, D. C.: The Arctic institute of north America.

Falk-Petersen, S., C. C. E. Hopkins and J. R. Sargent. 1990. Trophic relationships in the pelagic, arctic food web. *Trophic Relationships in the Marine Environment*: 315–333.

Grebmeier, J. M. and J. P. Barry. 1991. The influence of oceanographic process on the pelagic-benthic coupling in polar regions; a benthic perspective. *J. Mar. Syst.* 2: 495–518.

Gulliksen, B. and O. J. Lønne. 1991. Sea ice macrofauna in the Antarctic and the Arctic. *J. Mar. Syst.* 2: 53–61.

Hobson, K. A, W. G. Ambrose Jr. and P. E. Renaud. 1995. Sources of primary production, benthic-pelagic coupling, and trophic relationships within the northeast Water Polynya: insights from δ13C and δ15N analysis. *Mar. Ecol. Prog. Ser.* 128: 1–10.

Horner, R., E. Syvertsen, D. P. Thomas and C. Lange. 1988. Proposed therminology and reporting units for sea ice algal assemblages. *Pol. Biol.* 8: 249–253

Horner, R. A., S. F. Ackley, G. S. Dieckmann, B. Gulliksen, T. Hoshiai, L. Legendre, I. A. Melnikov, W. S. Reeburgh, M. Spindler and C. W. Sullivan. 1992. Ecology of sea ice biota. 1. Habitat, terminology and methodology. *Polar Biol.* 12: 417–27.

Lønne, O. J. 1992. Distribution, abundance and trophic role of the sympagic macrofauna in Svalbard waters. Dr. Sci. thesis. University of Tromsø.

Lønne, O. J. and G. W. Gabrielsen. 1992. Summer diet of seabirds feeding in sea-ice covered waters near Svalbard. *Pol. Biol.* 12: 685–92.

Lønne, O. J. and B. Gulliksen. 1989. Size, age and diet of polar cod, *Boreogadus saida* (Lepechin 1773), in ice covered waters. *Pol. Biol.* 9: 187–91.

Lønne, O. J. and B. Gulliksen. 1991a. On the distribution of sympagic macro-fauna in the seasonally ice covered Barents Sea. *Pol. Biol.* 11: 457–69.

Lønne, O. J. and B. Gulliksen. 1991b. Sympagic macro-fauna from multiyear sea-ice near Svalbard. *Pol. Biol.*: 471–77.

Legendre, L., S. F. Ackley, G. S. Dieckmann, B. Gulliksen, R. A. Horner, T. Hoshiai, I. A. Mel'nikov, W. S. Reeburgh, M. Spindler and C. W. Sullivan. 1992. Ecology of sea ice biota. 2. Global significance. *Pol. Biol.* 12: 429–44.

Loeng, H. 1991. Features of the physical oceanographic conditions of the Barents Sea. *Pol. Res.* 10, no. 1: 5–18.

Longhurst, A., S. Sathyendranath, T. Platt and Caverhill C. 1995. An estimate of global primary production in the ocean from satellite radiometer data. *J. Plankton Res.* 17, no. 6: 1245–71.

Lurin, B., S. I. Rasool, W. Cramer and B. Moore. 1994. Global terrestrial net primary production. *Glob. Change Newsl. (IGBP)* 19: 6–8.

Maykut, G. A. 1985. An Introduction to Ice in the Polar Oceans. *APL-UW 8510.*

Melnikov, I. A. 1997. The Arctic sea ice system. Gordon and Breach science publishers. Australia. pp. 204.

Nilssen, K. T., T Haug, V Potelov and Timoshenko. 1995. Feeding-habits of harp seals (*Phoca-groenlandica*) during early summer and autumn in the northern Barents Sea. *Pol. Biol.* 15, no. 7: 485–93.

Rigor, I and R Colony. 1997. Sea-ice production and transport of pollutants in the Laptev sea, 1979–1993. *Sci. Tot. Env.* 202, no. 1-3: 89–110.

Sakshaug, E and H. R. Skjoldal. 1989. Life at the ice edge. *Ambio* 18: 60–67.

Spindler, M. 1990. A comparison of Arctic and Antarctic sea ice and the effects of different properties on sea ice biota. *Geological History of the Polar Oceans* Arctic Versus Antarctic: 173–86.

Spindler, M. 1994. Notes on the Biology of Sea Ice in the Arctic and Antarctic. *Pol. Biol.* 14, no. 5: 319–24.

Syvertsen, E. 1991. Ice algae in the Barents Sea: types of assemblages, origin, fate and role in the ice edge phytoplankton bloom. *Pol. Res.* 10, no. 1: 277–89.

Untersteiner, N. and A. S. Thorndike. 1982. Arctic data buoy program. *Pol. Rec.* 21: 127–35.

Wheeler, P. A., M. Gosselin, E. Sherr, D. Thibault, D. L. Kirchmann, R. Benner and J. E. Whitledge. 1996. Active cycling of organic carbon in the central Arctic Ocean. *Nature* 380: 697–99.

Zwally, H. J., J. C. Comiso, C. L. Parkinson, W. J. Campell, F. D. Carsey and P. Gloersen. 1983. Antarctic sea ice, 1973–1976: Satellite passive-microwave observations. Washington, D.C.: NASA SP-459.

The Freezing of Soils: Ice in a Porous Medium and Its Environmental Significance

Peter J. Williams

Geotechnical Science Laboratories
Carleton University
1125 Colonel By Drive
Ottawa ON K1S 5B6 CANADA

1 Frost Heaving in Frozen Ground

1.1 Freezing of Water within Soils

Water in small spaces, like water in small droplets, does not freeze at 0°C. Neither does water that is very close to a mineral surface. Rather, depending on how small, and how close, the water freezes at temperatures that may be several degrees below 0°C. Thus as soils freeze, ice forms first in large openings and then gradually spreads into the smaller and smaller pores only as the temperature falls further.

1.2 Unfrozen Water in Frozen Soils

The (decreasing) quantity of water that remains unfrozen as the temperature falls, is the most important characteristic of a freezing soil. Examples are shown in Fig. 1.1. The curves are a function of the pore sizes (and thus of the particle sizes) and, at lower temperatures, of the area of mineral particle surface. In general, therefore, there is more unfrozen water the more fine-grained the soil.

The unfrozen water content and its dependence on the temperature and the type of soil, explains many important characteristics of freezing soil. These account for the long-term effects of freezing on natural terrain (Williams and Smith 1991) and for many of the geotechnical properties of freezing ground (Ladanyi and Andersland, 1994).

1.3 Soil Moisture Suction

Suction is the term describing the pressure state or hydraulic potential of water in soils, as the water content is decreased, for example, by evaporation (drying). Suction (or 'cryosuction) is also developed in the unfrozen water of a frozen soil. In this respect, the water which becomes ice in the freezing soil is analogous to that which evaporates in the drying soil.

NATO ASI Series, Vol. I 56
Ice Physics and the Natural Environment
Edited by John S. Wettlaufer, J. Gregory Dash and Norbert Untersteiner

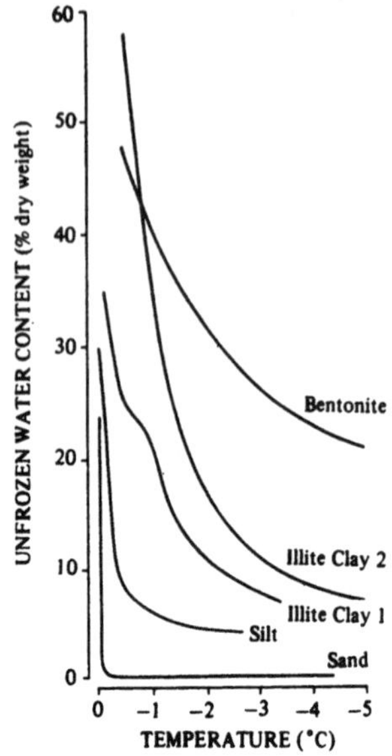

Fig. 1.1 The unfrozen water content of various soils at temperatures below 0°C (from Williams and Smith, 1991).

Plate 1.1 shows a soil (not frozen) prepared by grinding the rock mineral cryolite (Na_3AlF_6) into a fine powder of silt size particles. The soil is wet but not all the pores are filled with water. The refractive indices of water and cryolite are similar so the soil is transparent. Curved interfaces between the air in the pores and the water are seen. These are a result of surface tension and the small diameter of the capillary pores. These interfaces are associated with the development of the suction (the often sub-atmospheric, negative pressures) in the water. This suction is also the cause of 'capillary rise' in soils.

Suction in freezing soils is responsible for the movement of water towards the freezing zone and the accumulation of the layers of ice, Pl. 1.2a. It is these layers that result in frost heave, a volumetric expansion of the soil (Beskow, 1935). This has great significance in the development of terrain features as well as being problematic in many engineering constructions.

1.4 Frost Heave

Plate 1.2(b) shows a road during spring thaw forty years ago, when the melting ice layers caused the surface layers to almost totally lack bearing capacity. Roads are now usually built by using large quantities of coarse sand, crushed rock etc. in the foundation, because these materials do not show frost heaving.

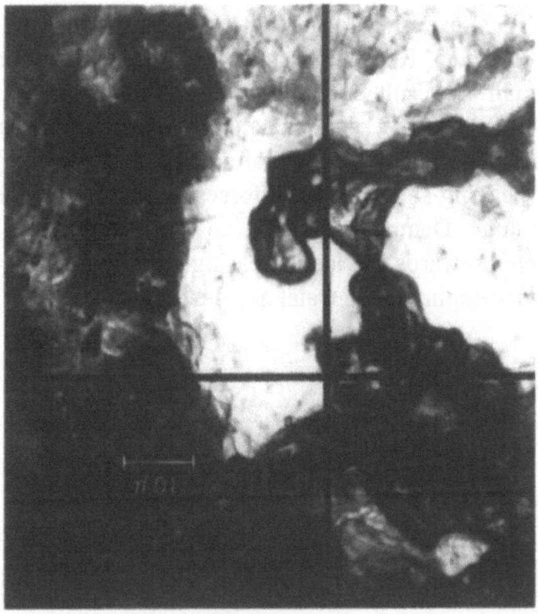

Pl. 1.1 Curved interfaces (capillary menisci) between air and water in the pores of a transparent soil of silt size particles (from Williams, 1966).

(a) (b)

Pl. 1.2 (a) Frozen silt soil with layers of ice. Sample is approximately 12 cm high (b) break-up of road following loss of strength due to thawing of ice layers giving excess moisture. (Photo: National Research Council of Canada).

Plate 1.3 (a) shows a house situated above permafrost (permafrost is frozen ground not thawing in the summer). Repeated, annual freezing and thawing of the 'active layer' above the permafrost, resulted in continuing soil displacements and destruction of the house. Plate 1.3 (b) shows a pingo, a naturally occurring hill, in this case, thirty or forty feet high, developed by the growth of a giant layer (or core) of ice. This extreme form of frost heaving shows that the forces involved, sufficient to lift such a weight of earth, are large. Demonstrably, the ice exerts a positive, lifting pressure, while water is attracted towards the ice increasing its volume. The water moves because of the suction developing in the water adjacent to the ice.

(a)

(b)

Pl. 1.3 Effects of permafrost and seasonal freezing and thawing. (a) deformation of building (b) pingo (note persons, on top)

1.5 Capillary Theory of Frost Heave

The capillary-porous nature of the soil is the basis of the capillary theory of frost heave. The thermodynamic relations are described well in Defay and Prigogine (1951). It is easy to come with objections to this theory. It seems improbable that the ice has curved interfaces with the water, like the air-water interfaces seen in Pl. 1.1. Stresses in the crystalline ice are not exactly comparable to the pressure of the air, and so on. But the capillary theory is compatible with a number of experimental observations.

Water rises up capillary tubes. In freezing soils, numerous experiments and the frost heave phenomenon itself demonstrate that water moves towards the interfaces of ice with water in the soil pores (capillaries). Furthermore, as water leaves the soil pores, moving into layers or other ice bodies larger than pore size, the suction is revealed by mechanical effects.

1.6 Consolidation and Aggregation Due to Freezing

The soil between ice layers becomes drawn together, in a manner similar to vacuum packed coffee. In the latter case, the removal of air lowers the intergranular pressure while the atmosphere presses on the pack of coffee grains to make it denser and stronger.

The freezing of a clay soil results in the formation of comparable, denser, stronger aggregations of clay particles, between the ice layers. In the newly thawed clay these are often visible as friable, schist-like fragments. Pusch (1977, 1979), Chamberlain and Gow (1979), Williams (1963) and others have pointed out this consolidation effect (to use the terminology of soil mechanics), that freezing has on clays and similar materials.

The suction becomes greater as the water content decreases. The suction of the water at a particular moisture content in the drying soil, is (when allowance is made for differences in surface tension) similar to that of the same amount of unfrozen water in that soil when frozen. It follows that the water in a frozen soil exerts a greater suction the lower the temperature. Secondly, the suction is really to be regarded as the difference in pressure between the ice and water, so that as the temperature is lowered, the soil may either exert a greater heaving pressure or a greater suction, or an element of both—depending on the externally imposed constraints (Williams 1967). The constraints are the weight of overburden or other restriction on heaving and the pressure of the water at large in the ground.

The size of the soil pores (the capillaries) and the pore size distribution are definitive. The observed curvature of the air/water interfaces in unfrozen soil (Pl. 1.1), is broadly in accordance with that predicted from the capillary equation:

$$h = \frac{2\sigma_{aw}}{r\rho g} \tag{1.1}$$

where h is the height of capillary rise, σ_{aw} = the surface tension air/water, r = the radius of curvature of the interface (meniscus), ρ = density of water, and the suction is:

$$p_a - p_w = \frac{2\sigma_{aw}}{r} \tag{1.2}$$

where p_a is the air pressure, and p_w is the water pressure.

The relation between the soil water suction and the water content reflects the distribution of pore sizes. Thus for a particular suction the interfaces air to water lie in pores of size r. Larger pores are empty, smaller ones are still filled with water.

The amount of water at temperatures below 0°C in the frozen soil likewise reflects the pore size distribution. The analogy between freezing and drying soils is quantitative as well as qualitative. Similar equations apply (substituting the surface tension ice/water) giving the pressure states of the ice and water. These pressures explain the magnitude of the consolidation of the soil between ice layers.

It is axiomatic in soil mechanics that the state of consolidation (the closeness of packing of the grains) is a function of the 'effective stress', that is, the difference between the pressure of the water in a soil and the external or total stress (Wu, 1981). In the freezing soil a large effective stress arises from the thermodynamics of the ice/water relationship, namely, the difference in pressure states of the ice and the water. In the next lecture, we shall see the significance of the pressure states for the microstructure of the soil.

1.7 The Mineral Surface Layer

At temperatures lower than about –1°C, most of the soil pores are filled with ice, such is their size, according to the equation:

$$T - T_o = -\frac{2\sigma_{iw}T_0}{\rho r_{iw}L} \tag{1.3}$$

where T = freezing point, K, of water in pores of radius r. T_0 = 'normal' freezing point of water, K. Thus, $T - T_0$ corresponds to the 'depression' of the freezing point, –C°.

The capillary model is no longer valid therefore, below about –1°C, yet in many soils there is still much unfrozen water. This must be in the form of layers around the particles (even though no discordance is seen in the dependence of suction on temperature, or of the unfrozen water content on temperature etc.). There is, for temperatures below about –1°C, a demonstrated correlation between the specific surface area of the soil (the area of particle surface) and the unfrozen water content (Anderson and Tice, 1972). The suction (or potential) in the unfrozen water (the difference between the pressure of the ice and that of the water) continues to increase as the temperature is lowered further.

Note too, that while the unfrozen water content as a function of temperature is dependent on the soil (Fig. 1.1), the suction or stress states are independent of the type of soil. Indeed, the stress states p_i and p_w are essentially described by equation 1.4, which is often known as the modified Clausius-Clapeyron equation It can indeed be derived from that purely general equation as Edlefsen and Anderson (1943) in their classic work showed (although interestingly, they concluded, wrongly, that a different variant of the equation would describe soil freezing):

$$P_i - P_w = \frac{(T - T_o)L\rho_w}{T_0} \tag{1.4}$$

The interpretation of frost heave and related phenomena in terms of capillarity and mineral surface effects is derived from the largely empirical work of agricultural soil scientists, and in particular their research into soil-water relations, in irrigation and similar topics. Interestingly, the extensive Russian studies into freezing soils have developed along rather similar lines (Yershov 1998), even though during Soviet times, there was notably little scientific exchange with the rest of the world. This followed from secretiveness concerning the cold regions and their political and economic significance.

Physicists on the other hand, have approached the questions of frost heave, more theoretically, on the basis of the existence of a layer of unfrozen ('premelted') water on ice (Wettlaufer and Dash, this volume). This has led to a second substantial literature and only rarely are the two approaches considered together (but see the review by Dash, 1995).

1.8 Importance of the Thermodynamic Relations for Macroscopic Properties of the Frozen Soil

1.8.1 Thermal Properties

The thermodynamic relations of the ice and water are imposed by the microstructure of the soil, the size of particles and pores in particular. These relations are, in turn, the cause of the particular values of the macroscopic properties.

The thermal conductivity and heat capacity of frozen soils are very temperature dependent. The latter is, more correctly, known as the effective or apparent heat capacity, because it is primarily latent heat of fusion that is exchanged as the proportions of ice and water change with temperature change. These proportions are soil dependent too (Fig. 1.1). The true heat capacity of the soil, the sum of the true heat capacity of the constituents, is by comparison, small and insignificant for most soils in the temperature range of greatest practical significance, 0 to –5°C.

The thermal properties are reflected in such terrain characteristics as the depth to which annual freezing or thawing occurs, the time and depth dependence of the temperature effects of climate change, the formation or degradation of permafrost and so on. The effects are also very significant in geotechnical design and are fundamental to the development of designs for pipelines or indeed most forms of construction.

But this is only the beginning. The changing proportions of ice and water and the concomitant changes in stress distribution are of great importance for the mechanical properties of frozen soils.

1.8.2 Thermal-rheologic Effects of Freezing in Natural Terrain

There are a variety of forms of terrain that are characteristic of the cold regions. There is patterned ground, which consists of repetitive arrangements of soil and/or rock fragments. The most striking are stone rings and similar features (Pl. 1.4a) in which boulders can be arranged in a pattern that suggests the work of man.

Other forms of patterned ground include hummocks and near circular patches, a metre or so across, covered with small stones. For this review we can only say that their formation is variously related to selective displacements of soil components, ascribed generally to the varied thermal properties and thermal regime of the ground components, the process of differential frost heave, and the related processes of creep. Although many of the forms are still not fully understood, there appears little need to resort to the more fanciful explanations found in the prolific literature. Troll (1944) listed some 1500 references to patterned ground and related phenomena and Washburn (1956) made an extensive review of the forms and theories. His book (Washburn 1979) is the classic work on the natural (or 'field') conditions and the distribution of these features of cold regions terrain.

(a) (b) (c)

Pl. 1.4 Patterned ground and solifluction. (a) stone polygons in which boulders are arranged in polygonal patterns (the reindeer antlers are about one metre wide (b) solifluction probably due to creep of ground when frozen. Each 'step' is about 20 cm high (c) similar solifluction 'step' or terrace front, but larger and on lower angle slope.

A second group of natural features are those occurring on sloping ground. These may be mudflows, landslides and other soil movements, similar to those occurring in temperate climates, except that the spring time conditions of poor drainage and strength loss causing them are a consequence of thawing. Quite different are a range of forms (Pl. 4b,c) in which displacement takes place by creep, apparently in the frozen state (Williams, 1988). The movements in this case are slow, just a few cm, sometimes less, per year, but continuing. They extend, (becoming smaller) to a depth of a metre or two and can also occur without surface expression.

Although the movements have been recorded in many field studies, there is still uncertainty as to when they occur. Probably this is early or late winter when the frozen ground is relatively warm. Temperature change (which occurs constantly in the near surface layers) is important.

Permafrost engineers are clear that frozen soil is prone to creep. This is not simply a matter of the ice present giving a creep component of deformation. Undoubtedly the composite material of soil minerals, ice and unfrozen water is more prone to creep than the soil when unfrozen. Of special importance is, surely, the change in proportions of the ice and water under changes of temperature or applied stresses. Furthermore associated with the changes in unfrozen water content are changes in pressure or stress in the ice phase and in soil moisture suction (or potential, in hydrological and soil science terminology, or free energy as pointed out by Schofield, 1935, 1938). Consequently the water tends to migrate. Local melting of ice can be followed by migration of water and its refreezing a short distance away. In this way the location of the ice bodies also changes. The displacements are slow and small but cumulatively lead to creep deformation of the soil mass as a whole.

Thus we see that the process of frost heave, the migration of water into ice segregations, and its attributes, occurs over a range of temperature of the frozen ground.

2 Contaminants and the Microstructure of Frozen Ground

2.1 Importance of Microstructure of Frozen Ground

The thermodynamic relations of ice forming in a porous medium are responsible for a wide range of properties and behaviour which distinguish freezing soils from soils more generally. It is useful to examine microscopically the structure of the materials both in the frozen and unfrozen states. Such examination reveals the origin of the thermal, mechanical and other properties of freezing soils as we know them from the historical development of the subject outlined above. Because these properties are now seen to be of such importance in the study of environmental effects in cold climates, whether interactions between constructions such as roads and buildings with the ground, or other disturbances of the terrain of anthropogenic or natural origin (such as climate change), more fundamental studies are timely. This is the practical importance of studying the microstructure of freezing soils—the microstructure determines their behaviour and properties.

A particularly topical geotechnical problem is that of soil contamination in cold regions (Popova, this volume). Whether due to spilled oil or industrial, perhaps military wastes, soil contamination can be a serious health hazard. It is an environmental disturbance in the cold regions just as in the temperate regions. Contaminants often pass through the soil even though it may be largely frozen, into the ground water and then may enter rivers, ultimately perhaps to spread in the world's oceans.

Only recently has it been realised how unique is the behaviour of freezing soils in this respect. Remediation is currently hugely expensive and technology developed for warmer climates is often inapplicable. In some cases, procedures may be ineffective or even harmful. The behaviour and effects of the contaminant as it moves into the ground depend on the microstructure which the contaminant itself modifies.

2.2 Microstructure of Soils

There is a tendency to regard a soil as simply a collection of mineral particles (grains) of easily defined sizes. The size distribution of these 'elementary particles' (to use a Russian term) leads, at least in engineering terminology, to the classification of a soil as a sand, silt or clay or some combination of these. It is also understood that the grain-size distribution largely determines the size of the pores, or, more precisely, the pore-size distribution. It is also fundamental, at least to the soils engineer, that when the particles are small, of clay size, they may be more or less densely packed, or consolidated (again, in engineering terminology). It is also known that the presence of certain substances in small quantity may effectively cement particles together—a state of affairs which, confusingly, geologists describe as 'consolidated'.

All this, however, is too simple a view, as pedologists have known for many years (Ugolini, this volume). In particular, elementary particles are often drawn together into aggregates which, in some respects, become the largest 'particles'. This aggregation necessarily will be associated with a large modification of the pore system. Instead of a pore-size distribution following directly from the size distribution of the elementary particles and the closeness of their packing, there is a pore size distribution that can be bimodal. There are then intra-aggregate pores and inter-aggregate pores. Within the two modes a range of pore sizes occurs.

2.2.1 Examining the Microstructure of the Unfrozen Soil

The microstructure of soils has been studied for decades, especially for agronomy and related topics. A widely used procedure is to impregnate a soil sample with resin such that thin sections of the strengthened material may be made. The thin sections can be examined with optical or electron microscopy. Examples of two soils which have never been frozen are shown in Pls. 2.1a and 2.2a. The same soils after freezing and thawing are shown in Pls. 2.1b, and 2.2b. The microstructure of soils varies strikingly with their lithological nature. The same is true of soils exposed to freezing.

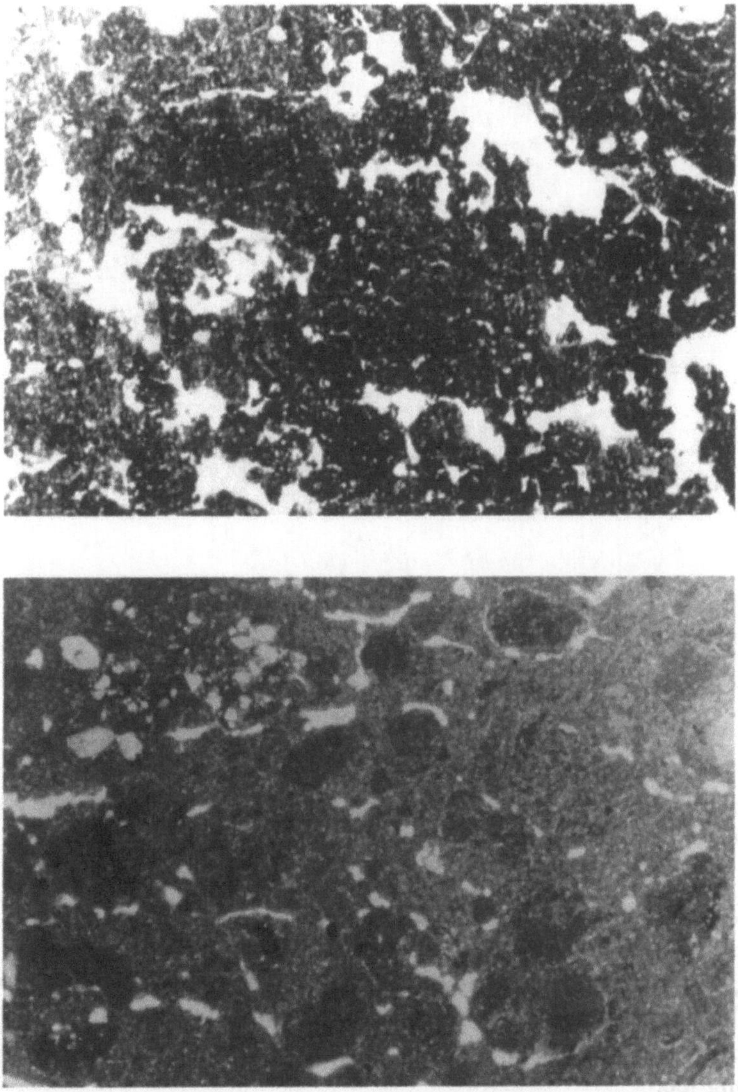

(a)

(b)

Pl. 2.1 Microstructure of Tourville silt (a) never frozen (b) after 1 freeze-thaw cycle. Frame length corresponds to 13.5 mm (White and Williams 1994, 1996)

2.2.2 Microstructure of Soils Exposed to Freezing

Effects of freezing on the microstructure of a soil were investigated, using samples that were taken after thawing, by Harris (1980), Bunting (1983), Van Vliet Lanöe (1985, 1987) and others. Until recently, however, the importance of these effects for properties of the frozen soil had not been analysed in detail. It has long been evident that freezing modifies geotechnical properties, but Pusch (1979) appears to be the first to have directly related microstructure of a soil (as observed in a soil after thawing) to the unfrozen water content. He observed (using an electron microscope) that freezing

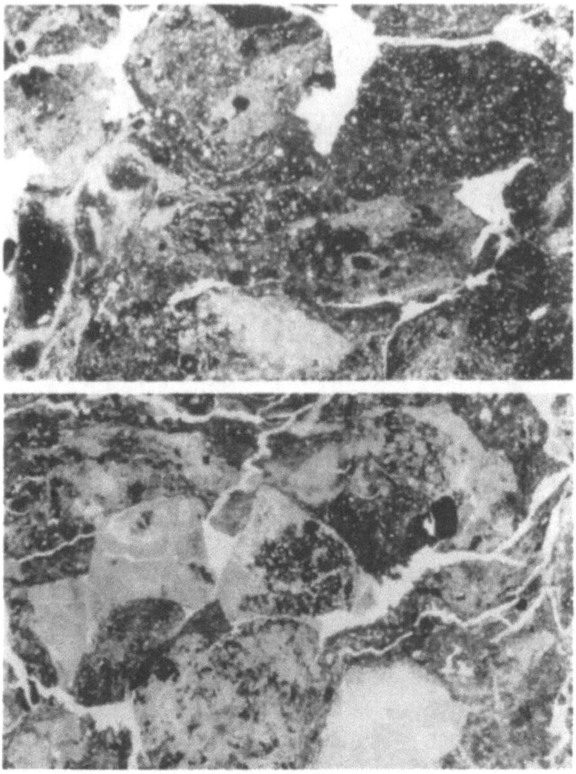

(a)

(b)

Pl. 2.2 Structure of Moley clay (from White and Williams 1996) (a) prior to any freezing (b) after one freeze thaw cycle. Frame length 13.5 mm.

of a remoulded clay (one in which the natural structure was destroyed by stirring) resulted in aggregation of the clay particles. The unfrozen water content was less, in second and subsequent freezings, over a range of temperature, and this can be understood from the reduction of total pore volume within the aggregates. The inter-aggregate pores on the other hand were enlarged with accumulating ice.

Over the last decade, Van Vliet Lanoe (1985, 1987) has shown how modifying the procedures allows not only the study of the structure of newly thawed soils but even of the structure of the soil while frozen. In the latter case, a sample taken in the frozen state is effectively freeze-dried and taking care to avoid disturbance, the impregnation with resin is carried out. The sample shown in Pl. 1.2a is impregnated in this way and is actually free of real ice.

The main difference between the newly thawed soil and the frozen soil is that on melting of the ice bodies larger than pore size, there is a closing of the space which they occupied as the excess water drains away. The sites of the ice segregations do not close completely (certainly not over a short time interval) and persist as discontinuities and elongated enlarged pores. The overall soil volume decreases, although as far as can be ascertained without any significant change in the intra-aggregate porosity during the thaw and consolidation.

The aggregates may themselves be formed, or enhanced, by the freezing process as noted (compare Pls. 2.1a and 2.1b) and the intra-aggregate porosity may continue to change with further freeze-thaw cycles (White and Williams 1994). Microstructural changes occur progressively although less marked for each succeeding cycle (compare Pls. 2.3a and 2.3b). After about ten cycles little change is seen in the microstructure. However Russian workers have observed that there can be breakdown of the elementary particles themselves over many cycles such that there is a gradual change in grain size distribution (Yershov, 1998).

Frequently there is a sorting of particles in the vicinity of ice accumulations with a closer packing of the particles towards the ice. The images, Pl. 2.4a,b, were obtained by scanning electron microscopy and accordingly are distinct in appearance (as well as in scale) from those obtained using transmitted light and optical microscopy. The packing of the particles is clearly seen.

When the Caen silt was saturated (that is, the soil pore space was essentially filled with water), a layered structure developed that did not directly correspond to the location of the ice layers. At least, the sorted layers of particles appeared to be more frequent than the ice layers even though the latter were numerous.

(a)

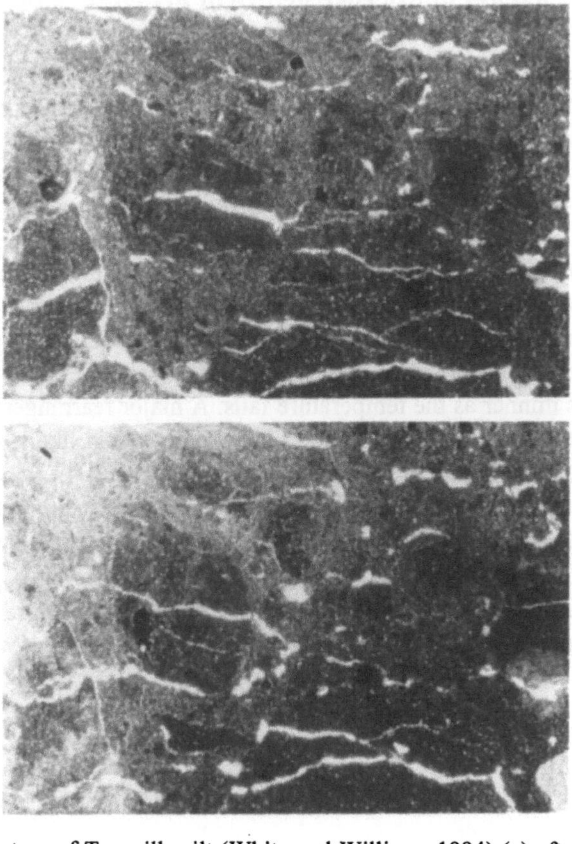

(b)

Pl. 2.3 Microstructure of Tourville silt (White and Williams 1994) (a) after five freeze-thaw cycles (b) after ten freeze-thaw cycles. Each frame is 13.5 mm.

(a)
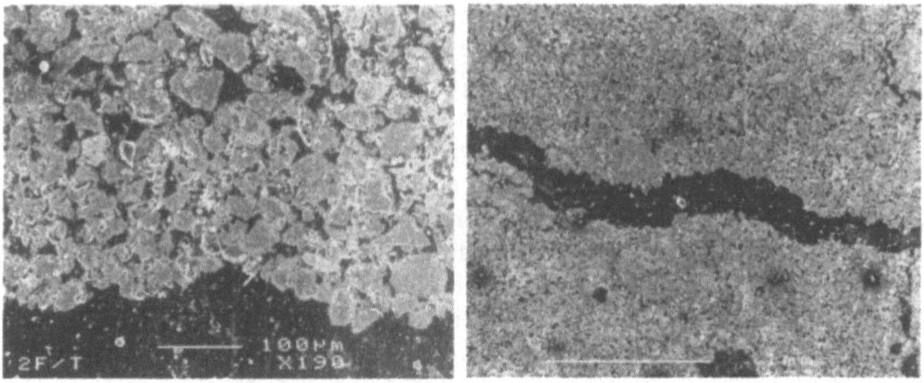

(b)

Pl. 2.4 Scanning electron micrographs showing closer packing of particles adjacent to ice layer (dark area) in Caen silt (a) frame length corresponds to 3 mm (b) higher magnification: frame length 0.59 mm (White and Williams 1997)

2.3 Effects of Changes in Microstructure in Freezing Soils

Changes in the geometry of pore space, such as is seen in the various micrographs, must influence the thermodynamic relations. Pores may be enlarged into macroscopic voids where ice accumulation occurs. Elsewhere, the suction forces generated in the water result in a decrease in pore (capillary) size—in most cases, in association with the formation of aggregates. The combined effects of enlargement and reduction of pore size do not, of course, balance one another out. The changes in pore size distribution will inevitably change the proportions of ice and water at each temperature, at least in a range of some degrees below 0°C. It is also reasonable to assume that there will be some modification of the effects of mineral surface on the content of unfrozen water in the mineral soil. The layer of unfrozen water associated with mineral particle surfaces becomes thinner as the temperature falls. A major rearrangement of particles and especially the close packing of particles in aggregates will likely reduce the available layer thickness of water to remain unfrozen.

Because of the dependence of the suction developed on the temperature, the lower the temperature that is reached, the greater is the consolidation of the aggregates (that is, the closer is the packing of the elementary particles in the aggregate). Thus any freeze-thaw cycle involving freezing to a colder temperature than previously is likely to have a more marked effect on that account. In the examples shown, the samples were cooled to approximately the same temperature in each freeze-thaw cycle. However, as the temperature in question was only –2°C and the dependence of unfrozen water content on temperature is large at that temperature (Fig. 1.1) any discrepancy in temperature could have an effect, although small.

These studies were carried out in a large-scale controlled environment facility as part of investigations of deformation of a buried pipeline by freeze-thaw (Williams, 1992). In the experiments it was observed that in successive freeze thaw cycles, even if the controlled parameters of temperature and ground water level remained the same, the magnitude of frost heave increased somewhat in each cycle. Other parameters, too, such as the temperature distribution in the ground over similar time periods, were found to be somewhat different. These differences may be ascribed, at least in large part, to the progressive changes in microstructure introduced by each freeze-thaw cycle.

2.4 The Effects of Changes in Environmental Conditions on Terrain

Soils which undergo freezing and thawing in natural conditions will generally have been subjected to many thousands of freeze thaw cycles already. Thus if temperature and other conditions remain essentially similar, the thermal, mechanical and hydraulic properties of the freezing, frozen and thawed soil, respectively, will remain the same one year to another. This does not mean that terrain conditions will be exactly comparable one year to the next. Changes in weather, a cold winter, or a wet autumn, may cause deeper winter penetration of freezing or greater frost heaving, for example. But these variations are part of a continuing pattern, in which there is little cumulative change.

If in contrast, there is a sharp and continuing change in moisture conditions, for example drainage or raising of the water table, on freezing there is a likelihood of lasting change of the microstructure of the soil. In turn, there will be an increase or decrease in the maximum depth of penetration of winter freezing and an increase or decrease in amount and possibly direction of frost heave, because of the changes in thermal properties brought about by microstructural alteration. Movements of boulders or stones and the sorting effects seen in patterned ground may accelerate or be retarded. Over a period of years there may be changes in drainage courses and changes in surface erosion. In areas of deep winter freezing or where permafrost occurs, there may be significant modifications to the stability of slopes, erosion by running water, local settlement (subsidence) of ground and associated modifications in vegetation cover.

Similar but often more marked changes occur after the addition of substances to the soil such as occurs after an oil spill or around waste dumps, or other areas where chemicals of anthropogenic origin are released onto or into the ground. It is reasonable to suppose that many contaminants will interact with the soil, after entering the pore system, in a manner that is quite fundamental. Studies of the microstructure of freezing soils before and after the addition of various chemicals that occur as contaminants show indeed very marked changes (White and Williams 1998).

2.5 Studies of the Effects of Contaminants on the Freezing of Soils

The literature on the effects of contaminants on freezing soils is still quite limited, unlike that for unfrozen soils where extensive research is underway (Winnicky, 1995, Can. Soc. Civil Engg., 1995). Because of the importance of freezing for the properties and behaviour of the soil, studies on unfrozen soils have only limited relevance.

Furthermore in the design of experiments with freezing of contaminated soil, it is desirable to have a facility in which the various parameters of the soil, the evolution of temperatures, water contents of materials and the availability of water during the freezing process can be monitored in detail. In the absence of such information it may be difficult to distinguish the particular effects of the contaminant.

The study of microstructure in the contaminated frozen soil reveals the fundamental nature of the effects of a contaminant. With such information it is possible to make predictions as to the effects of a contaminant on a variety of thermodynamic and mechanical properties, which in turn will determine the effects on terrain, and the rate and direction of spreading of the contaminant. Such information allows the development of rational and cost-effective remediation procedures.

Studies of microstructural changes in a silt soil in which contamination by various hydrocarbons was arranged, were carried out in the same large-scale controlled environment facility previously used for investigations of deformation of a pipeline by freeze-thaw effects (Williams, 1992).

Pl. 2.5 compares (with optical microscopy) the structure of contaminated soil with that of closely adjacent similar but uncontaminated soil. The most obvious effect is that the aggregates are smaller and the inter-aggregate voids more frequent, in the contaminated material, implying that the ice layers larger than pore size occur more frequently.

Plate 2.6, taken by scanning electron microscopy, shows the structure of contaminated material (10 parts per million of total mass) at two higher magnifications. In Pl. 2.6a, the structure is seen to be one of well-defined aggregations. In Pl. 2.6b an aggregation is seen to be a dense packing of 'elementary' silt particles. In Pl. 2.7 the structure of a similar material but with a 100 ppm concentration is shown. In this case the aggregates are seen to be rather larger and less linear. It also appears that, within an aggregate, the particles are even more tightly packed.

The relatively limited information we have of this kind does not yet allow us to define precisely and quantitatively, the changes for different kinds of soils that will be initiated by the presence of contaminants on freezing; nor does it allow us to establish the dependence of these changes on other factors in the local environment. The studies to date do establish the marked effects of contaminants on the behaviour of the freezing soil.

(a) 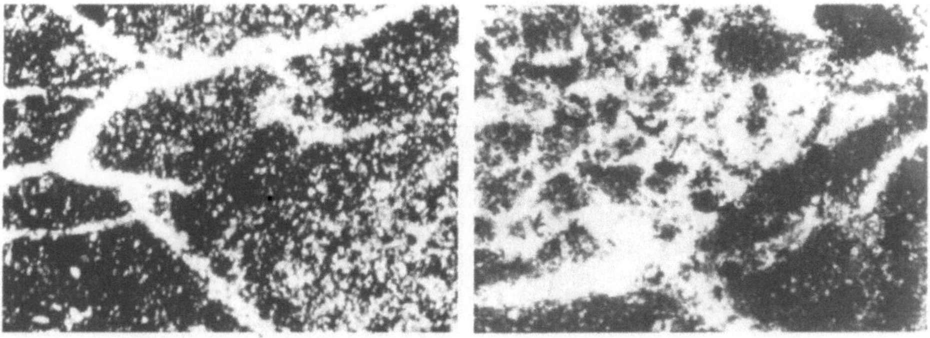 (b)

Pl. 2.5 Structure of Caen silt (a) after two freeze-thaw cycles, normal condition (b) after contamination by diesel fuel (10 parts per million) and two further freeze-thaw cycles (White and Coutard 1998). Each frame shows approx. 3 mm.

(a) 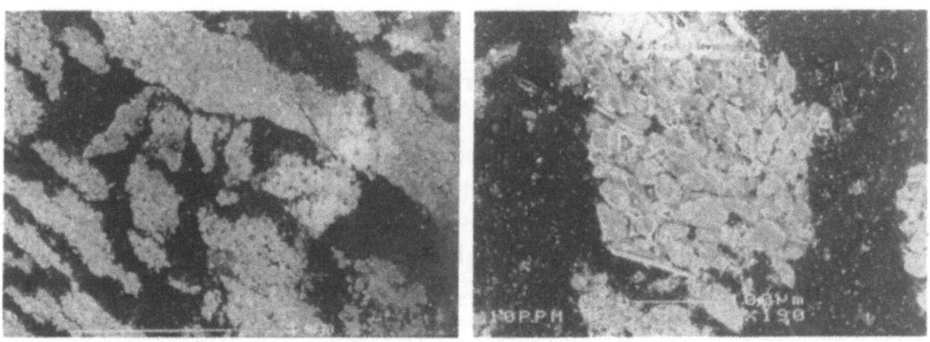 (b)

Pl. 2.6 Scanning electron micrographs. (a) Structure of Caen silt, contaminated at 10 ppm with diesel, showing aggregates following freezing. Frame length approx. 3 mm; (b) single aggregate, also visible, centre left, in (a), at greater magnification, frame length 0.56 mm (White and Coutard 1998).

Pl. 2.7 Caen silt, treated as 2.6b but contaminated at 100 ppm. Frame length approx. 3 mm (White and Coutard 1998).

2.6 Effects of Contaminants on Terrain

When contamination has occurred it is important to know what will happen in coming months and years. What will be the effect of the contamination on the terrain? Will there be increased thawing with subsidences and damage to roads and other structures? Will the contaminant travel through the ground, extending the contaminated area and perhaps finding its way into drinking water, or otherwise into the food chain of man or animals—even perhaps far distant from the original site? The exchanges of air and seawater between the cold and warm regions may spread the contaminants globally.

From the studies of microstructure described, and the conclusions one can draw as to the effects the contaminants will have on soil properties and thus on various processes in the ground and especially in the near ground surface layers, it appears that we have the possibility of predicting the course of events at a contamination site and thus of devising cost effective remediation procedures (Williams, 1996, 1998).

2.7 Role of Research in Contamination Remediation

Billions of dollars are being used in research and development for remediation of ground contamination in temperate areas. Quite large sums are being spent in efforts to correct contamination in the cold regions. Some of this expenditure is essentially wasted in that current procedures for remediation are often crude and ineffective under freezing conditions. 'Capping' the contaminated ground (burying it under additional soil) may be merely extending the problem if the contaminant can move in the ground, perhaps to more crucial locations. Excavation of the affected ground (increased by the added soil) may prove necessary later. Yet the disposal of excavated material can be an equally great problem. Sometimes incineration is suggested but this is awkward and expensive in remote locations. The use of frozen ground as a barrier requires, if leaks are not to occur, that the possibilities of the material travelling through the freezing material be thoroughly investigated (Dash, 1994, 1997, and this volume). Other methods, too, could be better assessed with a knowledge of the actual behaviour of frozen ground at the fundamental, microstructural scale.

Indeed, one can envisage the building up of a data bank of information on the conditions and processes associated with contamination in different types of soil. Working from a knowledge of the fundamental microstructure and the connection between this and the proportions of ice and water in the material, below 0°C, it will be possible to work out the effects of the contaminant on a range of soil properties. With such knowledge it will then be possible to predict the effects on the terrain over the near future and through longer periods. These effects may or may not be harmful; they may indicate future damage to the natural environment or man-made structures. More specifically it will be possible to predict eventual continuing movements of the contaminant—into water supplies, food chains (or even into more global fluxes) and to design the most cost-effective remedial actions.

Such an approach requires a large amount of research. But in relation to the enormous expenditures envisaged in response to contamination, the cost would be small—and certainly recovered later. Remote sensing of contaminated sites, particularly during the first years after contamination, will be a valuable adjunct to the recording of changes in terrain conditions. These changes will be a powerful diagnostic and prognostic tool when assessed together with the work of soil microstructure specialists, geomorphologists, physicists and others.

Acknowledgements

The references give some of the key works in the studies made over the past sixty years. My own involvement in the more recent work has been much aided by generations of graduate students. My thanks are especially due to my colleague (and former Ph. d. student) Dr. Les White, who is responsible for the soil microstructure studies, carried out over the last few years.

References

Anderson, D.M. and A.R.Tice, 1972, Predicting unfrozen water contents in frozen soils from surface area measurements. Highway Research Record, Vol. 393, 12–18.

Beskow G. 1935. Tjälbildningen och tjällyftningen med särskild hensyn til vägar och järnvägar. 242 pp. Statens väginstitut, Stockholm, Meddelande 48 (Also translated into English: Frost heaving with special reference to roads and railways. Tech. Inst. Northwestern Univ., Ill. 1947)

Bunting, B.T. 1983. High Arctic soils through the microscope: Prospect and Retrospect. Annals Amer. Assoc. Geographers, 73,4: 609–616.

Canadian Society for Civil Engineering. 1995. Proc. Wkshp. on Technologies and Techniques for Hydrocarbon Remedial in Cold and Arctic Climates. Cold Regions Res. Engg. Div. 264 pp.

Chamberlain, E.J. and A.J. Gow, 1979, Effect of freezing and thawing on the permeability and structure of soils. Engineering Geology, Vol. 13, 73–92.

Dash, J.G. 1994. Ground freezing for management of hazardous wastes. Proc. Seventh Intern. Symposium Ground Freezing. Nancy, France. Ground Freezing 94 (Ed. Frémond), Balkema, 351–354.

Dash, J.G., Fu, H.Y. and R. Leger 1997. Frozen soil barriers for hazardous waste confinement. Proc. Intern. Containment Technology Conf., St. Petersburg, Florida.

Dash, J.G., Fu, Haiying and J.S. Wettlaufer 1995. The premelting of ice and its environmental consequences. Reports on Progress in Physics, Vol. 58, 115–167.

Defay, R. and I. Prigogine, 1951. Tension superficielle et adsorption. Liége, Editions Desoer. 295 pp. Traité de Thermodynamique; conformement aux méthodes de Gibbs et de Donder.

Edlefsen, N.E. and A.B.C. Anderson 1943. Thermodynamics of soil moisture. Hilgardia, Vol. 15, 2. 298 pp.

Harris, C. 1980 Micromorphology of soils in soliflucted materials, Okstindan, Northern Norway. Geoderma, Vol. 23, 11–29

Ladanyi B. and O.B. Andersland, 1994. An Introduction to Frozen Ground Engineering. Chapman, New York. 352 pp.

Pusch, R. 1977. Ice formation in clays with special reference to their microstructural constitution. Proc. International Symposium Frost Action in Soils. Lulea, Sweden. Vol.1, 137–142.

Pusch, R. 1979. Unfrozen water as a function of clay microstructure. Engineering Geology, 13, 157–162

Schofield, R.K. 1935. The pF of the water in the soil. Third Intern. Congress Soil Science, vol. 2, 37–48.

Schofield R.K. and J. V. Botelho da Costa. 1938 The measurement of pF in soil by freezing point. Agric. Science, vol. 28, 645–653.

Troll, C. 1944 Strukturboden, Solifluktion und Frostklimat der Erde. Geol. Rundschau, 34, 545–694. (also as translation 43, Snow, Ice and Permafrost Res. Establ., U.S. Army: Structure soils, solifluction and frost climates of the earth. 121 pp.

Van Vliet Lanöe, B. 1985 Frost effects in soils. in Soils and Quaternary Landscape Evolution (Ed. Boardman) John Wiley & Sons, 117–158.

Van Vliet Lanöe, B. 1987 L'etude de la structure des sols gelés et dégelés: a) Méthodes de carottage et de préparation des échantillons b) Applications géotechniques. 2iéme Colloque Franco-Polonais, Gel et Génie Civil. Publ. Lab. Central des Ponts et Chaussées.

Washburn, A.L. 1956 Classification of patterned ground and review of suggested origins. Bull. Geol. Soc. Amer., 67, 327–402

Washburn, A.L. 1978 Geocryology. A survey of periglacial processes and environments. Arnold. 406 pp.

White, T.L. and P.J. Williams. 1994. Cryogenic alteration of frost susceptible soils. Proc. Seventh Intern. Symposium Ground Freezing. Nancy, France. Ground Freezing 94 (Ed. Frémond), Balkema. 17–24.

White, T.L. and P.J. Williams 1996. The role of soil microstructure in the geotechnical properties of freezing soils. Proc. Fifth Intern. Symp. Thermal Engg. and Sciences for Cold Regions.

White, T.L. and J-P. Coutard 1998. Modification of silt microstructure by hydrocarbon contamination in freezing ground. Polar Record, Cambridge Univ. Press (in press).

White, T.L. and P.J. Williams 1998. The influence of soil microstructure on hydraulic properties of hydrocarbon contaminated freezing ground. Polar Record, Cambridge Univ. Press (in press).

Williams, P.J., 1963. Suction and its effects in unfrozen water of frozen soils. Proc. Permafrost International Conference, N.A.C/N.R.C., Washington, D.C. 225–229.

Williams, P.J. 1966 Replacement of water in soil pores by air. The Engineer, Vol.223, 5796, 293–298.

Williams, P.J. 1967. Properties and behaviour of freezing soils. Norwegian Geotechnical Institute, Publn. 72, 120 pp.

Williams, P.J. 1988. Thermodynamic and mechanical conditions within frozen soils and their effects. Fifth Intern. Conf. Permafrost. Trondheim. Vol. 1, 633–638.

Williams, P.J. and M.W. Smith, 1989 (Pb. 1991). The Frozen Earth. Fundamentals of Geocryology. Cambridge University Press. 306 pp.

Williams, P.J., Riseborough, D.W. and M.W. Smith 1993. The France-Canada joint study of deformation of an experimental pipe line by differential frost heave. Intern. Jour. Offshore and Polar Engg.. Vol. 3, 1, 55–60.

Williams, P.J., Rees, W.G., Riseborough, D.W., White T.L. and K. Winnicky 1995. Research strategies for development of predictive and remedial measures for oil spills in permafrost regions. Proc. Wkshp. on Technologies and Techniques for Hydrocarbon Remediation in Cold and Arctic Climates. Cold Regions Res. Engg. Div., Can. Soc. Civ. Eng. 82–95.

Williams, P.J., Riseborough D.W., Rees, W.G. and T. L. White, 1998 (to be published) Models for Cold Regions Contaminant Hydrology. Conference Proc., Anchorage. Ann Arbor Press.

Winnicky, K. 1994. Study of Movement of Hydrocarbons through Freezing and Thawing Soils (Dew Line Cleanup). Final Report, Vol.1., Internal Report IR-67, Geotechnical Science Laboratories, Carleton University. 81 pp.

Wu, T.H., 1981. Soil Mechanics. T. H. Wu, Worthington, Ohio and Allyn and Bacon.

Yershov E. 1998 Transln. of Obschee Geokryologiia. Nedra. Moscow, 1990 550 pp., as: General Geocryology (Ed. Williams), Cambridge University Press, (forthcoming).

Ground Freezing Technology for Environmental Remediation

J.G.Dash

Department of Physics, University of Washington, Seattle WA 98195-1560

Abstract. Effluents from underground storage tanks and waste dumps of toxic and hazardous materials are entering or threatening aquifers which supply drinking water in many parts of the world. Cryogenic methods, using currently available technology, are suitable for the complete containment and immobilization of many of these sources of radioactive and chemical pollution. Thick frozen ground barriers can be installed as complete shields against downward as well as lateral migration, at sites as deep as hundreds of meters. Advantages include thermal and chemical stability, excellent resistance to diffusion, repairability *in situ*, economy of installation and long term maintenance, and ease of complete removal. The physical basis for these characteristics has been established by laboratory experiments and field demonstrations. On the basis of these tests the US Department of Energy has designated ground freezing among the top ten waste management technologies.

1 Introduction

Leachate from innumerable waste sites is threatening or already polluting the world's limited supplies of potable water. An idea of the magnitude of the problem can be gauged by the costs of remediation. In the U.S. alone, according to recent policy, cleanup cost estimates exceed $\$10^{12}$ U.S. [Russell 1991], although there is no existing technology that actually provides acceptable levels of cleanup within this figure [US DoE 1989]. The costs and infeasibility of massive cleanup is forcing a major reexamination and revision of US policy. A new emphasis is placed on risk assessment, which attempts to quantify the hazards to population and the environment, and to evaluate the costs and benefits of remediation. The assessments will help society establish sensible priorities for action where remediation is most urgent, and what level of cleanup is required. At the present time, the only available methods are expensive and of limited effectiveness. Ongoing research may eventually provide more economical and effective remediation techniques. But currently, while the hazards are being assessed and research explores new methods of remediation, pollutants are spreading. Therefore, containment of the leaking sites must be considered as an immediate temporary or long term remedy [National Research Council 1996]. Fortunately, there are several currently available technologies for containment by installing subsurface

NATO ASI Series, Vol. I 56
Ice Physics and the Natural Environment
Edited by John S. Wettlaufer, J. Gregory Dash and Norbert Untersteiner

barriers around the sites. One of the most effective types is composed of frozen ground.

Artificial ground freezing has been employed for many years in large scale engineering projects to bond soils for load-bearing strength during construction and excavation, and to seal tunnels and mine shafts against flooding from groundwater [Fremond 1994, Andersland 1995]. A newer application of artificial ground freezing is in hazardous waste management, to prevent the escape of liquid contaminants from waste sites [Krieg 1989, Dash 1989, Grant 1997, Iskandar 1997]. Studies have shown that the hydraulic conductivity of moist soils is reduced considerably at subzero temperatures [Horiguchi 1980, Nakano 1991]. Recent laboratory tests have shown that freezing techniques can make normally highly conductive ground impermeable to various liquids, including non aqueous liquids and brines. Subsequent to the laboratory tests, a field demonstration of a cryogenic barrier was carried out successfully for the US Department of Energy.

2 General Features of Cryogenic Barriers

2.1 Advantages

The use of thick ice barriers offers a number of advantages for confining hazardous wastes. A unique advantage of the cryogenic technology as compared to other techniques is its ability to achieve complete subsurface enclosure down to depths of several hundred meters [Jones 1982]. Ice is impervious to chemical attack, and it has negligible absorption of most chemicals. The permeability of a thick barrier at subzero temperatures is very low; the estimated leakage rate due to diffusion can be less than the detectable limit for all nuclear and chemical pollutants. The barriers are repairable *in situ*; if ground movement should fracture the barrier, the fissures can be filled and resealed quickly. Thick barriers have great thermal inertia, so that they can remain frozen without refrigeration for a long time; depending on the nature of the ground, and the thickness of the barrier, for months or years. The method is economically favorable at large sites, for intermediate and long term containment. Maintenance costs are low, allowing continued use for indefinitely extended periods. The barrier material is composed of pure water, and is environmentally benign. When the site is decontaminated, the ice walls can be allowed to melt, the pipes removed, and the area returned to its original state.

2.2 Technical Details

Fig.1 shows a schematic cross section of a possible design [Krieg 1989]. The overall scheme involves the installation of a boat-shaped rib cage of refrigeration pipes enclosing the contained volume of ground. The pipes are slanted, so that the volume can be enclosed on all sides and below. Alternative designs can involve ramp or cone geometries for the pipe arrays, or combinations of such features. Additional freedom of design can be obtained by the use of directional drilling. The double layer of pipes allows independent control over the thickness and temperature of the barrier. Additional pipes may be installed to inject water, and to monitor the integrity of the barrier. Containment of low freezing point fluids, such as hydrocarbons, may require barriers with zero pore space, so in such sites provision should be made for the injection of additional water in the barrier zone. The piping is a tube-within-a-tube design, the outer pipe is steel oil well casing, strong enough to permit installation by pile driving in most soils. The double array allows the entire intervening space to be frozen, producing a solid barrier of intermixed soil and ice. The injection and monitoring systems will require additional pipes, some of which must be perforated for access, yet clear of ice. This can be accomplished with internal heating and pumpout tubes, which are activated after the formation of the barrier, and then deactivated after they are cleared. The establishment of a complete barrier can be carried out quickly with large scale temporary refrigeration equipment, in a time as short as a few months, after which the ice can be maintained by a relatively modest refrigeration plant.

2.3 Economics

In most prior applications of ground freezing, there has been strong economic incentive to freeze quickly. However, in the case of hazardous waste containment, the typical priority is that seepage be halted in the long term. Then, significant economic savings can be obtained by slow freezedown, since the efficiency of the refrigeration process goes up significantly as the rate is slowed. An additional saving can be made by using low capacity refrigeration units for both initial freezedown and maintenance, thereby avoiding the expense of renting high capacity units for rapid freezing.

For periods longer than one year, the economic strategy dictates that the maintenance plant should be purchased. Construction should be oriented toward minimization of maintenance costs, and the piping geometry designed to allow total barrier formation to be accomplished slowly (in 4-12 months), rather than in a much shorter time. This permits greater spacing between the cooling pipes, extending from the typical construction spacing of 2m to as much as 10 m.

An added advantage of slower freezedown is the increased assurance of barrier integrity, due to the lower probability of instabilities and gaps in the frozen wall.

An analysis of the economics of a sample design was published some years ago [Dash 1989]. The major construction cost is the expense of drilling, which depends on the nature of the soil, the total length of piping and pipe diameter, with additional factors for hazardous working conditions, and the requirements of regulatory agencies. The size and shape of the site are important factors because the cost of containment increases as the area of barrier rather than the volume contained; thus the cost per unit volume contained tends to vary inversely as the characteristic linear dimension of the site. For sites in excess of several hectares, with soil conditions permitting pile driving, the installation cost is estimated between \$2.00 and \$4.00/m^3.

Maintenance costs are dependent on the size and depth of the site, the nature of the soil and the core temperature of the barrier. The heat load is determined by conduction and heat generated by radioactive sources. The flux from the ground decreases with time, as the cooled zone of surrounding soil becomes steadily extended. For a 4 hectare site, a barrier with 45° inclined walls and a core temperature of $-20°C$, the annual power cost for refrigeration one year after freezedown would be less than $20,000 U.S. After several years, the principal heat load will be due to conduction from the surface. Heating from radioactive sources will add an annual cost of about $3,000.U.S./MCi. The combined cost of installation and maintenance of a site of several hectares for 30 years would be less than $30 per cubic meter. The estimate compares favorably with the cost of alternative treatments. An analysis commissioned by the US Department of Energy estimated the cost of excavation and removal as $464/m^3 and of grout barriers as $76/m^3, not including the cost of removal [DoE 1989b].

2.4 Freezing of Water in Porous Media

A fundamental difference between water in bulk and water in a porous medium is that freezing is extended over a considerable temperature range below the normal transition temperature, which can lead to frost heave of the soil. Unfrozen water persists at the surfaces of the soil grains down to temperatures as low as -40°C. Frost heave involves the migration of this unfrozen water toward lower temperatures due to the system's tendency to lower the chemical potential. The dynamics of migration depend on the temperature, the impurity concentration, the temperature gradient and the hydraulic permeability of the soil. As the migration into a region proceeds to saturation, the local pressure will increase, and if it exceeds the external load the soil ruptures and the migrating water crystallizes out as an ice lens. The possibility of frost heave must be considered in the design of a barrier, but in many types of installations, the danger is minimal. In arid and coarse soils the rate is so slow that little motion will occur over long periods. In addition, since thermomolecular pressure drives water toward lower temperature, water migrates to the barrier from both sides, thus tending to balance the displacement forces on the barrier. The inward migration is beneficial since it increases the local moisture content within the barrier, tending to fill the pores. More extensive treatments of the subject of unfrozen water in porous media and frost heave are given in separate articles in this volume by Fukuda, Paterson, Wettlaufer and Williams [Wettlaufer 1998].

2.5 Leakage by Molecular Diffusion

Molecular diffusion remains as the dominant transport mechanism through the best barriers achievable [Daniel 1988]. The rate of transport is governed by several factors, including the inherent properties of the materials, thermodynamic factors, and barrier design. We estimate the diffusion through frozen ground, as follows.

Diffusion transport is described by Fick's law:

$$J = -D(dC/dx), \quad (1)$$

where J is the mass flux, D is the diffusion coefficient, C is the local concentration of the solution, and x is the direction of transport. The barrier penetration rate can be estimated from the time dependence of a one-dimensional model with simple boundary conditions: diffusion begins at $t = 0$ from a source of constant impurity concentration C_o into an initially pure semi infinite medium. After a time t the concentration at a distance x from the source is given by

$$C/C_o = erfc(x/2\sqrt{Dt}), \quad (2)$$

where $erfc$ is the complementary error function.

As discussed above, the cryogenic barriers will be composed of frozen ground containing one or more lenses of solid ice. The diffusion in each type of region will be treated separately.

The diffusion constants of light impurity molecules NaCl, HCl, and HNO_3 in crystalline ice at -15°C, are in the range of 4×10^{-9} cm^2/s [Barnaal 1983] . Diffusion constants generally decrease with temperature, and are typically smaller for heavier molecules, but for conservative estimates we base the calculations on the values above. Nevertheless, an estimate shows that transport is many orders of magnitude below detectable limits, even after geologically long intervals. According to Eq. (2), diffusion through a 5 m thick section of crystalline ice for a period of 10^4 years yields a concentration ratio $C/C_o < 10^{-21}$.

Molecules in frozen ground diffuse more slowly than in pure ice, due to the increased length of the diffusion path. A tortuosity factor reduces the diffusion constant from 0.01 to 0.8 of the original value. But for temperatures within ten or twenty degrees below 0° C an important counter effect is due to unfrozen water layers around the soil grains, since molecular diffusion in water is much more rapid than in ice. Typical diffusion constants of light impurities in bulk water are some 10^5 times the rate in ice. If a major fraction of the water in a cryogenic barrier were unfrozen the diffusion transport might be so great as to render the containment system ineffective. However, the liquid fraction decreases with temperature. For example, in a moderately fine soil with grain size 2 micrometers the cross sectional area of the liquid channels at $-20°$C is only about 0.1% that of the ice. In coarser soils and rock the unfrozen liquid films are still less important, because the specific area is less. The diffusion through such thin films is actually lower than in a channel of bulk water of the same thickness, due to a proximity effect, which decreases the mobility within a few molecular diameters of a solid boundary.

The thinness of the unfrozen liquid film at lower temperatures, tortuosity factor, and reduced diffusion coefficients bring the transport rate down into extremely low values. For temperature and soil grain size as discussed above, diffusion via the liquid film, through a 5 meter section for 10^4 years, will produce an estimated concentration ratio $C/C_o < 0.005$.

3 Laboratory Tests

The permeability of frozen ground barriers was directly measured in a recent laboratory project for the US Department of Energy. The soils were uncontaminated samples from DoE nuclear reservations: a coarse silty sand from Hanford, Washington, and a clay soil from Oak Ridge, Tennessee. The test fluids were dilute aqueous solutions of potassium chromate (K_2CrO_4), trichloroethylene (TCE) and radioactive cesium ($^{137}CsCl_2$), concentrated $CaCl_2$ brine, and pure decane. The choice of soils and chemicals provided a sampling of the characteristics of liquid contaminants at many of the DoE's leaking nuclear waste sites.

The measurements were carried out in small test cells provided with features enabling a variety of soil measurements over a range of temperatures and temperature gradients [Dash 1994]. Temperatures down to $-30°C$ were imposed on selected zones of the soil column by individually controlled thermoelectric cooling elements. The cells contained provisions for continuous monitoring of liquid movement, for injecting and removing liquids, testing the porosity by pressurized gases, profiling temperature and electrical conductivity along the soil column, and measuring the permeability of the frozen zone to various fluids under moderate hydrostatic pressure.

3.1 Procedures

To create conditions similar to an actual field installation, the central region of the soil column was cooled to subzero temperatures, while the portions above and below were maintained above freezing. In addition to imitating field conditions, the temperature gradients contributed toward achieving low permeabilities at moderate temperatures, particularly essential in the coarse sandy soil. The gradients produce a thermomolecular pressure inward, tending to cause migration toward the central lowest temperature. which produces a non porous frozen region. Such barrier regions can withstand great hydrostatic pressure gradients, for barrier core temperatures of only a few degrees below freezing can oppose hydrostatic pressures of many atmospheres. An additional important characteristic is that non porous frozen barriers are impenetrable to non aqueous liquids.

A simple technique was developed for creating non porous barriers, which is adaptable to field conditions. The region intended for the barrier was cooled below 0°C, while the space above was maintained above freezing. Water was then slowly added by injection into the soil immediately above the cooled region or by trickling down after flooding the surface. Water addition continued until the desired thickness of soil was saturated, after which the saturated region was frozen. The technique allows creation of non porous frozen regions of any desired thickness. It is adaptable to field conditions by installing a third layer of pipes within the outer two. To form the non porous barrier the outer layers are cooled below 0°C while the inner layer is maintained above freezing. Water is slowly injected or added at the surface, so that it freezes at the outer boundaries and fills the intervening space with liquid water. After the space is filled, the inner layer of piping is refrigerated, and the entire inner region is frozen. An added benefit of the technique is that it does not allow any water to escape outside the enclosed region, even during the formation of the barrier, and therefore does not liberate any solutes that may reside outside the region [Dash 1996].

3.2 Tests with Chromate and TCE Solutions

Tests of non porous barriers with potassium chromate and trichloroethylene solutions were performed over a period of several months. Hydraulic conductivities were calculated from the test parameters according to Darcy's law:

$$dV/dt = KA\Delta H/L, \tag{3}$$

where dV/dt is the volume flow rate through a column of cross sectional area A and thickness L, under a pressure head of depth ΔH. No leakage was detected in any of the tests, which were of limited duration due to contract deadline. Nevertheless, the test times were sufficiently long to show that the barriers were effectively *impermeable*, according to a standard classification scheme [Bear 1972], all K's being well below $10^{-9} cm/sec$.

3.3 Barriers Without Injected Water

Four experiments with chromate solution were conducted to test the effectiveness of cryogenic barriers with only the original moisture level and no water injection into the barrier zone. No leakage was detected in any of these tests, which can be attributed to the efficiency of the thermomolecular pressure in opposing the transport.

3.4 Radioactive Cesium

A special cell was designed for gamma ray logging measurements with ^{137}Cs. This isotope is a fission product in the manufacture of plutonium, and is one of the principal radioactive components in underground storage tanks at Hanford. The characteristic radiation is a highly penetrating gamma ray (662 keV). The radiation can be transmitted through the walls of the test cell with only slight attenuation, thus permitting continuous profiling of the isotope's distribution without removing the soil. The cell was similar to the others, except for a radiation enclosure of thick walled lead shields to protect personnel and reduce background activity. The cell and shields were mounted on a motor-driven frame, which enabled vertical scanning of the radioactivity along the soil column. Detection was obtained by a scintillation detector fitted with a narrow angle collimator, which provided a spatial resolution of a few millimeters along the soil column. As in the tests with chromate and TCE solutions, no leakage was detected in any of the cesium measurements.

3.5 Decane

Decane is useful for measurements of hydraulic conductivity in frozen and partially saturated soils [Andersland 1994], because decane remains liquid at low temperature (m.p.$-31°$C), and has a viscosity comparable to that of water. We carried out leakage measurements with decane on a 6 cm thick non porous region having a core temperature of $-25.5°$C., well within the range of decane's liquid phase. Pure decane was injected on top of the soil column, sufficient to produce a hydrostatic head of 28 cm. The liquid column height was monitored during the next two days. The results, if ascribed to leakage through the barrier, indicate an average conductivity $K = 5.3 \times 10^{-8}$ cm/s. However, all or part of the change may have been due to mixing of the decane with the unfrozen water above the freezing zone, rather than to actual penetration or transport through the barrier. Nevertheless, the measured rate, if ascribed to leakage, still designates the barrier as *impervious.*

3.6 Antifreeze

The most stringent test of a cryogenic barrier is the containment of brines and other antifreeze liquids, where the temperature of the barrier is higher than the freezing point of the mixture. Nevertheless, it is possible to contain such fluids for extended periods. When liquid contacts the barrier it induces melting whenever the freezing point of the brine is lower than the temperature of the ice. As melting proceeds the brine is diluted by the meltwater, thereby raising the local freezing point. Melting will continue as long as the supply of solute maintains a local freezing point below the temperature of the ice. The melt rate is limited by thermal conduction to supply the latent heat of fusion and by molecular diffusion of solute toward the interface. Since thermal conduction is typically much faster than molecular diffusion, the erosion rate will be governed primarily by the latter, and the rate will fall steeply as the diffusion path length increases. Thus, the lifetime of the barrier depends on its initial thickness, its temperature profile, the liquidus curve of the liquid, and the diffusion rate of the solute through the soil. A quantitative analysis and extensive testing would be required before reliable estimates could be obtained for the useful lifetime of barriers against brines and other antifreeze fluids. However, the qualitative trends described above indicate that the parameters of thickness, temperature, and soil permeability may be combined so that a barrier could remain intact for virtually any finite period.

A test with concentrated $CaCl_2$ brine was carried out in coarse Hanford soil, with a non porous barrier approximately 5.5 cm thick. The barrier was maintained such that the temperature increased monotonically in both directions from the center, from a minimum of $-25.5°$C at its core. Since the freezing point of the saturated solution was $-46°$C, contact with the ice caused immediate melting. The progress of melting was monitored by the height of the liquid column, changes of height being due to the volume change between ice and water (or solution). During the six day test the erosion was initially rapid, then fell non exponentially with time. The rate at the end of five days implied an effective hydraulic conductivity no greater than $2.8 \times 10^{-7} cm/sec$. However, it is to be emphasized that the monotonically decreasing rate implies that the brine was still contained within the frozen zone.

4 Field Demonstration

A full-scale cryogenic barrier was demonstrated for five months at a clean site in Oak Ridge, Tennessee [Scientific Ecology Group 1995]. Freeze pipes were installed in a double row, inclined at a 45° angle to establish an inverted vee-shaped subsurface barrier. The pipes were set on 2.5 m centers in each row, with rows 1.2 m apart.

After the installation of piping, refrigeration was supplied by two large mobile units delivering $CaCl_2$ heat exchange fluid circulating at $-20°C$. A complete subsurface enclosure, comprised of the frozen sides and bottom of the subsurface region, was obtained in about two weeks. By the end of the demonstration period the barrier wall thickness averaged more than two meters.

Diffusion transport was determined by chemical tracer tests before and after freezing. Because the barrier was installed in low permeability clay soil, it was necessary to prepare the area in order to obtain a critical test of the barrier's effectiveness. Before freezing a sand trench was dug transecting the barrier region, to provide a route for rapid transport of fluid, and therefore a stringent measurement of the effectiveness of freezing to block transport. Before freeze-down, substantial hydraulic conductivity through the sand trench was demonstrated by chemical tracer tests. Four hours after the injection of fluorescein in the center of the site it was detected in a monitoring well outside of the sand trench. After freezedown the site was injected with rhodamine solution, but no detectable rhodamine could be detected outside of the barrier for the duration of the demonstration.

Barrier integrity was also tested by measuring the ionic conductivity across the frozen zone, which showed negligible ionic transport through the frozen walls. Ground penetrating radar examination of the barrier wall after excavation of the interior unfrozen region showed a continuous frozen zone throughout the wall section.

The demonstration provided a practical example on which to base the cost of frozen ground barriers. After subtracting the special expenses for extra sensors and test support, the first year cost per cubic meter for installation and maintenance is estimated as \$200, and the average maintenance cost over a 15-year period is estimated as \$2.20 per cubic meter per year. The estimate of long term cost is based on calculated thermal losses in a typical soil, equipment amortization expense, power and maintenance costs [Dash 1989]. The cost per unit volume was relatively high because of the small size of the installation.

The bench tests and demonstration were deemed completely successful by the Office of Technical Development of the US Department of Energy [DoE 1995]. In 1996 the DoE designated frozen soil barriers as one of its top ten remediation technologies [Weapons 1996]. The DoE has recently awarded a contract for the installation of a cryogenic barrier around a radioactive site at the Oak Ridge National Laboratory.

References

Andersland OB and Ladanyi B 1994a. *Intro. to Frozen Ground Engineering*, Chapman and Hall.

Andersland OB Davies SH and Wiggert DC 1994b. Performance and formation of cryogenic containment barriers in dry soils: Barrier performance. Michigan State U. Rept to RUST Geotech, Inc., private comm.

Bear J 1972. *Dynamics of Fluids in Porous Media.* Elsevier NY.

Barnaal D and Slotfeld-Ellison D 1983. Pulsed nuclear resonance studies of doped ice. J Phys Chem **87**, 4321.

Dash, JG 1989. Ice technology for hazardous waste management. Waste Management **11**, 183.

Dash JG, Leger RJ, and Krieg, RK 1996. Nonporous Liquid Impermeable Cryogenic Barrier, US Patent 5,507,149.

Dash, JG, Fu H and Leger R, 1997. Frozen soil barriers for hazardous waste confinement, in *Ground Freezing 97*, S Knutsson, ed., AA Balkema Rotterdam, p.375

Dept. of Energy 1989a. Applied Research, Development, Demonstration, Testing, and Evaluation for Env. Restoration.

Dept of Energy 1989b, Basic Res. for Env. Restoration, Energy Research Office.

Dept of Energy, 1995. Frozen Soil Barrier Technology. Office of Innovative Tech. Summary Rpt.

Fremond M 1994 ed. *Ground Freezing 94*, Balkema publ.

IK Iskandar and FH Sayles 1997, Ground Freezing for Containment of Haz. Waste: Engineering Aspects. *Int. Symp. on Phys., Chem. and Ecol. of Seasonally Frozen Soils, Fairbanks Alaska 1997.* IK Iskandar et al. eds. CRREL Sp Rpt 97-10, 361-369. Jones, M.B., 1982. Ground Freezing Technology Advances with Speed. Tunnels and Tunneling, December.

Grant SA 1997, Artificially Frozen Ground as a Subsurface Barrier Technology,in *Barrier Technologies for Environmental Management*, D-153, Natl Res Council, Natl Academy Press, Wash DC.

Horiguchi K and Miller RD 1980. Experimental studies with frozen soil in an ice sandwich permeameter. Cold Regions Sci. and Technol. **3**, 177.

Iskandar I and Sayles FH 1997. Ground Freezing for Containment of Hazardous Waste; Engineering Aspects, in *Int Symp Physics, Chemistry and Ecology of Seasonally Frozen Soils*, ed. IK Iskandar et al, CRREL sp. Rpt 97-10, p.361.

Jones MB 1982. Ground freezing technology advances with speed. Tunnels and tunneling **14**, 31.

Krieg RK and Drumheller J 1989. A Closed Cryogenic Barrier for Containment of Hazardous Material Migration in the Earth. US Patent 4,860,544.

Nakano Y 1991. *Ground Freezing 91*, ed Y Xiang and W Changsheng, Balkema Rotterdam pp 65-70.

National Research Council 1996, *The Potential Role of Containment-in-Place in an Integrated Approach to the Hanford Reservation Site Environmental Remediation*, National Academy Press, Wash DC.

Russell, M., Colglazier, E.W., and English, M.R. 1991 *Hazardous Waste Remediation: The Task Ahead*, Waste Man. Res. and Ed. Inst., Univ. Tennessee, 1991.

Scientific Ecology Group, 1995. Demonstration of Ground Freezing Technology at SEG Facilities in Oak Ridge, TN, *Report to Martin Marietta Systems Inc. Hazardous Waste Remedial Actions Program.*

Weapons Complex Monitor: Waste Management and Cleanup 1996, *Top Ten Technologies Listed by DOE-Env. Man.*, Exchange/Monitor Publ Feb.13 p.6.
Wettlaufer JS, Dash JG, and Untersteiner N, eds *Ice Physics and the Natural Environment*, NATO ASI **58** *Series I*, (Springer-Verlag, Heidelberg 1998).

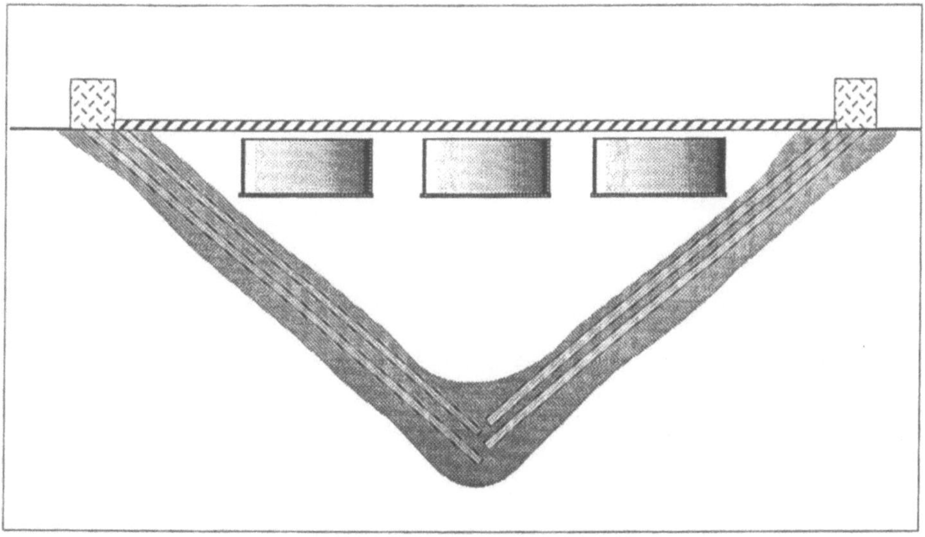

Fig.1. Cross section of a subsurface cryogenic barrier with a double array of refrigerating pipes.

Nuclear Contamination and Environmental Damage From Oil Spills in Polar Regions of FSU

Lydia Popova

Socio-Ecological Union
U.L. Krupskaya 8
Krop. 1, #187
117311 Moscow, Russia

1. Introduction

The Arctic region is usually defined as an area to the north of the Polar circle (66,5° N latitude). Climatologists consider July isotherm +10°C as the southern border of Arctic. A majority of experts define the Arctic as the area north of the southern limit of seasonal ocean ice, or in the case of land areas, north of the limit of forest. There are also other suggestions for latitude, geographic and climatic Arctic borders. In Russian legislation and in most of the social-economic investigations the term "Arctic" is used rather rarely. Meanwhile Russian statistics are based on administrative regioning of the country. In accordance with these 9 integrated administrative units (20 administrative regions) are related to the Russian Arctic (Fig. 1). Together with the marine economic zone and continental shelf the Russian Arctic constitutes more than 30% of the territory of the Russian Federation. Geographically in the south the Russian Arctic is bordered by the sparse growths of northern taiga and includes polar deserts, tundra, forest-tundra, and northern taiga. The main features of the Arctic are:

* very cold climate with the long winter (7–8 months) and low summer temperatures (11°C –13°C near the southern borders);
* short vegetation period;
* domination of precipitation over evaporation;
* specific hydrology—a lot of lakes and marshes, mouths of The Great Siberian Rivers;
* general prevalence of permafrost;
* seasonal differences in flows of solar radiation, modulated by alternation of polar days and nights;
* meager biodiversity (vegetation consists mainly of moss, lichen, grass and short shrubs), similar structure and functioning of ecosystems;
* relatively homogeneous distribution of indigenous people (more than 28 different groups of native people live in the European and Siberian North and in the Russian Far East) and old residents and concentration of recent residents in industrial centers;
* traditional economics—reindeer farming and fishing (primarily marine), scarce agriculture;
* domination of mining enterprises (which are located far apart from each other) in industry.

NATO ASI Series, Vol. I 56
Ice Physics and the Natural Environment
Edited by John S. Wettlaufer, J. Gregory Dash and Norbert Untersteiner

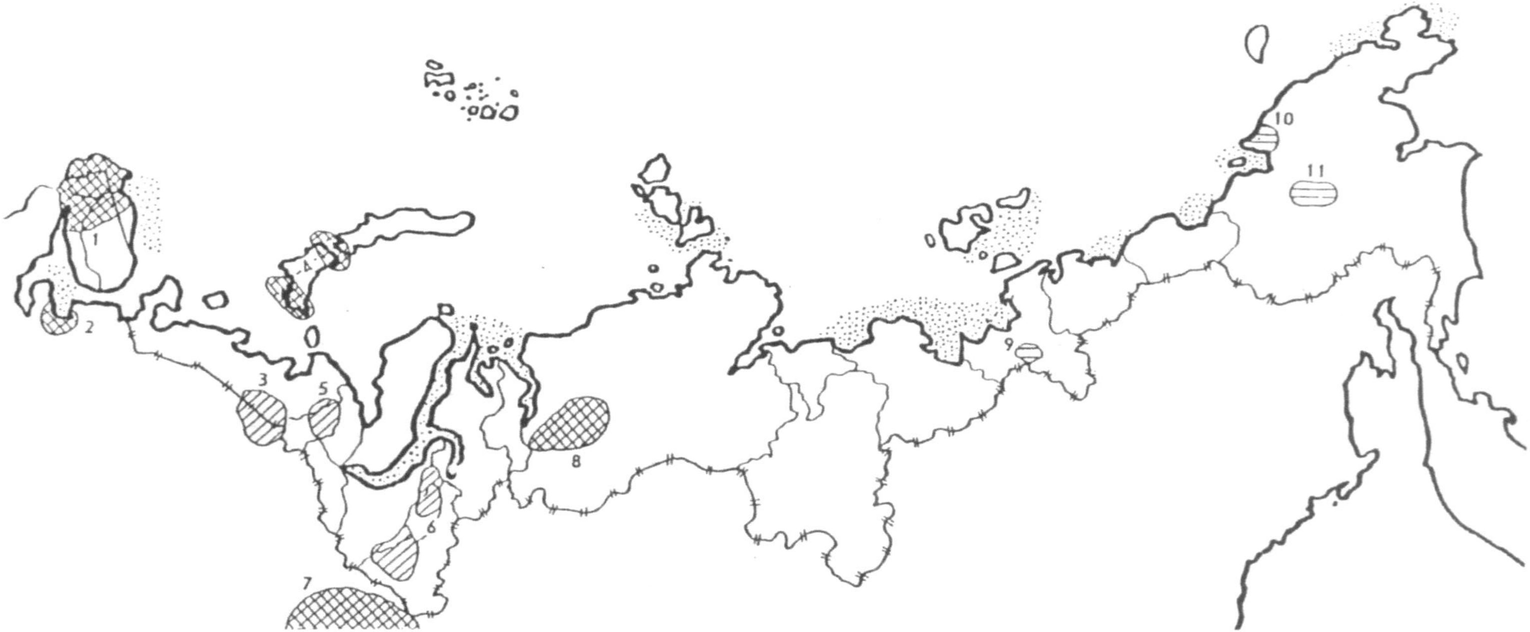

Fig. 1. **Impacted zones in the Arctic [Source: Arctic]**

⧺—⧺ **southern administrative border of the Arctic**

Situation:

crisis critical acute

Fig. 1. Impacted zones in the Arctic

The Arctic region plays crucial roles in the global environment: it is a place of origin of the global atmospheric processes and it serves as a filter for contaminated air flows. At the same time geographic location of the Arctic defines vulnerability of its ecosystems. The behavior of radionuclides and other pollutants is different from that in other regions. Restoration and self-cleaning of the environment are very slow due to the deceleration of all the biochemical processes in a cold climate. Restoration of destroyed soil, for example, takes 4–5 thousand years, and restoration of some kinds of vegetation takes hundreds of years.

The Russian Arctic is rich in natural resources—oil, gas, non-ferrous metals, gems, various chemicals used for production of fertilizers, etc. Oil and gas deposits (in oil equivalent) constitute more than 200 billion tonnes, and oil deposits on the sea shelf alone constitutes around 100 billion tonnes. 80% of fish and fish products in the Russian Federation come from the Arctic.

Industrial development in the Arctic started in the second decade of this century and was mainly focused on drilling and mining, construction and operation of smelters, construction of the cities. Simultaneously the Arctic was converted into a huge military base with its test-site on Novaya Zemlya and submarine fleet in the Kola Peninsula, and enormous GULAG empire. Industrial development in the Arctic was based on the labor of prisoners and soldiers, along with that of recruits.

Until recently a common perception of the Arctic was as an almost virgin land, not yet impacted by technological activity. Meanwhile a long term attitude to the Arctic as the resourceful colony of the Soviet Union resulted in an accumulation of liquid and solid waste and a heavy contamination of the environment. Official statistics indicate that total accumulation of the waste in the Arctic region is close to 100 million tonnes. 600,000 km^2, which constitute around 10% of the whole Russian territory, are contaminated by chemicals. Total or partial degradation of ecosystems can be observed on the territory of 80 km^2 (Arctic, p. 47]. Only Russian Arctic includes so called "impacted zones", where the environmental situation can be defined as acute or catastrophic (Fig. 1). The main polluters are Navies and militaries, mining industry, smelters and power plants. Transport and oil industry thus far have less impact on the Arctic environment.

The western part of the Arctic is subject to such specific pollution as the garbage brought by the Gulf Stream. The Eastern part of the sea now looks like a huge European dump. Woodwastes, plastic and metal garbage, oil and oil products film cover the sea surface near the eastern shores of the Barents and the Kara Seas along 4,000 kilometers. In some places this marine dump is 10 meters in width

Although data about contamination by various industrial pollutants have been gathered during the last forty years, information about the dumping by the Soviet Union of nuclear waste and submarine reactors into the ocean and rivers, which empty into the Arctic seas, fostered public interest in the Arctic environment.

Past dumping of nuclear waste into the Arctic seas and waters adjacent to the Arctic Ocean was partly revealed in a white paper released in 1993—the so called "Yablokov report" [Yablokov]. Since then a number of investigations have been conducted by the Russian, US, Norwegian, IAEA, and NATO research groups. Perhaps the most comprehensive evaluation of nuclear contamination of the Russian Arctic has been introduced in OTA study published in 1995 [OTA]. The following sources of radioactive contamination in the Arctic have been identified:

1) global fallout from nuclear testing;
2) "peaceful" nuclear explosions in USSR;
3) dumping of nuclear waste from submarine fleet, dumping of submarine reactors and fuel assemblies, accidents with nuclear submarines;
4) dumping of nuclear waste from reprocessing plants in Siberia into the rivers that empty into the Arctic Ocean;
5) dumping of nuclear waste from the reprocessing plants in the UK and France into the Atlantic seas;
6) fallout from the Chernobyl accident;
7) Kola and Bilibino nuclear power plants.

2 Nuclear Contamination in Polar Regions of FSU

2.1 Sources of Nuclear Contamination in the Arctic

2.1.1 Global Fallout From Nuclear Testing

Testing of nuclear weapons was the first and the largest source of contamination of the global environment by radionuclides. As the result of atmospheric tests by the Soviet Union, the United States, Britain, France and China 25 millions curies of ^{137}Cs, 16 million curies of ^{90}Sr and 6,5 billion curies of tritium were released into the global atmosphere, mostly between 1955 and 1966 [OTA. p. 34]. The Soviet Union conducted 715 nuclear tests on Semipalatinsk and Novaya Zemlya test-sites between 1949 and 1990.

Archipelago Novaya Zemlya, adjacent to the Arctic Ocean, has been used for testing nuclear weapons since 1954 by the order of the Soviet government. The first nuclear test took place on September 21, 1955. An explosion of a nuclear bomb was conducted in the bay of Chernaya Guba, to the South-West from Novaya Zemlya, at the depth of 100 feet [Bellona 1, pp. 120–121]. Between September 1955 and October 1990 132 tests were carried out at Novaya Zemlya: 87 in the atmosphere (between 1957 and 1962), 3 underwater (between 1955 and 1961) and 42 underground (between 1963 and 1990). At least 2 atmospheric nuclear explosions were carried out from military vessels in the Barents sea, to the west of the sound Matochkin Shar [Bellona 1, p. 122].

There are two periods in the history of Novaya Zemlya: one lasts from 1955 through 1962 and the other from 1965 through 1990. During the first period 87 atmospheric (out of which 4 were close to the earth surface) and 3 underwater explosions were carried out. The total amount of radioactivity released into the atmosphere at that time was 13 megacuries by ^{137}Cs [Arctic, p. 25]. From November 1958 through September 1961 there was no testing at Novaya Zemlya, but in September 1961 testing was resumed and between September 10 and November 4 1961 24 bombs with yields more than 1 Megaton each were tested.

Explosions were carried out every second day. Some of the yields amounted to 20–30 Megatons. The yield of the largest hydrogen bomb, which was exploded on October 30, 1961, was 58 megatons—6,000 times larger than that of the Hiroshima A-bomb [Bellona 1, pp. 123–124]. Although 94% of the yield of all Soviet nuclear tests has been released at Novaya Zemlya, the major part of radioactivity was distributed globally, as the largest explosions were carried out at the height of 1 km [OTA, p. 34]. General transport of radioactivity was to the east—that's why in Yakutiya contamination from nuclear testing is higher than on Novaya Zemlya. At places of underwater and above surface nuclear explosions in the Southern Island of Archipelago (Chernaya Guba) contamination of the shore and part of the marine shelf (area of a few hectares) is still rather high—up to 400 microrem./hour. The sea sediments are also contaminated by plutonium. This area is restricted for public access. [Materials, p. 246]. Since 1964 the Soviet Union conducted only underground nuclear tests at Novaya Zemlya—totalling 43 explosions. Underground testing had the goal of protecting the environment from radioactive releases. Nevertheless, every 5th Soviet underground test led to the release of radioactive gases into the atmosphere. After the explosion carried out on August 2, 1987, radioactive fallout was detected in all European countries. Above the site the dose exposure exceeded 500 rem/hour [Arctic, p. 25].

Levels of gamma radiation in most of the Arctic do not exceed natural background and are 2–3 microrem/hour. Average density of surface contamination is less than half in middle latitudes [Arctic, p. 25]. Nevertheless there are places of local contamination. Radionuclides are collected in moss and lichen and through the food-chain moss-reindeer-human get into human organisms. Accumulation of ^{90}Sr in bones of reindeer farmers in some cases is 60 times higher than that in bones of residents of middle latitudes [Arctic, p. 62].

2.1.2 "Peaceful" Nuclear Explosions (PNE) in Soviet Union

From January 1965 to September 1988 the Soviet Union carried out 116 "peaceful" nuclear explosions (PNE). 81 of the explosions took place in Russia. Explosions were used to conduct geological investigation, to extract oil and gas, to fight fires at oil wells, to build dams and canals, create underground storage cavities, primarily for gas, etc. Since early 1970s in Arctic and adjacent areas 29 or more PNEs were carried out [Arctic, 25]. There is almost no information about environmental impact of PNEs, but recent findings show that in some cases they have caused significant local contamination. Two PNEs in Yakutiya resulted in accidental releases of radionuclides into the environment. One PNE, named Kristall, was conducted on October 2, 1974 with the purpose of building a dam for the tailings from an enrichment plant. Right after the explosion the dose exposure on the surface of a cone was 65 microrem/hour, but at the depth of 0.4 meters it was 200 microrem/hour. After short-lived radionuclides decayed, and after clean-up in 1992 (18 years after the blast!) the dose exposure on the surface is 25 microrem/hour. But such long-lived radionuclides as ^{239}Pu, ^{240}Pu, ^{241}Am, ^{90}Sr, ^{137}Cs have been detected. At the distance of 100 meters from the epicenter of the explosion the content of plutonium in the soil exceeded that by a factor of

21,600 from global fallout. By this contamination an area can be compared with the region in South Australia, where Great Britain carried out its nuclear testing in 1955–1963 [NC, p. 17].

Another accidental release of radionuclides took place during PNE named Kraton-3, held on 24 August 1978 with the purpose of geologic investigation. This blast was 10 times more powerful than Kristall. As the result of this explosion, 100 hectares of taiga were made dead. It is supposed that radionuclides could be transferred to the distance of 150_250 km from the PNE. External dose exposure to gamma radiation was 500 rem at the distance of 3 km and 25 rem at the distance of 12.5 km. In 1990 levels of gamma radiation were 420 microrem/hour. High contents of ^{137}Cs, ^{90}Sr, ^{60}Co, ^{241}Am, ^{238,239,240}Pu have been detected in soil samples. Content of plutonium in some samples was 35,000 times greater than that from global fallout. 2.5 hectares around the well of PNE are contaminated by ^{137}Cs at the level more than 40 curies/km^2 (which is considered as the zone of ecological disaster) [NC, pp. 17–18].

Specific geographic conditions, which were not taken into account at the moment of these PNEs, were the cause of rather unsuccessful attempts to clean-up. Destruction of permafrost resulted in distribution of radionuclides in the environment [NC, p. 18].

2.1.3 Dumping of Nuclear Waste from Submarine Fleet, Dumping of Submarine Reactors and Fuel Assemblies, Accidents with Nuclear Submarines ("Komsomolets")

In 1965–1988 the Soviet Union dumped in Barents and Kara Seas near Novaya Zemlya and Kolguev Island 16 reactors of nuclear submarines, 7 of them with unloaded fuel cores with the total activity 2,300,000 curies, and a fuel assembly from the icebreaker "Lenin" [Materials, p. 245]. This dumping was first described in 1993 [Yablokov]. In addition to this dumping 17 vessels with solid radioactive waste aboard were sunk in Kara and Barents Seas by the Northern Fleet [Bellona 1, p. 98]. Besides, the report reveals dumping of liquid radioactive waste from 1959 to November 1991 at five different areas in Barents and Kara Seas and six accidental releases in bays and elsewhere (Fig. 2). This dumping amounted to almost 200,000 m^3 of liquid waste with more than 20,500 curies of radioactivity. Leaks from the storage and an accident aboard a nuclear submarine also contributed radioactive contamination. The largest dump site is Abrosimov Fjord where 1.2 million curies was deposited and East Novaya Zemlya Trough, into which 800,000 curies was dumped. In 1960s, the first years of dumping, some of the containers did not sink and were sunk by gunfire [Bellona1, p. 103].

On April 7, 1989 the Soviet nuclear submarine Komsomolets sank in the Norwegian Sea approximately 480 km off the Norwegian coast. The submarine lies at a depth of about 5,000 feet [OTA, p. 33]. The submarine has a nuclear reactor with spent fuel and two torpedoes with nuclear warheads. Recent estimates by the Russian scientists of the inventory of radionuclides in the reactor core of the Komsomolets

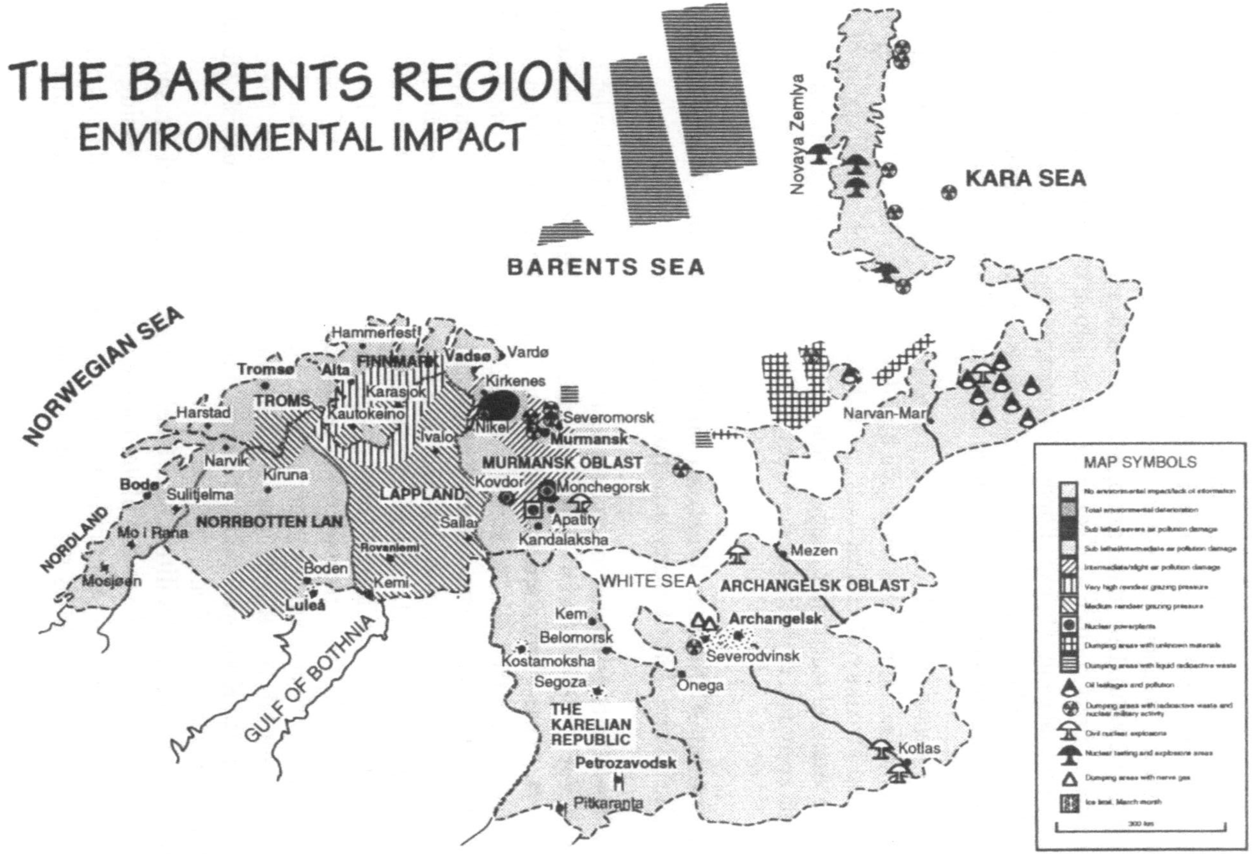

Fig. 2. The Barents region: Environmental impact

specified about 76,000 curies of 90Sr and 84,000 curies of 137Cs [OTA, p. 35]. It has been reported that the reactor is in a stable mode, the reactor compartment appears adequate and water exchange in its region is limited. More concern arises from two nuclear warheads, weapons plutonium activity of which has been estimated at 430 curies (96% of 239Pu). Due to the damages of the warheads' outer shells nuclear material is now in direct contact with seawater. Investigation of samples of water, sediments and biota taken during expeditions to Komsomolets do not indicate significant release of warhead plutonium into the environment. Nevertheless Russians have made attempts to seal the torpedo section to minimize the rate of warheads corrosion.

Most of the wastes dumped was low- and intermediate-level, but spent fuel and reactors from nuclear submarines introduce high-level wastes. Since the time of the dumping total inventory of radioactivity has reduced due to the natural decay of radionuclides. Some of calculations suggest that less than 130,000 curies remains in the reactors and spent fuel deposited in the Arctic [OTA, p. 49].

2.1.4 Dumping of Nuclear Waste From the Reprocessing Plants in Siberia into the Rivers That Empty into the Arctic Ocean

Russia inherited a huge nuclear military-industrial complex from the Soviet Union. Three facilities for production of fissile materials for nuclear bombs are located in Siberia, on the banks of big rivers or their tributaries. A plutonium production facility "Mayak" was built in the Urals, 70 km away from Chelyabinsk. The city which hosted the Mayak working force was named Chelyabinsk-65. Recently it was renamed «Ozyorsk». Mayak is surrounded by lakes and bogs, from one of which the river Techa emergest. Through other rivers it is connected with the Great Siberian River Ob'. Another facility was built near Tomsk, to the east from Chelyabinsk. The closed city had the name Tomsk-7, recently it was renamed «Seversk». Tomsk-7 is located on the bank of the river Tom', a tributary of Ob'. The third facility is located on the bank of the Great Siberian River Enisey, 70 km away from Krasnoyarsk. The closed city was first named Krasnoyarsk-26 and recently renamed to Zheleznogorsk. Ob' and Enisey empty into the Kara Sea.

Among other plants, facilities near Chelyabinsk, Tomsk and Krasnoyarsk also have reprocessing plants for plutonium separation. Reprocessing plants became major contributors into radioactive contamination of the environment not only around the facilities themselves, but also along the rivers thousands miles away.

In 1957 a tank with high-level waste from the reprocessing plant at Mayak exploded due to the failure of the cooling system. This accident resulted in release of 20 million curies of radioactivity, mainly ^{137}Cs and ^{90}Sr. Most of radionuclides fell near the tank, but 10% were carried away and contaminated 23,000 km^2 of land northeast of Mayak. From 1948 to 1953 Mayak discharged high-level waste directly into the Techa river. When radioactivity was discovered in the mouth of the Ob', discharges into the river were stopped, and radioactive waste was dumped into the lake Karachay, located on the Mayak site. Cesium and strontium contained in the lake water, leaked through the bottom of the lake into the ground water. A radioactive plume of the ground water under Karachay is migrating at a speed of 80 meters a year towards the river Mishelyak. There is a concern that radioactivity can break into the open hydrological system and contaminate Ob'. Another concern comes from the reservoirs on the Techa River that had been built in order to keep radioactivity. These reservoirs are separated from the river by a series of dams. If the water of the last reservoir overflows the dam, contaminated water will get into Techa and then to Ob'. Thus far groundwater and water in reservoirs is being monitored. Different proposals exist to prevent distribution of radioactivity, including construction of breeder reactors on the Mayak site. It is supposed that reactors will use the river water for cooling, which would lower the level of water in reservoirs due to the water heating and evaporation. In 1967 the summer drought dried out the lake Karachay and a tornado spewed out the radioactive dust from the lake. 600 curies of radioactivity were carried 75 km away from the lake and contaminated over 2,700 km^2. This plume partly covered the previous plume from the 1957 accident. Now the lake Karachay is being filled in with hollow cement blocks in order to trap the sediments and to limit further release of radioactivity. Meanwhile Mayak continues to discharge liquid waste into the Karachay [OTA, p. 40].

Liquid radioactive waste from Tomsk-7 and Krasnoyarsk-26 facilities have been stored in tanks and injected into deep geologic formations. Through-way reactors in Tomsk-7 and Krasnoyarsk-26, which discharged their coolants into Tom' and Enisey, were shut down a few years ago.

Cesium, strontium and plutonium have been detected in sediments of Ob' and Enisey downstream as far as the mouths of the rivers and in Enisey floodlands. In some places contamination was higher than that from global fallout. Although nuclear military facilities in Siberia could be potential contributors to the radioactive contamination of the Arctic, global fallout onto the land seems predominant. Some part of radionuclides is washed into the rivers and ultimately transported to the Kara Sea. As the catchment area of the river Enisey is smaller than that of Ob', its contribution into the radioactive contamination of the Arctic would be smaller [OTA, p. 45]. Potential contributions of the weapons complex to Arctic pollution depends on many factors, and permanent monitoring of the weapons production sites is necessary.

2.1.5 Dumping of Nuclear Waste From the Reprocessing Plants in the UK and France into the Atlantic seas

Radioactive releases from the reprocessing plants in Sellafield (UK) and at La Hague (France) also contribute to contamination of the Barents and Kara Seas. ^{137}Cs and ^{90}Sr are transported from the Irish Sea and around the coast of Scotland into the North Sea. From there, part of the radioactive flow travels through the Norwegian Sea to the east straight into the Barents Sea, from which contaminants get into the Kara Sea. The remainder travels with the West Spitsbergen Current up through the Fram Strait into the Nansen Basin. It takes 5–6 years for ^{137}Cs to travel from the Irish Sea to the Barents Sea, and 3–4 years to travel from La Hague. Most of the plutonium released from Sellafield stays in the Irish Sea, but some forms of plutonium oxides stay dissolved in the sea water and can be transported to the Barents Sea. Composition of plutonium isotopes in water samples allows conclusion that plutonium from Sellafield can travel as far as the Barents Sea.

2.1.6 Fallout from the Chernobyl Accident

At a result of the Chernobyl accident significant amounts of radioactivity were released into the environment, but most of radionuclides fell out not in the Arctic region. Nevertheless some ^{137}Cs has been deposited and transferred there. By some estimates, Chernobyl fallout contributed 1–2% of the total 1991 ^{137}Cs concentration in the Arctic basin [OTA, p. 37]. Contribution of ^{137}Cs as a product of industrial activity (including fallout from the Chernobyl accident) into the dose exposure from the ground surface, is less than 10–15% [Arctic, p. 24].

2.1.7 Kola and Bilibino Nuclear Power Plants

As reported by the State Committee on Hydrometeorology of the Russian Federation, Kola and Bilibino nuclear power plants do not have significant radioactive impact on the environment. It should be taken into consideration, that 6 reactors of these plants started operation more than 20 years ago and must be decommissioned soon. Potential risk of radioactive contamination comes from the storage facilities for solid and liquid radioactive waste at these plants. At the Kola plant, 5,000 m^3 of solid and 5,000 m^3 of liquid waste have been collected, and at the Bilibino plant—2400 m^3 and 700 m^3, correspondingly [Arctic, p. 26].

1.2 Environmental and Health Effects of Nuclear Contamination in the Arctic.

Despite the fact that global effects of the past dumping are unlikely to be catastrophic, specific features of the Arctic environment define vulnerability of its ecosystems and health effects for the population. Findings of a few expeditions to the Barents and Kara Seas indicate that concentration of radionuclides in sediment organisms (mollusks, crustacea, echinodermata) and in marine fish is relatively low—lower than in Baltic and Northern Seas, for example [Arctic, p. 34]. On land radionuclides are collected in moss and lichen and through the food chains get into the human organism. The most hazardous is ^{90}Sr, which replaces calcium in tissues. Accumulation of strontium in lichen is 115 mg/kg of dry material while background accumulation is 1.6 mg/kg, in moss—140 mg/kg with the background 3.0 mg/kg [Arctic, pp. 54–55]. Transport of cesium and strontium through the food chain is two orders higher than in more southern latitudes. Strontium intake for reindeer farmers is 3–4 times higher and cesium intake 100 times higher than for residents of middle latitudes [Arctic, p. 122]. Specific activity of ^{90}Sr in bones of reindeer farmers is 60 times higher than that in bones of people busy in other sectors of economics [Arctic, p. 62]. Self-clean-up of organisms from ^{137}Cs in the North is 5–10 times slower than in the same food-chains in southern regions of Russia [Arctic, p. 122]. Correlation between morbidity rates due to cancer and radiation level has been identified {Arctic, p. 62]. One should bear in mind, that the immune system of indigenous people in the North is not adapted to many infectious diseases that are widely spread in middle latitudes. This factor would make them more sensitive to the diseases that are caused by environmental pollution. Still more information is necessary to better understand the range of risks for individuals and local communities.

2.3 Potential Future Contamination

Potential risk of further contamination of the Arctic comes both from the previous dumping and operation of military sites, primarily Northern Fleet and nuclear test-sites on Novaya Zemlya.

The Russian active nuclear fleet in the north consists of about 88 submarines. Besides, there are 2 nuclear cruisers and 7 nuclear icebreakers (currently active 5 icebreakers), operated by the Murmansk Shipping Company [OTA, p. 119].

Annual yield of liquid radioactive wastes by the nuclear fleet (200 reactors) is 6–7 thousand cubic meters and of solid radioactive wastes 4.5 thousand cubic meters. 30% of liquid waste is being produced in the White Sea region, and 70% in the Barents Sea region [Materials, p. 245].

Dismantlement of nuclear submarines decommissioned in the process of disarmament is a serious problem. More than 1 million curies of solid radioactive waste and more than 7,000 cubic meters of liquid waste have already been accumulated on Kola Peninsula. Containment of liquid radioactive waste is not adequate. Construction of a new plant was stopped because of the lack of funding. Storage sites for solid waste operate since 1986 and many of them are 90% full. Significant amounts of spent fuel have been stored in on-land and floating waste facilities: more than 100 reactor cores (21,000 fuel assemblies) in a good condition and 700 damaged fuel assemblies that should be isolated from the environment. All storage facilities for spent fuel are 100% full. More than 100 reactor cores are still in reactors of decommissioned nuclear submarines that are currently based in the vicinity of residential areas [Bisovka, pp. 2–3].

The service ship Lepse operated by the Murmansk Shipping Company, which is a home for 643 assemblies with the spent fuel, requires urgent remediation. About 90% of the spent fuel aboard the Lepse is either damaged or unextractable, as it has been encased in concrete for prevention of radioactive releases.

Maintenance of the nuclear fleet results in annual releases into the atmosphere of 100,000 cubic meters of radioactive aerosols, basically ^{85}Kr and ^{133}Xe.

As already mentioned, Komsomolets presents a potential risk of contamination of the environment by radionuclides from nuclear warheads aboard the submarine. Nevertheless, in accordance with the estimates by the Russian experts, an upper-bound release rate of ^{137}Cs is about 13.5 curies per year, and release rates of other radionuclides are likely to be an order lower [OTA, p. 38]. There is rather limited water exchange between the deeper layers (below 1,000 meters), which reduces risk of transportation of radioactivity into the marine environment. Further efforts are necessary to seal more holes near the nuclear warheads or even to encase the submarine into a protective shell (containment) and to continue monitoring.

It is important to elaborate appropriate measures for the prevention of the future radioactive releases. Urgent measures include solution of the problems of the Northern Nuclear Fleet. Construction of storage facilities for the spent fuel and solid radioactive waste, and a plant for utilization of liquid radioactive waste are the first priorities in the Russian Arctic.

3 Oil Spills and Environmental Damage in Polar Regions of FSU

3.1 State of Oil Industry in Russia

The situation in oil industry in the Russian Federation at the moment can be described as critical. It is especially urgent for oil transporting pipelines. More than 70% of pipelines are 15–35 years old. 550 thousand kilometers of pipelines are high risk constructions [Materials, p. 284]. Aging of oil fields results in higher chemical activity of oil and sewage, which causes intense corrosion of pipes and increased numbers of accidental leaks of oil. Low quality of pipes and construction works and mechanical damage also lead to oil spills.

By some estimates, 3.5% of all oil produced (including by-pass gas, calculated into the oil equivalent) is lost in oil fields due to the lack of incentives for rational use of oil. A lot of oil is lost in collectors, from reservoirs, and storage tanks, and during transportation of oil by the railroads [Materials, p. 272].

Abandoned oil fields are also high risk units because of possible spontaneous fountaining of oil from the bore holes and wells. In Russia there are thousands of such constructions—9,000 in Tyumen' region only, and 6,000 in Komi. Fires also happen in operating wells and usually are caused due to the human error or lack of discipline. Oil fountaining and fires in wells and bore holes require intervention of specialized troops [Materials, p. 283].

Accidents with the oil containers and tank-cars in the railroads are another cause of oil spills in Russia. 30% of tank-cars should be replaced immediately [Materials, p. 283].

A recent problem is thefts of oil from the pipelines. Very often after these thefts pipes stay damaged for a while and leak. In 1993 642 cases of thefts of oil from the pipelines were revealed. But this is mainly a problem in the regions with unstable social and political situation, such as the North Caucasus [555 cases out of 642 in 1993].

3.2 Oil Contamination in the Arctic

It is rather difficult to get statistics about oil spills in the Arctic. The most infamous is 1994 Usinsky accident in Komi, when approximately 100,000 tonnes (estimates vary from 14,000 tonnes by official data to 200,000 tonnes by the data from environmental organizations) of oil leaked onto the soil from a broken pipeline [Tupikin]. Basins of rivers Kolva and Usa, tributaries of the river Pechora, were the most damaged regions at the result of the accident. In some places thickness of the oil film near the bank was from 10 to 25 cm. More than 900 tonnes of oil have been collected from only one spot of 700 square meters. As reported by the Usinsk Committee for Environmental protection, this accident was preceded by 14 smaller scale accidents at the same pipeline in August 1994.

Currently only large spills, when territories over 1 hectare get contaminated, are statistically accountable. Investigation conducted in some oil production regions, however indicate that small-scale accidents, when contaminated territories are smaller than 1 hectare, dominate and constitute 80–95% of all spills, or 40–60% of contaminated land [Chizhov, p. 37].

During the navigation period 5,000 river vessels operating in the Arctic produce 90 thousand cubic meters of sewage contaminated by oil products. Annually every fishing boat discharges 10–20 tonnes of oil products with its sewage [Materials, p. 247].

Arctic seas are subject to contamination by oil and oil products [Figs. 3 and 4]. Barents Sea is mostly contaminated by the oil products brought there from the Atlantic by the ocean. The highest concentrations of oil were observed in the western part of the sea. The main contributors of oil to the Kara Sea and other seas to the east are Siberian rivers—Ob', Pur, Enisey, Lena and others. Concentrations of oil in the mouths of the rivers and in bays are higher than in open parts of seas.

Currently there are 29 military bases in the Arctic North [Arctic, p. 247]. Usually they are lacking installations for waste treatment and utilization. Leakages of oil products from military storage of oil products and fuel cause contamination of both soil and seawater. After the reduction of military troops in general and in the North in particular, these bases were abandoned and it is not clear who will take responsibility for clean-up.

3.3 Impact of Contamination by Oil Products on the Arctic Environment and Health of the Population

3.3.1 Mechanical Destruction of Soil, Permafrost and Vegetation

One of the major specific features of the Arctic is permafrost, which comprises 60% of the Russian territory [Arctic, p. 80]. Any human activity in the Arctic causes damage to permafrost and leads to consequences harmful to the environment. With the beginning of geologic investigation a tremendous number of bore holes have been drilled in the Arctic, in late 1980s—hundreds of them.

Arctic permafrost is very diverse. That's why in every region changes in permafrost depend on its specific features, and environmental response to technological intervention is different.

The most dangerous is mechanical destruction of permafrost. Drilling of bore-holes and construction of wells, pumping out underground water and sewage leads to thermokarst settlings around the well and sometimes to its total destruction. Construction of roads also destroys the upper layer of soil and changes snow density and thickness.

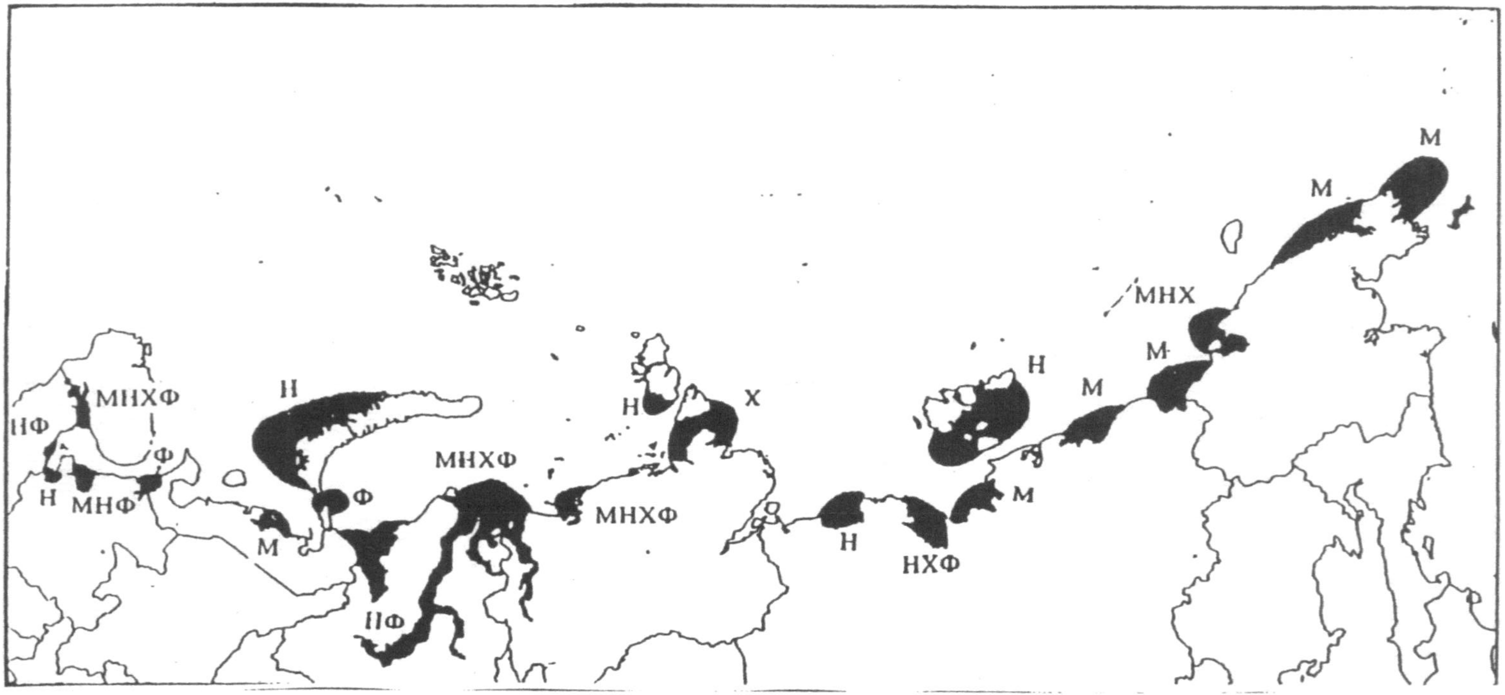

H - contamination by oil
M - contamination by metals
X - contamination by chemicals
Φ - contamination by phenols

Fig. 3. Places with the highest level of the sea water pollution

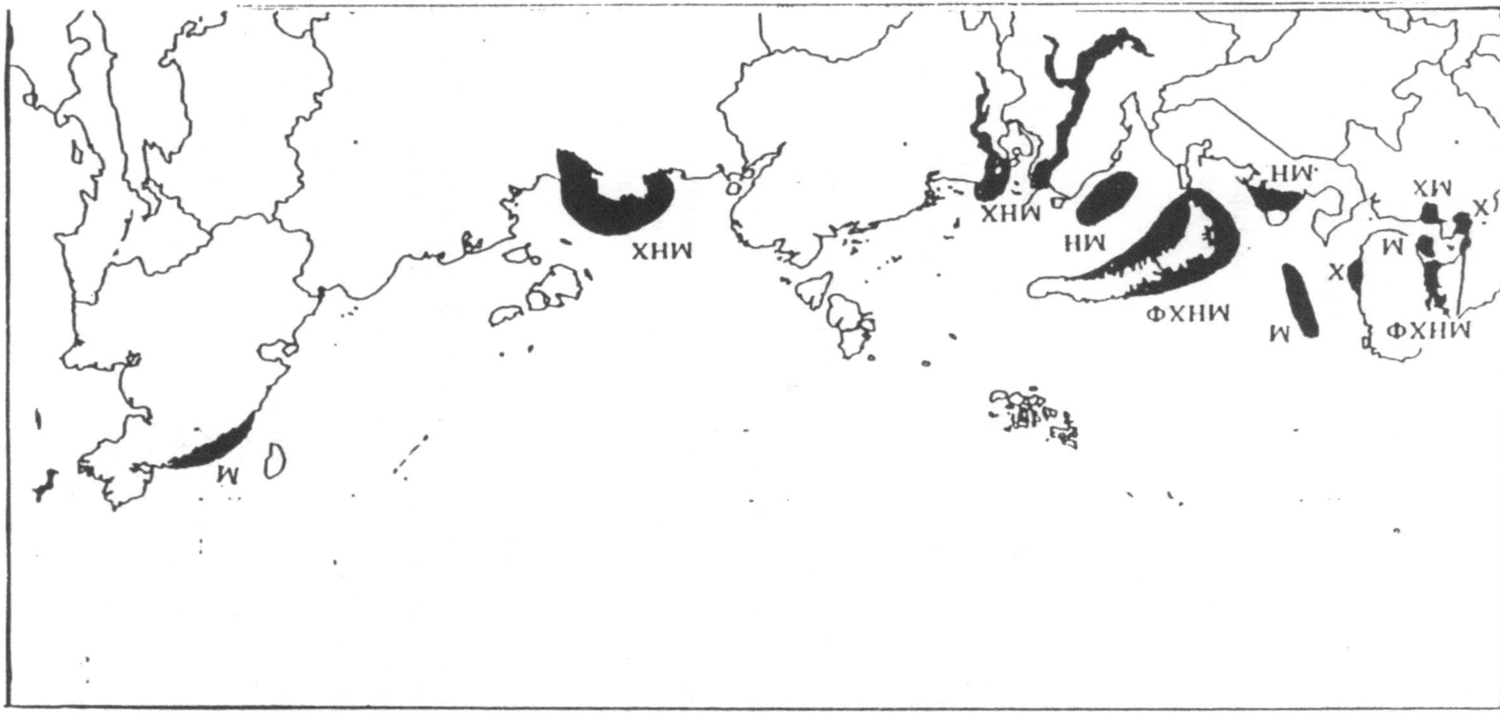

Н - contamination by oil
М - contamination by metals
Х - contamination by chemicals
Ф - contamination by phenols

Fig. 4. Places with the highest level of the sediments pollution

On the northern shore of Chukotka, for example, in the 1960s–1970s during the summer transportation of bore-hole constructions on heavy platforms, the upper layer of soil was totally stripped off. As permafrost in that region is marked by a high content of ice (70–80% of bed-rock), the whole area became a huge wet quick ground, a non-passable badland [Arctic, p. 83]. In other places construction of roads led to total or partial destruction of vegetation. In its turn, 60% of the accidents in pipelines are due to specifics of permafrost [Ibid., p. 83].

Oil spills sometimes cause irreversible changes in soil and impact vegetation and biological productivity of soil. As the result, soil is subject to erosion, degradation, and other destructive processes.

3.3.2 Impact on Fauna

Oil spills are deadly dangerous for Arctic fauna. A lot of animals die in oil-traps (collection of oil in low-lying areas). Birds die in lakes and seas due to contamination of feathers and violation of air-exchange processes. As oil contamination of egg shell kills germinating birds, reproduction of birds in most of the oil-fields in the Arctic became impossible. Oil film on the sea surface is even more dangerous for biota than oil diluted in seawater.

Oil spills are especially dangerous for biota during the warm season, with the maximum of solar radiation. In this case oil becomes more toxic, which is extremely harmful in the Arctic environment, where self-cleaning processes are slow [Arctic, p. 39].

Penetration of toxic oil products underground causes contamination of ground waters, primarily in the southern part of the Arctic. In the regions where ground water is used for drinking, this contamination poses certain threat to the health of local population.

3.3.3 Impact on Local Communities

Industrial development in the Arctic in general has a negative impact on the Native people. It interferes with traditional life-styles in the most harmful way. Use of reindeer pastures for industrial construction, destruction of landscape, contamination of water, massive logging, destruction of spawning fields, etc.—all this leads to the destruction of traditional economics and sometimes provokes active protests from the local population. Only in Yamal 6 million hectares of pastures have been destroyed [Ecology, p. 23]. Apparently in the process of industrial development protective measures should be undertaken. For example, some kinds of traditional activity, such as fishing and hunting should be allowed for the Native people only and prohibited for industrial workers. At the same time models of neotraditional lifestyles for indigenous people do not look realistic. It is necessary to include local population into the modern types of activity, to secure for them working places at enterprises, to build for them houses in towns and villages, etc.

3.3.4 Prognosis of the Environmental Situation in the Arctic due to the Development of Oil and Gas Fields

The Russian government has ambitious plans to develop oil and gas fields in the Arctic North, basically with the purpose of exporting crude oil and gas. Moreover, the North is projected to become a main exporting area [Strategy, p. 302]. Oil production will be intensively developed in two main regions: in Komi (Southern region) and in Arkhangelsk oblast, Nenets Autonomous County and on Pechora Sea shelf. The main gas deposit proposed for the development is the Shtockman deposit in the Barents Sea. Annually 6.5 million tonnes of oil will be produced at and shipped from the Prirazlomny oil deposit only. It is evident that oil and gas development in the Barents and Kara Seas will interfere with the traditional economics such as the fishery industry [Shtockman]. At the moment a decision about the route for oil transportation in the North has not yet been made. There are two options: either oil will be directed to the southern system of mainstream pipelines or a new terminal will be built to export oil directly by tankers. In the final choice a terminal will be built either in the sea 60 km north from the town Varandey, or on the river Indiga, or on the north shore of Kolguev Island, or east from this island [Strategy, pp. 302–303]. In any case, all oil produced will be exported as there are no oil refineries in the North, and the regional demand will be covered by oil supplies from the Central and North-Western regions.

The Barents and Kara Seas are the most contaminated in the Arctic (Fig. 2). Deterioration of the environmental situation in the Barents, Kara and Chukotsky Seas regions is very likely. Future oil spills will threaten most of waterfowl in Western Arctic. All year international navigation by the Northern Sea Route will contribute to oil contamination of the Arctic.

4 Legal Protection of the Arctic

Thus far there are no laws or regulation for the protection of the Arctic in Russia. Nor there are amendments to already active legislation. At the same time Russia is one of eight Arctic region countries (United States, Canada, Denmark, Finland, Iceland, Norway, Russia, and Sweden) who approved in June 1991 and signed at the recent conference in Norwegian city of Alt the Declaration on Arctic Environmental Protection. The Declaration is based on the Arctic Environmental Protection Strategy (AEPS), jointly elaborated by the experts from the participant countries, which include Russia. In the framework of AEPS, a Working Group on Emergency Prevention, Preparedness and Response will be addressing such problems as nuclear accidents and oil spills.

Unfortunately one can hardly expect that the Declaration can protect the Russian Arctic and secure rights of the Native people, as it has mostly recommendatory character. Attitudes toward the Arctic as a resource-rich colony, lack of environmental legislation and regulation with the introduction of market economy will create a destructive climate in relation to the Arctic, will lead to irreversible changes in the Arctic environment and will damage its ecosystems for ever.

Special protective measures and relevant legislation are needed to ensure the compliance of economic development in the Arctic with environmental protection.

5 Conclusion

The Arctic region has a unique environment of global importance. Previous activities of the Soviet Union, particularly military, led to contamination of the Russian Arctic by radioactivity and different chemicals. Although data collected by international experts do not indicate significant immediate risk for the Arctic and global environment and for the public health, permanent monitoring and elaboration of appropriate measures are necessary to prevent further contamination and reduce possible negative effects in the future.

It is important to identify possible sources of the future releases of contaminants. Urgent measures include solution of the problems of the Northern Nuclear Fleet. Other potential sources of contamination must be monitored for the earliest prevention activities to be undertaken, if necessary. International cooperation in this field is crucial. Without it Russia in its current economic situation is unlikely to take all responsibility for sound environmental and safety policy in the region which has a global value.

It is also important to recognize that to make international agreements and foreign assistance for solution of environmental problems in the Arctic effective, they must be enhanced by the relevant legislation and regulation still to be elaborated in Russia.

Acknowledgments

The author wishes to thank Dr. S. Knutsson for providing the map reproduced as Fig. 2. through the effort of Lisa Haugen of the Applied Physics Laboratory, University of Washington.

References

1. Government Commission on Matters Related to Radioactive Waste Disposal at Sea ("Yablokov Commission"), created by Decree No. 613 of the Russian Federation President, Oct. 24, 1992, Facts and Problems Related to Radioactive Waste Disposal in Seas Adjacent to the Territory of the Russian Federation (Moscow, 1993) (quoted as Yablokov).
2. Nuclear Waste in the Arctic: An Analysis of Arctic and Other Regional Impacts from Soviet Nuclear Contamination, OTA-ENV-623 (Washington, DC: U.S. Government Printing Office, September 1995) (quoted as OTA).
3. Environmental Security of Russia, No. 2: Materials of Interministerial Commission on Environmental Security (September 1994–October 1995), Security Council of the Russian Federation (Moscow, Judicial Literature, 1996) (quoted as Materials).

4. On Safeguarding Environmental Security in Arkhangelsk and Murmansk Regions in Connection with The Problems of Nuclear Submarines Utilization: Material prepared by the working group, headed by the adviser to The Chairman of The Committee on Supervision over Nuclear and Radiation Security N.Z.Bisovka and established by the order of The Secretary of The Security Council of The Russian Federation, issued on February 26, 1997, No. 6-rsb (MVKe-2-1/1) (quoted as Bisovka)
5. Sources of Radioactive Contamination in Murmansk and Arkhangelsk Regions, by Thomas Nilsen and Nils Boemer (Bellona Report, Version I, 1994) (quoted as Bellona 1).
6. Northern Fleet: Potential Risk of Radioactive Contamination of the Region, by Thomas Nilsen, Igor Kudrik and Alexander Nikitin (Bellona Report, No. 2, 1996).
7. Radionuclides in Ecosystems of the Barents and Kara Seas Region, by G. Matishov et al. RAN, Apatity, 1994.
8. Econord: Environmental Magazine for the Barents Region, No. 1/94.
9. Econord: Environmental Magazine for the Barents Region, No. 3/94.
10. Proposals on Environmental Investment Projects in the Russian part of the Barents Region: Report/Stage 1. Vol. 2. Radioactive Contamination (Environmental Financial Corporation of Northern Countries, December 1995).
11. Russian Arctic on The Edge of Catastrophe, by L. Bogoslovskaya et al. (Center for Environmental Policy in Russia, Moscow, 1996) (quoted as Arctic).
12. Environmental Consequences of Underground Nuclear Explosions, by V. Menschikov and B. Golubov (Nuclear Control, October 1995: Center for Political Studies, Moscow) (quoted as NC).
13. New Russian Energy Policy: editor-in-chief Yu. Shafranic (Energoatomizdat, Moscow, 1995) (quoted as Startegy].
14. Ecology. Oil and gas. By A. Gritsenko et al. (Nauka, Moscow, 1997). (quoted as Ecology).
15. Scientific and Methodological Approaches to the Estimation of the Gas Condensate Extraction Impact on the Arctic Seas Ecosystems (on the example of the Shtockman gas-condensate deposit): editors-in-chief G. Matishov and B. Nikitin (Kola Science Center, Apatity, 1997) (quoted as Shtockman).
16. Theoretical Bases of Critical Limit Values of Antropogeneous Loads on Subarctic Surface Waters. By T. Moiseenko (Kola Science Center, Apatity, 1997).
17. Oil and Gas Production Impact on Forest and Ecosystems of Middle Ob' Basin, by B. Chizhov (NDI Working Papers, Nizhnevartovsk, 1994) (quoted as Chizhov).
18. Black oil will make us bankrupts. By V. Tupikin, Nezavisimaya Gazeta, No. 213, 9 November 1994 (quoted as Tupikin).

Lecture Notes on Water in Ice: Microscopic and Geophysical Scales

John F. Nye

H.H. Wills Physics Laboratory, University of Bristol, Bristol BS8 1TL, United Kingdom

I: Microscopic Scale

1 Geometry

Polycrystalline ice (I_h) near its melting point is observed to contain microscopic veins of water, 10-100 μm in width, which follow the three-grain intersections. The veins form a connected network. Their existence and their detailed geometry are explained by the relative surface energies of the liquid-solid interfaces and the grain boundaries. The veins meet in fours at the four-grain intersection points, forming tetrahedra with open corners and non-spherical faces, whose shapes have been computed.

2 Effect of Stress

When stress is applied to ice further internal melting occurs at favourably oriented grain boundaries, with the formation of lenses of water. These may be made to focus light. The resulting diffraction pattern has been observed and also computed, and yielded a value (25±1°) for the dihedral angle of water at a grain boundary. However, it is known that the angle varies widely from one boundary to another, depending on the relative orientation of the two crystals involved.

3 The Freezing Point Depends on Impurity Concentration and on Interface Curvature

The veins can contain a large proportion of the soluble impurities of a complete specimen and therefore are very saline, even in specimens frozen from very pure water. Consequently, the water in them remains unfrozen even many degrees below the normal freezing point. The freezing point, and therefore the temperature of the whole specimen, depends both on the impurity concentration in the veins and on the microscopic curvature of the vein walls. As the temperature is lowered and the veins shrink, the concentration of impurities in them increases, and the curvature effect also in-

NATO ASI Series, Vol. I 56
Ice Physics and the Natural Environment
Edited by John S. Wettlaufer, J. Gregory Dash and Norbert Untersteiner

creases. The predicted relation between vein size and temperature has been verified by laboratory experiments in which both were continuously measured.

4 The Thermal Behaviour of Polycrystalline Ice

The above considerations account for a specimen's thermal behaviour. A continuum theory, in which the veins are imagined spread uniformly through a specimen, will be set up; it leads to differential equations, given below, that describe the diffusion of both heat and impurities along the veins; diffusion of impurities proves to be an essential feature. Because of the latent heat of the vein water, the ice behaves macroscopically as a continuum with an anomalous specific heat capacity that depends strongly on temperature.

5 Equations of the Continuum Model

We consider three space dimensions, and time t.

u = temperature depression below 0°C
A, K = constants
c = concentration of impurities in the veins
r = radius of equivalent cross-section of a vein
λ = length of vein per unit volume of ice
L = latent heat of melting per unit mass
σ = heat capacity of ice per unit mass
D_{th} = thermal diffusivity of ice
D_c = diffusivity of salt in water.

Melting point dependence

$$u = \frac{A}{r} + Kc,$$

Conservation of energy

$$\frac{\partial u}{\partial t} - \frac{\lambda L}{\sigma}\frac{\partial(\pi r^2)}{\partial t} = D_{\text{th}}\nabla^2 u,$$

Conservation of mass of solute

$$\frac{\partial}{\partial t}\left(r^2 c\right) = \frac{1}{3} D_c \operatorname{div}\left(r^2 \operatorname{grad} c\right)$$

6 Rotting of Polycrystalline Ice

When a specimen of polycrystalline ice is held in water close to 0°C, the veins become visible to the naked eye, and over a period of days or weeks they grow in thickness to a few tenths of a millimetre. The continuum theory explains both the time scale for this process and the size of the veins. It shows that asymptotically ($t \rightarrow \infty$) the vein size is proportional to $t^{1/3}$ (t = time). For example, in a specimen 40 mm in diameter, with grain size 5 mm, veins are predicted to be 0.3 mm thick after 10 days and 0.7 mm thick after 100 days. The asymptotic formula for vein size has no adjustable constants, containing merely material constants, grain size and specimen size; it is independent of the impurity concentration in the veins. The same continuum theory explains quantitatively how the cross-section of a vein flares out as it approaches the surface.

7 Polythermal Glaciers

From this point of view a polythermal glacier is a single medium with continuously varying properties, rather than consisting of distinct phases, namely cold (below the melting point) and temperate (at the melting point) with sharp boundaries between them, as is usually assumed. The equations provide a physical model for the transition region between the 'cold' and the 'temperate' ice, which turns out to be quite narrow.

References

Harrison, W. D. 1972. Temperature of a temperate glacier. *J. Glaciol.* **11**, No. 61, 15-29.

Harrison, W. D. and Raymond, C. F. 1976. Impurities and their distribution in temperate glacier ice. *J. Glaciol.* **16**, No. 74, 173-181.

Lliboutry, L. 1971. Permeability, brine content and temperature of temperate ice. *J. Glaciol.* **10**, No. 58, 15-29.

Mader, H. M. 1992a. Observations of the water-vein system in polycrystalline ice. *J. Glaciol.* **38**, No. 130, 333-347.

Mader H. M. 1992b. The thermal behaviour of the water-vein system in polycrystalline ice. *J. Glaciol.* **38**, No. 130, 379-394.

Nye, J. F. 1989. The geometry of water veins and nodes in polycrystalline ice. *Journal of Glaciology*, **35**, No. 119, 17-22.

Nye, J. F. 1991a. Diffraction by a small unstopped lens. *Journal of Modern Optics*, **38**, No. 4, 743-754.

Nye, J. F. 1991b. Thermal behaviour of glacier and laboratory ice. *J. Glaciol.* **37**, No. 127, 401-413.

Nye, J. F. 1991c. The rotting of temperate ice. *Journal of Crystal Growth*, **113**, 465-476.

Nye, J. F. 1992. Water veins and lenses in polycrystalline ice. In *Physics and Chemistry of Ice*, edited by N. Maeno and T. Hondoh, Hokkaido University Press, 1992, Sapporo, pp. 200-205.

Nye, J. F. and Frank, F. C. 1973. Hydrology of the intergranular veins in a temperate glacier. *International Association of Scientific Hydrology Publication 95* (Symposium at Cambridge 1969—Hydrology of Glaciers), 157-161.

Nye, J. F. and Mae, S. 1972. The effect of non-hydrostatic stress on intergranular water veins and lenses in ice. *J. Glaciol.* **11**, No. 61, 81-101.

Raymond, C. F. and Harrison W. D. 1975 Some observations on the behavior of the liquid and gas phases in temperate glacier ice. *J. Glaciol.* **14**, No. 71, 213-233.

Walford, M. E. R. and Nye, J. F. 1991. Measuring the dihedral angle of water at a grain boundary in ice by an optical diffraction method. *J. Glaciol.* **37**, No. 125, 107-112.

II: Geophysical Scale

1 Jökulhlaups

Temperate glaciers contain water on many different scales, ranging from intergranular veins on a scale of μm (lecture 1) to tunnels on a scale of metres. This lecture will deal mostly with one of the most spectacular examples of subglacial water flow, namely the periodic catastrophic outbursts of water (called jökulhlaups) from the subglacial lake Grímsvötn, Vatnajökull, Iceland. Apart from the exceptional size of this flood many features of its mechanism are typical of other glacial outbursts. The draining of the lake in the centre of the icecap, which used to happen about every 10 years and now happens about every 5 years, is accompanied by an enormous outburst of water from beneath the edge of the ice, some 50 km away from the lake, into the sea. Any road along the south coast of Iceland has to cross the outwash plain and for many years none was built because of the periodic jökulhlaups. The 1996 outburst was particularly spectacular because it coincided with an eruption of the Grímsvötn volcano. It caused great damage to the road and bridges.

The mechanism by which the floods occur has been studied in some detail. The lake is sealed until it reaches a critical level which enables it to lift the glacier. The resulting outflow tunnel is simultaneously melted open by frictional heat and closed by plastic deformation under the weight of the overlying ice. The differential equations for non-steady water flow in a subglacial tunnel (see below) will be derived and applied to the 1972 outburst from the lake. The discharge: time relation observed during the growth stage, and the abrupt ending of the flood, are accounted for in a way that is insensitive to the details of the subglacial tunnel system, although some difficulties will be referred to later.

2 Differential Equations

The equations for non-steady flow in a glacier conduit are:

Geometry and flow of ice:

$$\frac{\partial S}{\partial t} = v_i m - K_0 S(p_i - p)^n, \qquad (p < p_i)$$

Continuity:

$$\frac{\partial S}{\partial t} = v_w m - \frac{\partial Q}{\partial s}$$

Flow of water:

$$\rho_w g_s - \frac{\partial p}{\partial s} = \frac{NQ^2}{S^{8/3}}$$

Energy:

$$Q\left(\rho_w g_s - \frac{\partial p}{\partial s}\right) - \rho_w \sigma S \frac{d\theta_w}{dt} = mL + m\sigma(\theta_w - \theta_i)$$

Heat transfer:

$$0.205\left(\frac{2Q\rho_w}{\pi^{\frac{1}{2}} S^{\frac{1}{2}} \eta}\right)^{0.8} \kappa(\theta_w - \theta_i) = mL + m\sigma(\theta_w - \theta_i)$$

where

s = position in tunnel
t = time
S = tunnel cross-section
v_i = specific volume of ice
v_w = specific volume of water
θ_i = temperature of the ice
θ_w = temperature of the water
p_i = ice overburden pressure
p = water pressure

m = mass melted per unit length per unit time
K_0 = constant
σ = specific heat capacity of water
L = latent heat per unit mass
$n \approx 3$
Q = water discharge (volume per unit time)
g_s = s component of gravity
$N = (S/R^2)^{2/3} \rho_w g n'^2$ = constant
n' = Manning's roughness coefficient.

3 Stability

The equations may be applied to a fundamental question of stability. If a subglacial tunnel can grow unstably, as in the growth phase of a jökulhlaup, one may ask why intergranular veins in a glacier apparently do no such thing. The answer is found by studying the steady state in which growth by frictional melting is just balanced by plastic closure. *The steady state may be stable or unstable according to the conditions imposed at the ends.* By considering a complete system in which there are reservoirs at both ends of the channel, one finds the following: if the *outflow* end is held at a fixed pressure, the system is unstable when the *inflow* end is held at fixed pressure, but stable when it is fed from a reservoir of small area. This explains why veins develop just the right underpressure to balance contraction by plastic flow against expansion by melting; this is merely a consequence of the tendency to seek a steady (stable) state.

A related question is: why can ice-dammed lakes exist? Why do they not simply drain away through the intergranular vein system? The answer is found by considering the direction of the potential gradient.

4 Difficulties

There are certain difficulties with these theories. Notably, it is rather crude to assume that the outflow tunnel for a jökulhlaup is circular in cross-section. Computer simulations suggest that this assumption does not account too well for the falling limb of the hydrograph. It seems more likely that the cross-section is more like a shallow arch, but the difficulty here is to find a good way of calculating how it collapses. Another problem is that the flood sometimes starts somewhat before the lake level has risen to the level necessary for it to lift off the glacier. This is an important matter but the reason is uncertain. The temperature of the lake water is also an issue.

References

Bjornsson, H. 1975 Explanation of jökulhlaups from Grímsvötn, Vatnajökull, Iceland *Jökull*, **24**, 1-26.

Nye, J. F. 1976 Water flow in glaciers: jökulhlaups, tunnels and veins *J. Glaciol.* **17**, No. 76, 181-207.

Rothlisburger, H. 1972 Water pressure in intra- and subglacial channels. *J. Glaciol.* **11**, No. 62, 177-203.

Shreve, R. L. 1972 Movement of water in glaciers *J. Glaciol.* **11**, No. 62, 205-214.

Thorarinsson, S. 1953 Some new aspects of the Grímsvötn problem. *J. Glaciol.* **2**, No. 14, 267-275.

Thorarinsson, S. 1974 *Vötnin strid: saga Skeidarárhlaupa og Grímsvatnagosa.* Reykjavik, Bókaútgáfa Menningarsjóds.. *J. Glaciol.* **2**, No. 14, 267-275.

Weertman, J. 1972 General theory of water flow at the base of a glacier or icesheet. *Reviews of Geophysics and Space Physics*, **10**, No. 1, 287-333.

Local Ice Deformation Under the Influence of Natural Forces. Field Observations and Analyses of Cyclic Oscillations

Yevgeny Aksenov and Peter Wadhams

Scott Polar Research Institute, University of Cambridge, Lensfield Road, Cambridge, England CB2 1ER

Keywords. Sea ice, local deformation, field observations, SIMI

Introduction

Studies of deformations of an ice cover, its failure, natural ridging and rafting are of vital importance to ice forecasting, navigation and the safety of offshore structures. Waves as a mechanism of transfer of deformation energy play a part on the local scale and mesoscale. In this paper, we discuss the spectral structure of cyclic deformations of the ice floe. We mainly relied on the Sea Ice Mechanics Initiative (SIMI) field experiment, the Beaufort Sea, 1993–94. Of special interest were the data from the West Fall/Winter camp (75°N, 142°W) set up on a 2.1 m thick multi-year ice floe. Tiltmeters, BP-Delta strainmeters and three-axis inertial accelerometers were applied in keeping the record (Fig 1).

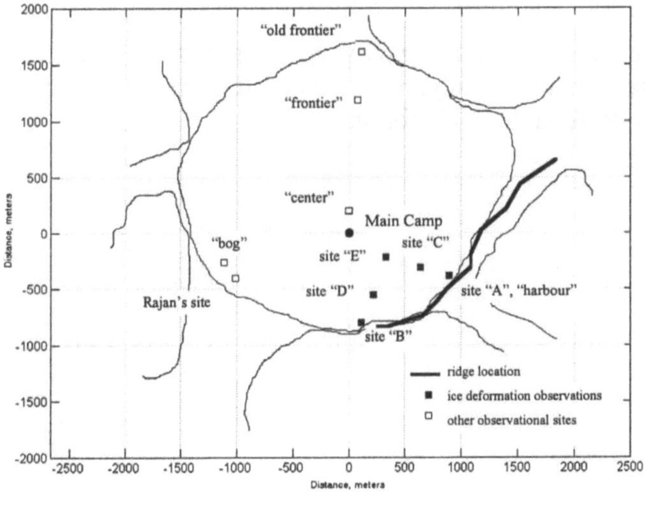

Fig. 1. Layout of sensors on experimental floe. *A, B, C, D* and *E* are sites of deformations measurement. *Harbour, center, bog, frontier, old frontier* are sites of CRREL stress gauges. Thick solid line shows location of ridge formed late in September 1993

NATO ASI Series, Vol. I 56
Ice Physics and the Natural Environment
Edited by John S. Wettlaufer, J. Gregory Dash and Norbert Untersteiner

1 Spectral Structure of Ice Deformations

1.1 Short-Period Waves

In addition to the SIMI experiment, some other campaigns, such as the MIZEX-84, CEAREX-88/89, AIMEX-93/94, Drifting Stations *North Pole*-23 and -28, etc., carried out the similar short-period wave (2–4 s period) observations (Martin and Drucker, 1991; Wadhams and Wells, 1995). This type of wave was reported as of frequent occurrence with a significant level of power.

In the SIMI experiment, the short waves were observed at the two rather distant sites *A* and *B*, near the forming ridge (Fig. 1). They occurred in several separate wave packets with a frequency of 0.32–0.5 Hz (variations in frequency of about ± 0.01 Hz) and each having a duration of 5–20 min, amounting to 50 min altogether (Fig 2a). Their power spectral density was significantly greater than that of flexural gravity waves (for the tilt the order is 1–2 × 10^{-2} arc degree^2/Hz). The waves were found to be vertically polarized, propagating from the 20 m wide source, beside the ridge (Fig 1). The ice cover bobbing induced by the unbalanced pushing upwards or downwards of newly fractured ice blocks adding to the growing ridge has been considered as a cause of oscillations (Wadhams and Wells, 1995).

The simplest model of bobbing oscillations of an ice floe with a draft *h*, which comprises a free non-damped oscillation under the hydrostatic vertical restoring force, gives a resonant frequency:

$$\omega \cong \sqrt{g\rho_w / h\rho_i}\,, \tag{1}$$

where g is acceleration due to gravity, *h* is mean draft of ice floe, and ρ_w and ρ_I are water and ice density, respectively. It gives an estimate of the bobbing frequency of about 0.33 Hz for an ice draft of 2,5 m.

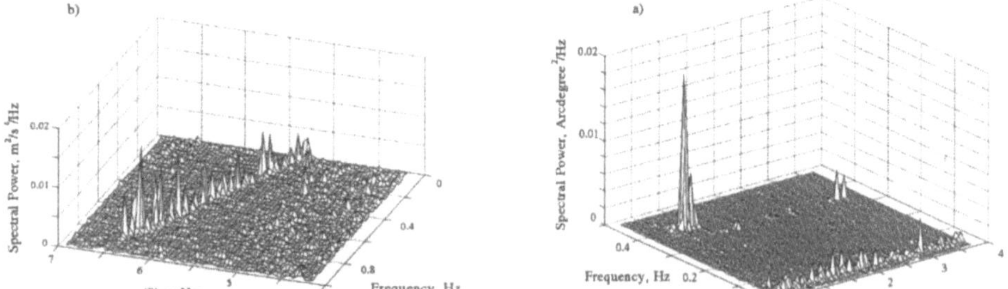

Fig 2. Time structure of spectrum for waves emitted during ice ridging (a) and ice cracking (b). Swell waves also are presented (a).

Ridge building laboratory experiments (Tuhkuri and Lensu, 1997) show periodic oscillations of the ice sheet with a well-pronounced peak of 9 Hz. The same mechanisms may be responsible for the oscillations observed in the field.

There were other attempts to monitor seismo-acoustic emission due to ridging and cracking, based on geophone/hydrophone arrays (Farmer and Xie, 1995; Dempsey and Adamson, 1995). Because these observations and the local ice deformation measurements were not taken concurrently, we cannot perform an intercomparison.

1.2 Long-Period Waves

Flexural gravity waves with a frequency of 0.036 Hz and a relatively small variability of 0.014 Hz occurred extensively (Fig 2a). It is believed that they are caused by long period swell penetrating from the open water area of the Greenland or Chukchi Seas.

Sporadic oscillations with the frequency of 0.14–0.16 Hz and the average length of 10–20 min (variability in frequency = 0.015 Hz). These are locally generated by turbulent atmospheric pressure microfluctuations. The mechanism has resonant character, and the favourable conditions for them are when the mean wind speed is nearly equal to or higher than the minimum phase speed of flexural gravity waves (Wadhams, 1986).

Oscillations with the period of 11–12 hrs were associated with semidiurnal tidal waves which cause tilting of the ice floe. And their propagation direction has been found to coincide with that of the tidal waves in this region of the Beaufort Sea (Kowalik and Matthews, 1982).

Internal waves observed in the SIMI experiment had a period of about 40 min which is longer than those from the CEAREX being of 20–30 min (Czipott et al. 1991). In both cases, the maximal tilt of the ice floe was about 10 arcs.

Infragravity waves with a period of about 1 min having a duration about 2 hrs and a non-regular occurrence pattern were seen during the SIMI experiment. Such a spectral structure leads us to assume that the origin of the waves is more likely to be related to non-regular events such as oscillations of the seabed due to seismic activity or to non-linear interactions between swell waves of short periods.

1.3 Events with Variations in Frequency

Ice cover oscillations with regular variations in frequency were noticed both near the border of the floe and in the central area (Fig. 2b).

The Doppler shift of frequency due to motion of the wave source relative to the point of observation can be an explanation of the phenomenon observed. A theoretical prediction of wave emission from the tip of a crack penetrating through ice floe was made by Slepyan (1993). The asymptotic solution presumes the emission of shear bending waves with frequency of about 0.58 Hz, wave length 38.5 m and phase velocity equal to 22.4 m/s for the crack moving faster than Kelvin's phase velocity. Smirnov and Shushlebin (1988) observed variations in frequency of seismic waves related, they assume, to movement of the crack tip.

2 Conclusions

The preliminary analysis shows that cyclic deformations in the ice cover have very sophisticated spectral structure. Besides the well-studied tidal and internal waves, long-period swell and flexural bending, short-period vertical shear waves related to ice ridging and oscillations with varying frequency have been found.

It is hoped that further analysis of short-period cyclic deformations together with seismo-acoustic observations will provide the more detailed outline of energy emission during ice failure within a wide frequency band. We also plan to study the strain and stress data obtained by J. Richter-Menge, CRREL (Fig 1), in order to relate stress release and cyclic deformations during ice failure.

References

Czipott, P.V., M.D. Levine, C.A. Paulson, D. Menemenlis, D.M. Farmer, R.G. Williams, Ice flexure forced by internal wave packets in the Arctic Ocean, *Science*, 254, 832–835, 1991.

Dempsey, J.P. and R. Adamson, Scale Effects on the Fracture and Constitutive Behavior of Sea Ice, *Proc. of the Sea Ice Mechanics and Arctic Modeling Workshop*, North West Research Assoc., Inc., Bellevue, WA, vol. 1, 84–99, 1995.

Farmer, D. and Xie Yu., Acoustic and Seismic Studies of ice Mechanics, *Proc. of the Sea Ice Mechanics and Arctic Modeling Workshop*, , North West Research Assoc., Inc., Bellevue, WA, vol. 1, 1–8, 1995.

Kowalik, Z. and J.B. Matthews, The M_2 tide in the Beaufort and Chukchi Seas, *J. Phys. Ocean.*, 12(7), 743–746, 1982.

Martin, S and R. Drucker, Observation of short-period ice floe accelerations during Leg-II of Polarbjorn drift, *J. Geophys. Res.*, 96 (C6), 10567–10580, 1991.

Slepyan, L.I., Crack dynamics in a plate under bending, in *Ice mechanics—1993*, edited by J.P. Dempsey, Z.P. Bazant, Y.D. Rajapakse, S.S. Sunder, pp. 119–129, ASME, New York, 1993.

Smirnov, V.N. and A.I. Shushlebin, The results of the studies of natural deformations of sea ice field, *Proc. of the First European Offshore Mechanics Symposium,* Norway, 1990, pp. 512–516.

Tuhkuri, J. and M. Lensu, Ice tank tests on ridging of an ice sheet, A preliminary report, Helsinki University of Technology, Ship Laboratory, Helsinki, 1997.

Wadhams, P., The seasonal ice zone, in *The Geophysics of Sea Ice*, edited by N. Untersteiner, University of Washington, pp. 825–991, Plenum Press, New York and London, 1986.

Wadhams, P. and S. Wells, Ice surface oscillation measurements on SIMI using strain, heave and tilt sensors, *Proc. of the Sea Ice Mechanics and Arctic Modeling Workshop*, North West Research Assoc., Inc., Bellevue, WA, vol. 2, 176–189, 1995.

Corrugations of the Sea-Ice–Ocean Interface Caused by Ocean Shear

Daniel L. Feltham[1] and M. Grae Worster[2]

[1] Department of Mathematics, University of Dundee, 23 Perth Road, Dundee, DD1 4HN, UK

[2] Institute of Theoretical Geophysics, Department of Applied Mathematics and Theoretical Physics, University of Cambridge, Silver Street, Cambridge, CB3 9EW, UK

Keywords. sea ice, corrugations, ocean, Bernoulli effect

The topography of the underside of sea ice is dominated by ice keels (often tens of metres deep) separated by expanses of relatively flat ice extending hundreds of metres horizontally (Davis & Wadhams 1995). However, there is evidence of some local variability in the thickness of sea ice in the flat regions (Wettlaufer 1991) of several centimetres over distances of about a metre. Such variations contribute to the drag exerted by sea ice on the ocean and enhance gravity drainage of brine from its interstices. We describe here a mechanism that can generate corrugations on the underside of sea ice in the presence of ocean shear, caused predominantly by wind-driven drift of the ice.

The mechanism we describe relies on the fact that sea ice is not solid but forms a porous medium of ice crystals through which interstitial brine can flow. Such two-phase media of growing (or melting) crystals and residual melt are often referred to as "mushy layers" and are common in metallic alloys during casting. A potent mechanism for brine drainage from sea ice is buoyancy-driven convection (gravity drainage) within the mushy layer, as described, for example, by Weeks and Ackley (1986) and by Wettlaufer et al. (1997). These observations provide evidence for the idea that the pores between the ice crystals form an interconnected network and that flow through it can be described by Darcy's equation, which relates the interstitial fluid velocity to the local pressure gradient. In the present context, pressure gradients are generated by the ocean flowing beneath incipient corrugations on the underside of the sea ice, as described below.

Consider the flow of ocean past perturbations to the underside of sea ice, as illustrated in Fig. 1a. The effect of this flow on the temperature field is shown in Fig. 1b. Within the ocean, the isotherms are compressed on the upstream face of each corrugation and rarefied on the downstream face, resulting in a net migration of the corrugation downstream but with no amplification. There is experimental evidence of such downstream migration given by Gilpin et al. (1980), who investigated the effects of fluid flowing past corrugations in a layer of solid (not mushy) ice.

The flow in the ocean is faster past the crests (marked c) than past the troughs (marked t). As a result of the Bernoulli effect, the pressure is lower where the flow is faster and higher where the flow is slower. This pressure variation has no effect at a solid boundary but adjacent to the porous sea ice its effect is to suck brine up through

NATO ASI Series, Vol. I 56
Ice Physics and the Natural Environment
Edited by John S. Wettlaufer, J. Gregory Dash and Norbert Untersteiner

the troughs and out of the crests, as shown in Fig. 1a. The strength of the induced flow in the sea ice increases with the speed of the flow past the sea ice and varies with the wavelength of the corrugations. If the wavelength of the corrugations is very large, the interface is nearly flat and the Bernoulli effect is weak. If the wavelength of the corrugations is much less than the thickness of the viscous boundary layer in the ocean adjacent to the sea ice then the pressure is governed by viscous stresses and the induced flow is again weak. Thus there is an optimal range of corrugation wavelengths for large induced flows.

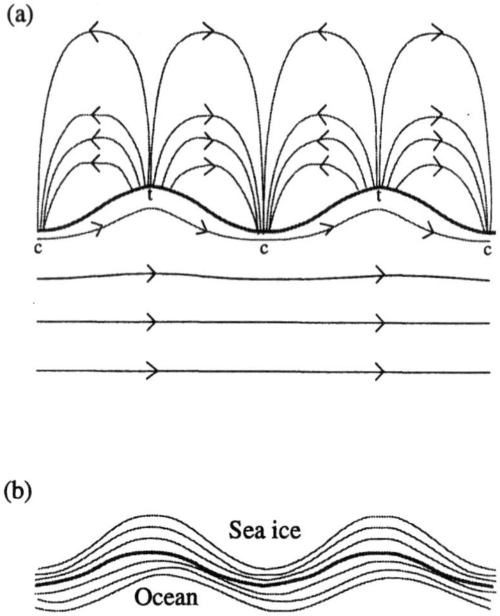

Fig. 1. (a) Flow within the sea ice (above) and ocean (below). The sea-ice–ocean interface is shown by the heavier line. (b) Effect of flow on isotherms. Note that the isotherms are compressed on the upstream face of each crest (marked 'c') in the ocean and are compressed at the crest within the sea ice. The latter enhances solidification at the crests and may therefore cause corrugations to grow.

Within the sea ice, the induced flow compresses the isotherms near the crests and rarefies those near the troughs. Thus the upwards heat transfer is enhanced at the crests allowing more rapid removal of latent heat and promoting solidification there. If the flow is too weak, however, the isotherms are insufficiently deformed and a random perturbation will decay due to the stabilising effect of thermal diffusion.

The physical effects described above can be analysed quantitatively using a linear stability analysis of the governing equations (Feltham & Worster 1998). For a given flow rate past the sea ice, we can calculate how porous the sea ice must be in order for our mechanism to become important. A measure of the porosity of the sea ice is its permeability, which is proportional to the square of the spacing of the ice platelets. In Fig. 2, we plot the variation of the critical permeability required for our mechanism to become important versus the wavenumber of the incipient corrugations for a freezing and an ablating sea ice layer. If the permeability is greater than the values given by Fig. 2 then perturbations of the given wavenumber will grow. Thus the minima of the curves give the minimum permeability for which a random perturbation can be expected to grow. Only those perturbations with wavelengths around the minima will grow so that one would expect to see corrugations of a particular wavelength with crests aligned transverse to the direction of the flow relative to the sea ice.

Naturally, the precise conditions required for incipient corrugations to grow depends upon the parameter values assumed for the ocean and sea ice. For a flow rate of

0.2 m s^{-1} (Omstedt and Wettlaufer 1992), a growth or ablation rate of 10^{-6} m s^{-1} (Wettlaufer et al. 1997) and a sea ice depth of 0.1 m, we predict corrugations if the sea ice permeability is approximately 10^{-10} m^2 or greater. This corresponds to a platelet spacing greater than about 10^{-4} m, which is fairly typical. It is conceivable that turbulence (which is not considered in our analysis) may act to decrease the speed of the flow required in the ocean.

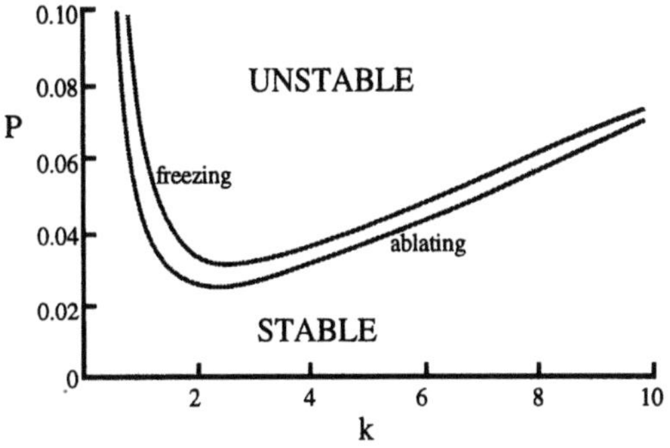

Fig. 2. Dimensionless curves of critical permeability P versus wavenumber k. Corrugations will appear in the region marked 'UNSTABLE', where the permeability is sufficiently large. Equivalently, for a given permeability there is a critical flow speed, or shear rate, at which corrugations will form. Ablating sea ice is more unstable than advancing sea ice.

In this brief communication, we have described the physical mechanisms that can cause the sea-ice–ocean interface to become wavy in the presence of flow. A theoretical analysis of this mechanism suggests that corrugations may form and grow given flows of magnitudes within observed ranges, it will be interesting to see if these predictions are borne out by future field observations.

References

Davis, N. R. & Wadhams, P. 1995 A statistical analysis of Arctic pressure ridge morphology
J. Geophys. Res. 100, 10915–10925.

Feltham, D. L. & Worster, M. G. 1998 Flow-induced instability of a mushy layer *J. Fluid Mech.* (under consideration).

Gilpin, R. R., Hirata, T. & Cheng, K. C. (1980) Wave formation and heat transfer at an ice-water interface in the presence of a turbulent flow *J. Fluid Mech.* **99**, 619–640.

Omstedt, A. & Wettlaufer, J. S. 1992 Ice Growth and Oceanic Heat Flux: Models and Measurements. *J. Geophys. Res.* **97**, 9383–9390.

Weeks, W. F. & Ackley, S. F. 1986 *The Geophysics of Sea Ice* NATO Advanced Study Institute Series B, Physics, vol. 146. New York: Plenum. Chap 1.

Wettlaufer, J. S. 1991 Heat Flux at the Ice-Ocean Interface *J. Geophys. Res.* **96**, 7215–7236.

Wettlaufer, J. S., Worster, M. G. & Huppert, H. E. 1997 The phase evolution of young sea ice *Geophys. Res. Lett.* **24**, 1251–1254.

Climatic Changes in the Mountain Glacier Area of Pamir

Alexander Finaev

Laboratory of Glaciology and Meteorology. Institute of Geology, Academy of Sciences
Dushanbe, Tajikistan

1 Introduction

Atmospheric changes in Central Asia during the 20th century [i, ii] are influenced by the amount of dust storms [iii] and other climatic parameters. The arid zones generate enormous quantities of dust particles which undergo transformation, transport and sedimentation, significantly affecting the climate and other aspects of the natural environment.

The high mountain ranges of Tyang-Syang and Pamiro-alai form an orographic basin open to air flow from the southwest. Mountain ranges rising 5000 to 6000 m above sea level in the eastern and northern parts of Central Asia form a barrier preventing southwestern air masses from penetrating further. These conditions are typical for this area of Tajikistan. Air masses coming into this region are formed over the western part of the continent.

When dust storms develop in the deserts of Middle Asia and break into the mountain country of Pamiro-Alai the dust load in the convective layer of the troposphere increases. Meteorological stations record dust haze during these episodes (Fig. 1).

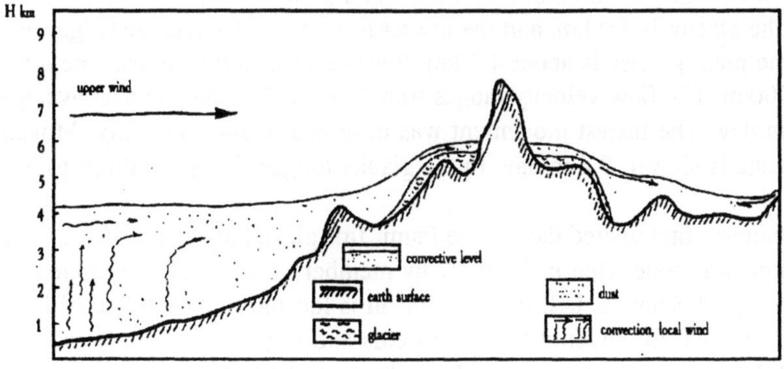

Fig. 1. Transport of aerosol.

NATO ASI Series, Vol. I 56
Ice Physics and the Natural Environment
Edited by John S. Wettlaufer, J. Gregory Dash and Norbert Untersteiner

Haze caused by dust storms is often transported over large distances through the lower troposphere. The vertical extent of the haze layer may reach 3 to 4 km. Particle number densities at 1.5 m may reach 8.4 cm^{-3}.[iv] Haze penetrates far to the east along the river valleys, but it is rarely observed in the Pamir Mountains (3–6 days/yr).[v] Annual trends show that the highest number of days with haze occurs during July to October and the lowest during winter. While the average duration of a haze episodes does not exceed 40 hr/yr in the north, that figure increases to 200–400 hr/yr in the Central Valley, and it grows to 500–600 hr/yr in the extreme south. The annual duration of haze in Pamir varies from 4 to 40 hr/yr.

2 Background: Glacial Geography of Pamir

Glaciers and snow cover are a major component of the natural environment of Tajikistan. The general area of glaciation covers more than 8000 km^2, or 6% of Tajikistan. More than 400 km^3 of fresh water are stored in the glaciers, eight times the annual flow of all rivers in Tajikistan. Over 1000 glaciers have an surface area of more than 1.5 km^2 each. Sixteen glaciers are more than 16 km long. Central-Pamir is the largest glaciated region of Tajikistan. The general area of glaciers reaches 2500 km^2, or about one third of the entire glaciation of Tajikistan. Glaciers of the Central-Pamir have different sizes. The longest are Fedchenko and Grumm-Grdgimailo. Another major glacial region is Zeravshan, covering an area more than 556 km^2.

Fedchenko, the longest mountain glacier in the world, is located in the northern part of Pamir and was discovered in 1878 by V.F. Oshanin. The Gorbunov Meteorological station is located in the middle part of the Fedchenko glacier, at a height of 4169 m. The glacier covers an area of 651.7 km^2; the length of the main hanger is 77 km; and the volume of ice in the glacier is 136.5 km^3. Fedchenko glacier is feed by 49 tributary glaciers of various sizes. The region surrounding the Fedchenko glacier contains 100 glaciers, with a general area of 824.1 km^2, and a volume of 144.4 km^3. The highest point of the glacier is 7.0 km, and the lowest is 2.9 km. The average height of the firn line on the main glacier is about 4.7 km. Seismic measurements indicate a thickness of 800–900 m. The flow velocity ranges from 0.6 to 0.9 m/day, with an average speed of 0.73 m/day. The fastest movement was observed in June and July. Movement in the firn zone is slower, 0.6 m/day. In the glacier tongue the speed drops to 0.4 m/day or less.

Mountaineers first visited the remote Pamir firn-glacial plateau in 1957. Glaciological research was conducted in 1976–77 by members of the Institute of Geography of the Academy of Science, USSR. The plateau is located in an interval of heights between 5400–6000 m, covered by snow fields and high-altitude glaciers, reaching a height of 7495 m at Communism Peak. In the central part of the plateau the thickness of firn and ice can reach 300 m[vii]. The discharge of ice from the northern boundary of the plateau occurs by "calving" of ice blocks every 15 to 20 minutes. Blocks as large as 100 tons break off once every 2 to 3 days. The resulting ice avalanches flow downward to the ablation area of Fortambek glacier, where huge avalanche cones form. These avalanches constitute the total loss of ice mass for the plateau.

The Zeravshan glacier belongs to the largest glaciers of Central Asia. It is not connected to other significant units of glaciation of Pamir-Alay. The system of the Zeravshan glacier consists of 21 glaciers with an area of 132.6 km^2 and an ice volume of about 15 km^3. The area of the main flow is 38.7 km^2; the length is 27.8 km, with an average width of 1.6 km. The average slope of the surface is about 8 degrees. Seismic data from 1962 show an average thickness of ice of 200 m. At an elevation of 2810 m the tongue of the Zeravshan glacier is the lowest in that region. The average height of the firn line is 4000 m. The highest point is Piramidalniy peak with an altitude of 5510 m. [viii]

Regular surveys of the terminus of the Zeravshan glacier, taken by members of hydrographical service of Tajikistan, show that in the last years glacier decay has been rapid. In some years the glacial tongue receded by up to 100 m.

3 Climate Change

In Tajikistan's mountainous region, long-term meteorological data were recorded at stations situated 300 to 4200 m above sea level. From these data we can estimate climatic change in the middle troposphere. About 80 meteorological stations (60–70 years of observations) are situated on the mountain slopes. Kurgan-Tube (426 m above sea level), Dushanbe (803 m above sea level) and the Fedchenko Glacier stations (Gorbunova, 4169 m above sea level) are located in the valleys, which are open to air flow from the west. The Kharamkul station (2800 m above sea level) has its own orographic conditions and is sometimes influenced by northerly air flow. Data from these stations provide long periods of vertical profiles.

4 Air Temperature

An analysis of the annual air temperature data from 1929 to 1993 shows a positive temperature trend at all levels (Fig. 2). The Kurgan-Tube temperature increase is 1.9°C/100 years. At 803 m above sea level the temperature increase is 1.72°C/100 years. Data from the Kharamkul station show a temperature increase of 1.33°C/100 years. A temperature increase of 0.7°C/100 years is found at 4169 m above sea level. The data show that troposphere warming decreases with elevation, leading to increased vertical gradients of temperature (Table 1). The vertical gradient increases by 3.3% per 70 years between levels 1 and 4. This should reduce the stability of the atmosphere and result in a stronger vertical mixing.

Monthly temperature changes are different throughout the year, but a warming tendency is shown in every month except in February (Fig. 3). The highest temperature increase occurred in the last half of the year at the Kurgen-Tube and Dushanbe stations. In Dushanbe, the greatest warming occurred in November and reached 7.3°C. In addition, Duarte Maria Luz (1993)[vi] showed that, in the summer period (1962–1965), air temperature in the troposphere has an upward trend at tropical and middle

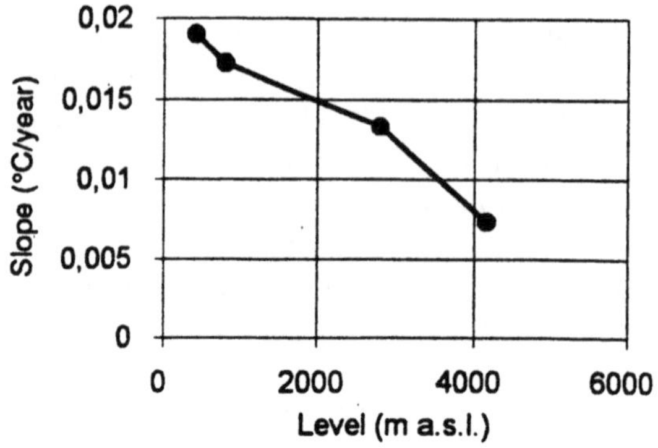

Fig. 2. Trend of annual air temperature.

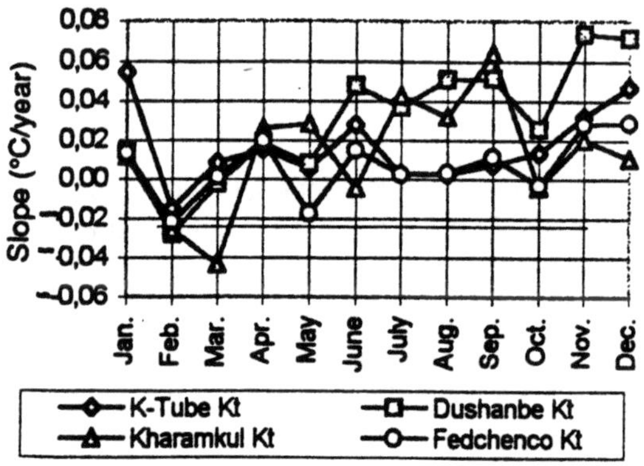

Fig. 3. Trend of monthly air temperature

Table 1. Vertical temperature gradient between different levels.

Station	H(m)	No of level	Delta of level	Gradient (°C/100 m)		
				1926 year	1996 year	%
Kurgan-Tube	426	1	1-2	-0,31	-0,34	9,7%
Dushanbe	803	2	2-3	-0,60	-0,62	3,3%
Kharamkul	2800	3	3-4	-0,68	-0,71	4,4%
Fedchenco	4169	4	1-4	-0,60	-0,62	3,3%

latitudes. Warming was very slight at the high-mountain stations in summer, but in November and December the temperature rose by 2.8°C.

5 Precipitation

A positive tendency is also found in annual precipitation. The trend increases with elevation above sea level (Fig. 4). The rate of precipitation increase at the lowest station (Kurgen-Tube) was 4.6 mm/100 years; at the second station the increase is 64.7 mm/100 years, at Kharamkul it was 181.6 mm/100 years, and at the highest

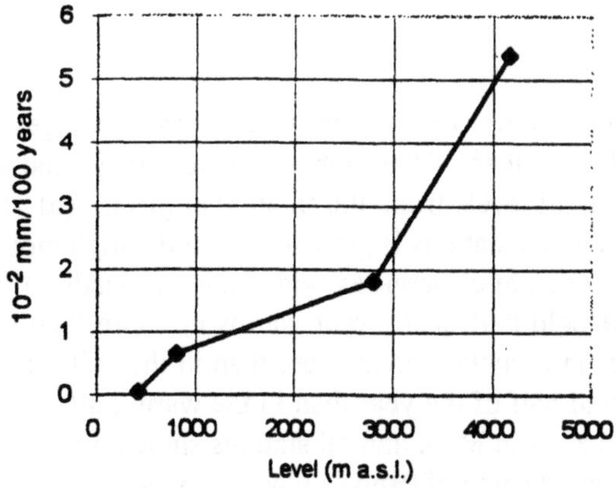

Fig. 4. Annual precipitation trend.

station (Fedchenko) the precipitation increase was 536.8 mm/100 years (Table 2). The percentage increase of precipitation was greater at the high elevations than in the low valleys.

Monthly changes of precipitation varied throughout the year. The monthly increase was lowest for all stations in June (Fig. 5).

Table 2. Trend of annual precipitation change per 100 years on different levels.

Station	H(m)	No of level	Precipitation (mm)			
			1900 year	2000 year	%	Delta P (mm)
Kurgan-Tube	426	1	287,8	292,5	1,6%	4,7
Dushanbe	803	2	632,0	696,7	10,2%	64,7
Kharamkul	2800	3	1 267,9	1 449,5	14,3%	181,6
Fedchenco	4169	4	1 027,6	1 564,4	52,2%	536,8

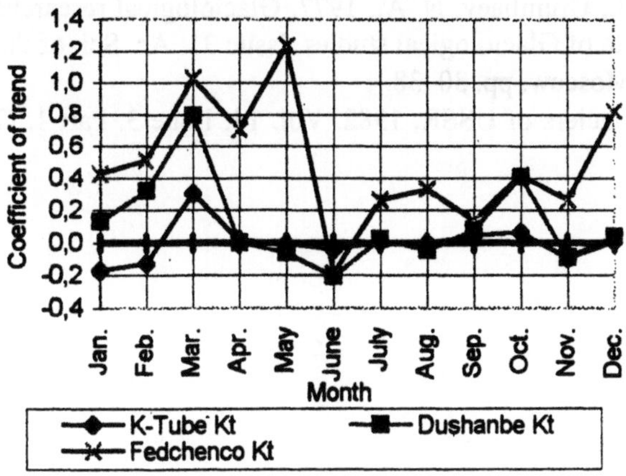

Fig. 5. Trend of monthly precipitation.

6 Conclusion

Analysis of these data allows us to draw some conclusions about climatic changes in Tajikistan and their connection to the changes of the global climate. (a) Annual air temperatures show a steady increase at all levels from 400 to 4000 m (a.s.l.). (b) The rate of air temperature increase on the plain stations is greater than in the high mountains. (c) The vertical gradient of temperature increases with time and leads to increased atmospheric mixing. (d) The cold half of the year warms up more than the warm half. (e) Precipitation increases in mountain areas more than in the valleys. (f) Precipitation increases more in the cold half of the year than in the warm half. (g) In valleys, a reduction in precipitation occurs in May, and all stations show a reduction in June. (h)Climate warming has caused of some of Pamir's glaciers to recede.

References

[i] Molosnova T.N. and Ilinyak E. P. 1991. Climate variations of total atmosphere circulation and types of synoptical processes above Middle Asia. Proc. Middle Asia Scientific Research Hydro-Meteorological Institute. Issue 141 (222). P. 3–11.

[ii] Subbotina O.N. and Chevichalova T.M. 1991. Particularities of long term change of atmosphere circulation on the territory of Middle Asia. Proc. Middle Asia Scientific Research Hydro-Meteorological Institute. Issue 141 (222). P. 12–21.

[iii] Zolotokrylin A.N. 1996. Sand storms on Turan lowland.

[iv] Ivanov V.A., Finaev A.F. Concentration and size distribution of aerosols in a mountain valley. Proc. Main Geophysical Observatory in Leningrad. 1987. Issue 507. P. 121–125.

[v] Climates of the USSR. Handbook, 1970. Issue 31. Vol. V. Gydrometeoizdat, Leningrad. P. 214.

[vi] Duarte Maria Luz. 1993. Temporal variations of the vertical profile of temperatures of Argentina. Int. J. Climatol. 13.4.437–445.

[vii] Durgerov, M. B., Urumbaev, N. A., 1977. Glaciological researches of the Pamir's firn plateau. Data of Glaciological studies. Issue 31. Ac. Sci. of the USSR. Institute of Geography. Moscow, pp. 30–38.

[viii] Catalogue of glaciers of USSR, 1982. Vol. 14. Issue 3. Part 1. Gydrometeoizdat, Leningrad.

Geometric Selection in Ice Polycrystals: Concavity, Faceting, and Kinetics

V.A. Hodgkin[1], J.S. Wettlaufer[1,2], and J.G. Dash[1]

[1] Dept. of Physics, Box 351560, University of Washington, Seattle 98195

[2] Applied Physics Laboratory, Box 355640, University of Washington, Seattle 98195

Abstract. Ice crystallization dynamics are observed with crossed polarizers in a Hele Shaw cell. Major features include the evolution of two thin ice sheets with parallel crystallite structures. Typical grain boundary grooves are seen at the junctions of most crystallites and water, but occasional giant faceted grooves are observed. The faces of the giant grooves are formed by basal planes, which are nucleated at the lower temperatures at the clefts. The grooves are asymmetric, yet track with the direction of the advancing interface.

1 Introduction

The morphology of ice crystallization is a distinguishing feature in the different textures of ice occurring in Nature, e.g. frost on windows, frozen lakes, glaciers, and sea ice. Current studies attempt to predict the textures that result from freezing under a given set of conditions, such as temperature gradient, convection, and solute concentration. Recent work describes how a matrix of ice and brine forms when salt solutions are frozen from above (Worster 1997, Wettlaufer 1997). In current experiments we are studying the crystallite texture resulting from freezing along temperature gradients in confined, quasi two-dimensional geometry.

2 Experimental Arrangement

The experiments are performed in a vertically oriented thin Hele Shaw cell having transparent 75 mm × 25 mm sapphire windows separated by a 3 mm thick plastic frame. The cell is cooled by thermoelectric elements in contact with a window face at its upper and lower ends. Observations are made at low power magnification through crossed polarizers. In the preliminary studies reported here, the cell ends were cooled at fixed power levels; the lower end was cooled sufficiently to nucleate ice locally and the ice was then observed as it advanced to the upper end of the cell. The water was deionized, and the cell was open to the atmosphere for pressure relief.

NATO ASI Series, Vol. I 56
Ice Physics and the Natural Environment
Edited by John S. Wettlaufer, J. Gregory Dash and Norbert Untersteiner

3 Observations

The course of freezing in several runs was recorded on video tape and subsequently studied in detail. All of the runs showed the following principal features. Nucleation was followed by rapid growth of a thin film of ice on the lower region of the cooled window. The ice developed needle-like structure and grew as feathery dendrites. As freezing continued, a more oriented structure developed, with grain boundaries parallel to the temperature gradient. The ice grew as two thin superimposed sheets, the `lower' sheet lying against the cooled window advancing ahead of the `upper' sheet. The two sheets had a combined thickness estimated to be less than 0.5 mm. A particularly striking feature is discussed below..

4 Giant Faceted Grain Boundary Grooves

Figure 1 shows a video image of a giant faceted grain boundary groove. At least one such groove occurred on one or both ice sheets in each of the runs made to date. In most cases the grooves were carried along by the advancing interface throughout the growth toward the top end of the cell.

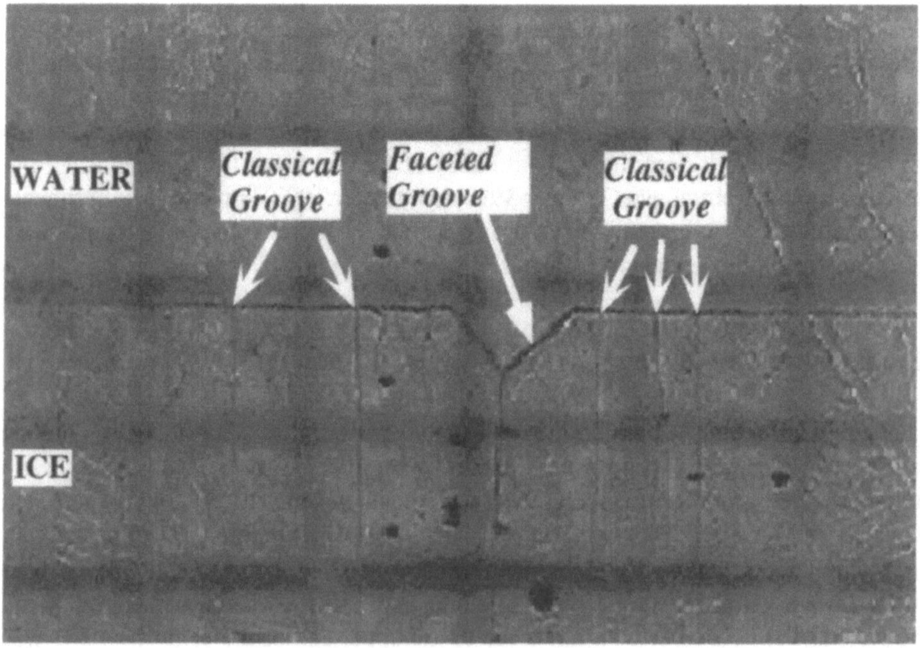

Fig. 1. Video image of a giant faceted grain boundary. Size can be judged by the millimeter scale background. Ice is growing upward and the mean ice/water interface is the dark horizontal line along which $T = T_m$ and where the classical grain boundary groove's terminate. The giant facet exposes a region of supercooled water.

Giant faceted grain boundary grooves were first discussed by Wilen (1995). Grain boundary grooves at a solid-melt interface are analogous to the meniscus of a liquid that wets the wall of a vessel. The liquid rises to a certain height L to reduce the surface energy by an amount proportional to σL, where σ is the difference between the surface energies of the wall when dry and wet. The rise is opposed by gravitational energy. In the case of a solid-melt interface, the groove is caused by $\Delta\gamma$, the difference between the solid-liquid interfacial energy and the grain boundary energy between the two crystallites. In place of gravitational energy, the depth is opposed by the latent heat of the supercooled liquid within the groove. Typical grooves have rounded faces and depth L given by

$$L \approx \left(\frac{\Delta\gamma T_m}{q_m \nabla T} \right)^{1/2}, \tag{1}$$

where T_m is the melting temperature, q_m is the latent heat of fusion and ∇T is the gradient within the liquid filling the groove. In Wilen's (1995) study of ice freezing in pure water, most of the grooves were of typical rounded shape, with depths consistent with estimates based on Eq.(1). However, a fraction were some 30 times deeper, with planar facets. They were shown to be formed by basal plane facets, nucleated at the cleft at δT below T_m The grooves were studied as they grew over short distances, limited by the small size of the cell. In the current cell we can observe the grooves as they advance over 50 mm or more. Several facts are concluded from the observations. (i) The facets are dislocation-free. (ii) The facets grow by individual monolayers nucleated independently on each facet. (iii) The grooves are typically anisotropic, the facets being inclined at different angles with respect to the gradient. (iv) The anisotropy tends to cause the growth to deviate to one side, yet the grooves track along the gradient. (v) The rectilinear track is due to the stiffness of the grain boundary.

Continuing study will examine the conditions for the development of the giant grooves, and further measurements of their properties.

Acknowledgments. We acknowledge extremely useful conversations with L. A. Wilen. This research is supported by Army Research Office grants DAAH34-96-1-0433 and DAAH004-96-1-0290 and NSF grant OPP95-23513.

References

Wettlaufer, J. S., Worster, M. G., & Huppert, H. E., *J. Fluid Mech.* **344**, 291–316 (1997).

Wilen, L.A. & Dash, J.G., *Science* **270**, 1184–86 (1995).

Worster, M.G. & Wettlaufer, J. S., *J. Phys. Chem. B* **101**, 6132–36 (1997).

Snowpack Accumulation Trends in California

Tammy Johnson[1], Jeff Dozier[2] and Joel Michaelsen[3]

[1] Department of Geography, University of California, Santa Barbara, CA 93106-4060, USA
[2] Donald Bren School of Environmental Science and Management, University of California, Santa Barbara, CA 93106, USA
[3] Department of Geography, University of California, Santa Barbara, CA 93106-4060, USA

Keywords. Climate change, snow, California, water resources, Sierra Nevada

1 Introduction

Mountainous areas, particularly in the western United States, provide a large fraction of the fresh water supply. This reserve, which supplies most of California's growing water needs, is vulnerable to changes in climate. Regional precipitation patterns, especially snow, which is a sensitive indicator of temperature and precipitation changes, are predicted to vary according to future climate models.

Utilizing 3 General Circulation Models (GCMs), monthly California temperatures were predicted to increase an average 3.8°C in a doubled CO_2 atmosphere (Lettenmaier and Gan, 1990). A site at 2813m in the Sequoia National Forest was modeled using these GCM outputs, coupled with an energy-based snowmelt model. Predicted snowmelt runoff hydrograph changes ranged from 19 to 93 days earlier (Tsuang and Dracup, 1991).

Analyses of climatic and hydrologic time series in the Sierra Nevada indicate recent trends toward increased winter temperatures, increased March fractional flows to annual totals (Dettinger and Cayan, 1995), reduced April through July fractions (Roos, 1991) and unchanged April through July total flows (Wahl, 1991). These findings suggest earlier snowmelt runoff and possibly increased annual precipitation. However, specific changes are unclear since streamflow integrates snow accumulation and melt throughout the season and across all elevations of entire river basins. This study uses a statistical model which links snow water equivalent (SWE) measurements over a 50-year time-series to analyze spatial snow accumulation trends in the Sierra Nevada at different elevations.

NATO ASI Series, Vol. I 56
Ice Physics and the Natural Environment
Edited by John S. Wettlaufer, J. Gregory Dash and Norbert Untersteiner

2 Procedure

2.1 Snow Course Data

We utilized all available California Cooperative Snow Survey data, which included more than 50,000 measurements spanning 81 years from 393 snow course stations extending over 3450 meters in elevation, 9(latitude and 7(longitude. Since these data were not collected for time-series analyses, there were several data difficulties including inconsistent monthly sampling, added and removed stations, possibly moved or otherwise altered snow courses and earlier sample collection timing.

Stations were selected for time-series analyses that consisted of at least 10 years of measurements within a period of at least 30 years. In addition, monthly analyses required 4 observations per station-month throughout the record while seasonal maximum analyses required at least 3 measurements per station-year. All measurements were normalized to station averages.

2.2 Inconsistent Sampling

Analyses were separated into 100-meter elevation bins to both diminish station sampling bias and to exploit the spatial snow course clarity that is missing from the river data. To determine the effects of a monthly and irregular sampling schedule we analyzed daily snow sensor data from 47 snow pillows that spanned 25 years. To check the effects of sample timing in days from the first of the month, we compared each day within all the station-years' maximum months to the remaining station-years' single monthly SWE measurements. The maximum SWE value and month of melt were again determined and differences from the first-of-the-month results were plotted by days from the first.

2.3 Discontinuous Stations

To find possibly inconsistent stations, we generally followed a procedure for detecting discontinuities in historic temperature data (Easterling and Peterson, 1995). Stations were grouped by river basins and elevation bins with at least 5 nearby, highly correlated reference stations. Group averages were subtracted from station data to create standardized variances from group means. For every station-year, the variances before and after that year were summed and the difference was calculated. These differences were plotted for every station to look for "jumps" which indicate that something about the measurements for that station changed when neighboring stations did not.

Potential problem stations were identified if there was both a trending as well as an extreme difference in the mean, that was defined to be a jump at least two standard deviations from the average jump. 33 stations meeting these criteria are distributed throughout the range at all elevation bins and the variances from the group means trend positively about as often as negatively.

3 Results

3.1 Sample Timing Effects

Monthly sampling causes a false earlier maximum SWE timing trend of 0.18 days over 50 years. Although 80% of the monthly sample days identify the correct maximum SWE month, monthly sampling creates melt timing errors with 3.4 times more erroneous earlier melt months than false later ones. The primary reason is that additional SWE melts before the next monthly measurement is taken.

Monthly sample timing has very little effect on measured maximum SWE. The average erroneously reduced amount is 0.3% per day sampled after the first. However, there is a greater effect on the calculated timing of maximum SWE. An erroneous 0.7 day earlier melt trend over the 50-year, 2.7 day earlier sampling, is due to the changing sample schedule. These monthly sampling effects total to an average false trend of 0.9 days. However, when broken into elevation bins, the greatest schedule change of 7.2 days amounts to a false earlier melt of 2.1 days.

3.2 Range-Wide Trends

We found progressively larger decreasing SWE trends for the lower elevations throughout the season, whereas March 1 and April 1 trends are increasing at higher elevations (Figures 1 and 2). Areas below 2400m are accumulating an average 11% less snow and melting 8 days earlier whereas higher elevations have gained 9% more snow over 50 years. Local variations in snowmelt timing at higher elevations result in no average change. Since seasonal maximums typically occur near the April 1 measurement, the maximum SWE trends graph looks similar (Figures 3 and 4). There are tighter confidence intervals for the April 1 results because there are more data.

All analyses of elevation bins below 2400 m indicate a strong elevational component to SWE accumulation, indicating that lower elevation snow losses are probably a result of warmer winter temperatures and rain-on-snow events. Increased April 1 and maximum SWE values at higher elevations suggest increased precipitation in these areas. This could be the result of warmer air masses having higher moisture contents.

4 Acknowledgements

T. Johnson wishes to thank NASA's Earth Observing System for funding this research.

5 References

Dettinger, M.D., and D.R. Cayan, 1995: Large-scale atmospheric forcing of recent trends toward early snowmelt runoff in California, Journal of Climate, 8, 606–623.

Easterling, D.R., and T.C. Peterson, 1995: A New Method of Detecting Undocumented Discontinuities in Climatological Time Series, International Journal of Climatology, 15, 369–377.

Lettenmaier D.P. and T.Y. Gan, 1990: Hydrologic Sensitivities of the Sacramento-San Joaquin River Basin, California, to Global Warming, Water Resources Research, 26 (1), 69–86.

Roos, M., 1991: A Trend of Decreasing Snowmelt Runoff in Northern California, Proceedings of the 59th Western Snow Conference, Juneau, Alaska, 29–36.

Tsuang, B.J. and J.A. Dracup, 1991: Effect of Global Warming on Sierra Nevada Mountain Snow Storage, Proceedings of the Western Snow Conference, Juneau, Alaska, April 12–15, 1991, Fifty-Ninth Annual Meeting, 17–27.

Wahl, K.L., 1991: Is April to July Runoff Really Decreasing in the Western United States?, Proceedings, 59th Annual Western Snow Conference, April 12–15, 1991, Juneau, Alaska, 67–78.

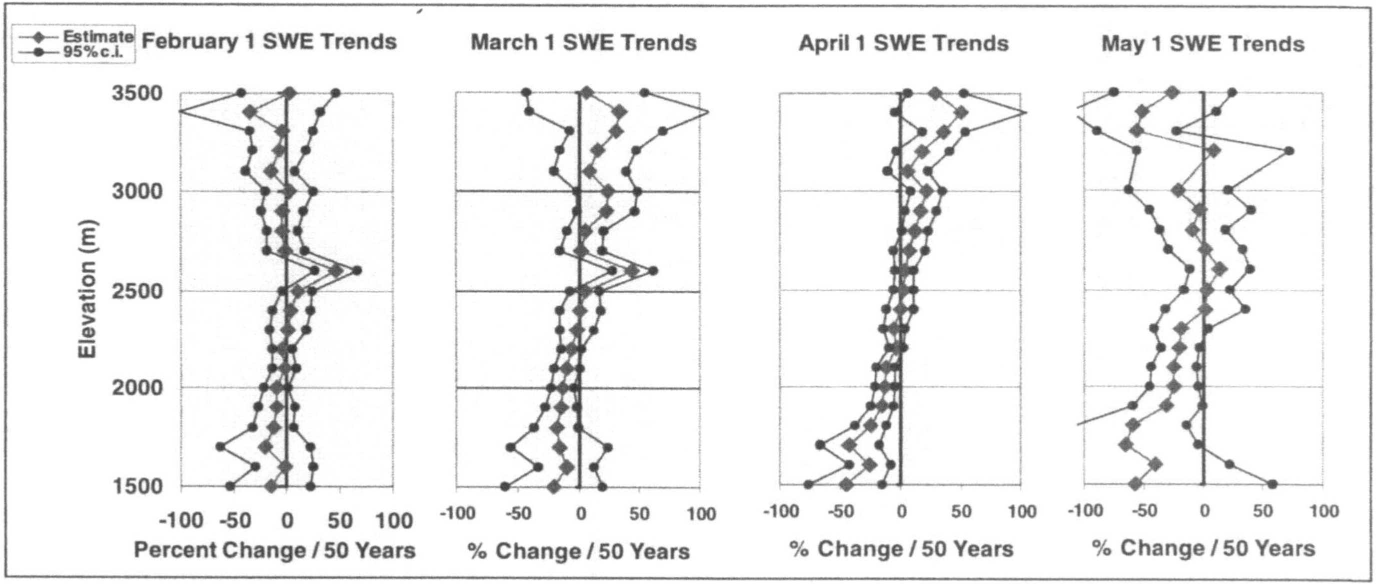

Fig. 1. Snow accumulation trends depend on elevation and month.

SWE % Change Per 50 Years		
Month	**<2400m**	**>2400m**
February	-7	0
March	-11	18
April	-19	18
May	-34	-13

Fig. 2. Average monthly SWE percent changes above and below 2400 meters per 50 years.

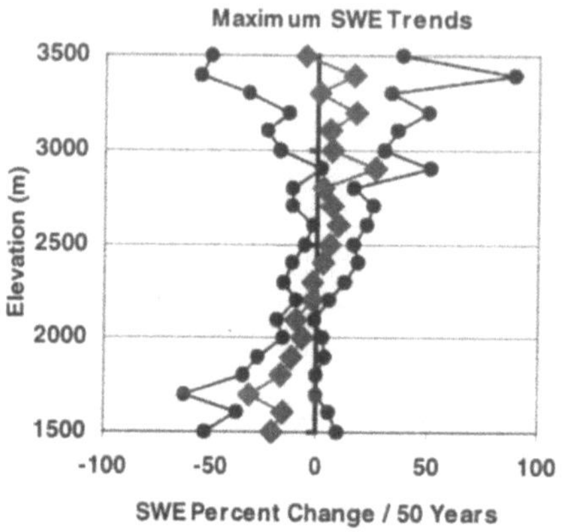

Fig. 3. Maximum SWE trends.

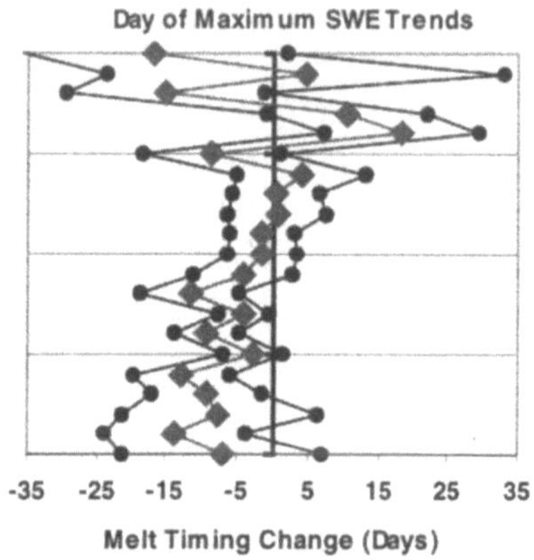

Fig. 4. Snowmelt timing trends.

Neutron Spectroscopy of Vapour Deposited Amorphous Ice

A.I. Kolesnikov and J.-C. Li

Department of Physics, UMIST, PO Box 88, Manchester, M60 1QD, UK

Abstract. Neutron scattering data for low-density amorphous ice, produced by low-temperature and low-rate vapour deposition, shows that the dynamical properties of the vapour deposited amorphous ice are different from other low-density ices.

Keywords. Amorphous ice, vibrational spectra, neutron scattering

1 Introduction

The properties of amorphous ice (am-ice) are important for the study of ice forms in astrophysics, as well as for better understanding of behaviour of water in its liquid state. High-density amorphous (hda) ice (1.31 g/cm^3) can be produced by pressurising ice-Ih above 10 kbar at T<140 K [1]. It can be recovered after releasing the pressure at 77 K (1.17 g/cm^3). The recovered high-pressure phase ice-VIII can be also transformed to hda-ice by warming to ~130 K [2]. By further heating, the hda-ice transforms to low-density amorphous ice (lda-ice, 0.94 g/cm^3) [1]. Other ways of making the am-ice are by vapour deposition (dep-ice) onto a cold surface [3-6] or by hyperquenching of m-size water droplets [7]. The deposition usually gives rise to the lda-ice [5,8,9]; however, hda-ice was also detected [3,6]. The hda-ice produced by deposition was only observed by X-rays [3] and electron diffraction [6], while the lda-ice was observed by neutron diffraction (ND) [4,5,8]. Electron diffraction and X-rays see mainly surface of samples, while ND sees the bulk material. We supposed in [8] that the hda-ice, observed in [3,6], represented only the surface properties of the sample.

We present now the inelastic neutron scattering (INS) studies of dep-ice in wide range of neutron energy transfers, 0-500 meV.

2 Experiment and Results

To prepare the sample, water vapour (doubly distilled H_2O and 99% D_2O) was condensed on the surface of an aluminum sample-can at T<15 K. The deposition was carried out with a flow-rate of 3 and 6 m/h for two H_2O samples (for INS measurements at TFXA and HET spectrometers [10], respectively), and of 9 m/h for D_2O. The

NATO ASI Series, Vol. I 56
Ice Physics and the Natural Environment
Edited by John S. Wettlaufer, J. Gregory Dash and Norbert Untersteiner

energy resolution of the spectrometers is ~2%. HET can also perform ND, which allowed us to measure the structure and vibrational spectra of the samples.

Fig. 1 shows the ND patterns for the H_2O and D_2O dep-ice, and after annealing to 185 and 255 K (for D_2O sample). The data represents the superposition of structure factors from the ice and aluminum sample-can (maximum d-spacing for Al is 2.34 Å). The position of the right-most peak for dep-ice sample is equal to d3.7 Å, which is in good agreement with the literature data for lda-ices [1,2-5,7].

The characteristic value of INS spectrum for ices is the position of its low-energy cut-off for the librational band. In the present spectra (Fig. 2) this position and also the whole librational band of the dep-ice is shifted by ~3.5 meV to lower energy compare to ice-Ih. Below the first acoustic peak, an additional low-energy excitations is seen (Fig. 3a) for dep-ice compare to ice-Ih (after annealing to 215 K). The first peak is observed at about the same energy, ~7 meV, but the peaks at around 20, 28 and 37 meV in ice-Ih are missing for the dep-ice (curve 1 in Fig. 3a). Absence of long range order in the molecular packing of the dep-ice is supposed to be present, this results in featureless behaviour of this part of the spectrum.

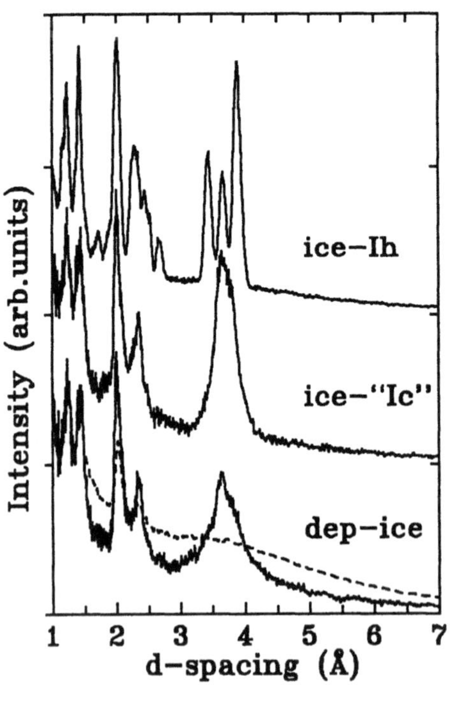

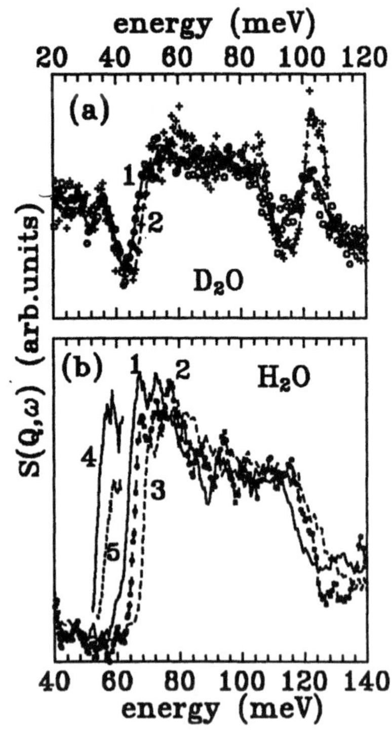

Fig. 1 (left). ND patterns (HET, T=5 K) of the D_2O (solid curves) and H_2O (dashed curve) ice.

Fig. 2 (right). The librational part of INS spectra (T=5 K) for: (a) D_2O dep-ice (solid curve-1 with points) and ice-Ih (dashed curve-2 with crosses) measured on HET; and (b) H_2O dep-ice (solid curve-1) and after annealing to 170 K (long-dashed curve-2 with points) measured on TFXA; the spectra of ice-Ih (short-dashed curve-3) and part spectra of the high-pressure phase ice-VI (solid curve-4) and hda-ice (short-dashed curve-5) [11,12] are shown for comparison.

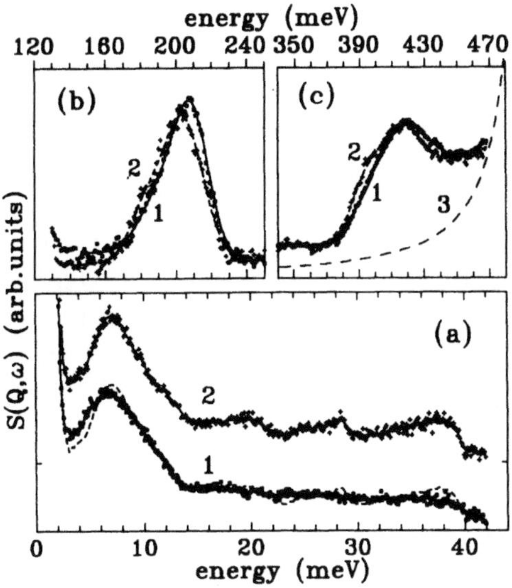

Fig. 3. The INS spectra (HET, T=5 K) for H_2O dep-ice (solid curve-1 with points) and ice-Ih (dashed curve-2 with crosses) in the range of: (a) inter-molecular translational vibrations (the plotted spectra are shifted along the y-axis, for better comparison the spectrum 2 is also plotted over the spectrum 1 as a dashed curve); (b) bending modes; and (c) stretching modes (an estimated time-independent background is shown as a long-dashed curve-3).

The bending mode peak (Fig. 3b) for dep-ice is observed at higher energy by 5.9 meV, compares to ice-Ih. The average position of the stretching mode peak (Fig. 3c) for ice-Ih is at 414.1 meV, that is ~4.4 meV softer than for the dep-ice. This is ~1.6 times larger than the similar shift, observed in the Raman spectra [9]. The shift can be attributed to difference in oxygen-oxygen separation, $R_{O\text{-}O}$. Using an empirical relation between the energy of stretching mode and $R_{O\text{-}O}$ value in ice [13], it is found that the $R_{O\text{-}O}$ in the dep-ice is by 0.03 Å longer than in the ice-Ih ($R_{O\text{-}O}$ 1.79 Å).

These data clearly indicate that the low-temperature and low-rate water vapour deposition has produced a lda-ice, which differs from ice-Ih and hence from another lda-ice, obtained by heating to ~130 K the hda-ice (their INS spectra are similar [14]).

References

1. O. Mishima, L.D. Calvert, and E. Whalley, *Nature* **310** (1984) 393; **314** (1985) 76.
2. A.M. Balagurov, O.I. Barkalov, A.I. Kolesnikov *et al.*, *JETP Lett.* **53** (1991) 30.
3. A.H. Narten, C.G. Venkatesh, and S.A. Rice, *J. Chem. Phys.* **64** (1976) 1106.
4. J. Wenzel, C.U. Linderstrom-Lang, and S.A. Rice, *Science* **187** (1975) 428.
5. M.R. Chowdhury, J.C. Dore, and J.T. Wenzel, *J. Non-Cryst. Solids* **53** (1982) 247.
6. P. Jenniskens and D.F. Blake, *Science* **265** (1994) 753.
7. A. Hallbrucker, E. Mayer, L.P. O'Mard, *et al.*, *Physics Letters* A **159** (1991) 406.
8. A.I. Kolesnikov, J-C. Li, S. Dong *et al.*, *Phys. Rev. Lett.*(1997) in press.
9. T.C. Sivakumar, D. Schuh, M.G. Sceats, and S.A. Rice, *Chem. Phys. Lett.* **48** (1977) 212.
10. 1992 ISIS User Guide to Experimental Facilities, SERC, ed. B. Boland, S. Wapham, 75 pp.
11. J.C. Li, J.D. Londono, D.K. Ross et al., J. Chem. Phys. **94** (1991) 6770.
12. J. Li, J. Chem. Phys. **105** (1996) 6733.
13. D.D. Klug and E. Whalley, J. Chem. Phys. 81 (1984) 1220.
14. J.-C. Li and D.K. Ross, Nature **365** (1993) 327.

Comet 46P/Wirtanen: The Influence of Grain Sintering on the Evolution Layer

Konrad J. Kossacki[1] and Sławomira Szutowicz[2]

[1] Institute of Geophysics of Warsaw University, Pasteura 7, 02-093 Warsaw, Poland, E-mail: kjkossac@fuw.edu.pl

[2] Space Research Center of Polish Academy of Sciences, Bartycka 18a, Warsaw, Poland

Abstract. Cometary nuclei are composed mostly of water ice and mineral grains. When a comet passes near the Sun, several processes modify the subsurface layer of the nucleus which, according to theory, produce more cohesion and stratification. Presently a Rosetta mission is being prepared to comet 46P/Wirtanen. Plans are to land on the nucleus and penetrate its subsurface layer with a mechanical tool (experiment Mupus). Thus, understanding of the processes responsible for the evolution of a cometary nucleus is of key importance for the success of the mission and for interpretation of the results. This work is intended to estimate how quickly the process of grain sintering could modify the outer part of nucleus. The numerical model is adopted from previous work by Kossacki [1]; however, at present the evolution of the cometary orbit is included.

1 Introduction

At large distances from the Sun, a cometary nucleus is cold and can evolve only owing to absorbtion of cosmic radiation. However, as the comet approaches the Sun, the surface of its nucleus is warmed, and various processes start modifying the properties of the cometary matter. Some of them tend to increase the cohesiveness of the material, but others have the opposite result. Cometary simulation experiments predict that the metamorphic processes increasing the strength of granular ice should dominate over those that reduce its cohesiveness [3].

Strengthening of porous cometary ice can be caused by the migration of vapor in a thermal gradient (pore migration [4]). When the temperature at the comet's surface is higher than deeper in the nucleus, this process leads to a gradual growth of density in the subsurface layer. This effect is enhanced by the presence of a dust mantle, which increases temperature and vapor pressure.

Another important process responsible for hardening of the cometary ice is sintering of the ice grains [5] leading to the continuous growth of the grain-to-grain contact area. It has been already demonstrated [5], [2] that this process has a significant influence on the evolution of cometary-like material under laboratory conditions. According to [2] the sintering of ice grains can be much

NATO ASI Series, Vol. I 56
Ice Physics and the Natural Environment
Edited by John S. Wettlaufer, J. Gregory Dash and Norbert Untersteiner

more significant than vapor migration in strengthening cometary matter. Therefore, it is interesting to analyze the influence of grain sintering on the evolution of a real cometary nucleus. The modeling [1] assuming a non-evolving orbit predicts the formation of a cohesive subsurface layer up to about 10 meters thick. The calculations accounting for the evolution of the cometary orbit are presented here.

2 Description of the Model

The cometary nucleus is composed of materials of different volatility. Thus, when the surface of the nucleus is heated, its outer layer becomes depleted of the volatile components. This is because of sublimation followed by partial escape to space and partial migration toward the centre of the nucleus. Since the less volatile component of a cometary matter is mineral dust, the stratification process leads to the formation of a possibly very porous [6], [7], [2] dust cover on the surface of the nucleus. The dust mantle evolves in a very complex way. The sublimation of volatiles removes nonmineral components from the cometary matter below the dust layer, which makes the dust cover grow from the bottom. However, the sublimated vapor flowing out to space blows out the mineral grains. This results in the continuous loss of the mantle mass. Altogether, the escape of ice and mineral components of the cometary nucleus leads to the gradual decrease of its radius.

The aim of this work is to analyze how quickly the process of grain sintering can modify the cometary nucleus and to what depth below the dust mantle the increase of cohesiveness can extend. The performed analysis is similar to that presented in [1]; however, in the present work, no assumption of a non-evolving orbit is made. The orbital motion of the comet is now calculated according to the forced precession model of the cometary nucleus. Nonlinear variations in the direction of the spin axis with time are caused by changing reactive force acting on the oblate nucleus [8]. The forced precessional model adopted to orbital calculation allows [9] determination of parameters describing the precession rate simultaneously with corrections to the orbital elements obtained from the observational equations by the least-squares method in an iterative process of orbit improvement. To find the orbital solution describing 50 years of the comet's motion, 107 positional observations were used. The equation of motion was numerically integrated using a method of recurrent power series. It should be noted that two close approaches of the comet to Jupiter in 1972 and 1984 substantially diminished the perihelion distance.

The evolution of the material texture and temperature distribution below the dust layer is simulated using a self-consistent model of the thermal and structural evolution of an initially homogeneous ice/dust mixture. This model was originally designed for interpretation of the laboratory experiments and has been found to be successful in simulating the observed evolution of cometary-like materials (water ice, water ice + CO_2 ice, water ice + CO_2 ice + mineral dust) in a vacuum chamber [2]. The same model is used In the present work.

It is assumed that the dust layer is eroding at the same rate as it is growing owing to the movement of its bottom (sublimation front of water ice) toward the center of cometary nucleus. The thickness of the dust mantle is therefore constant. The surface temperature is calculated according to the energy balance equation accounting for the absorption of solar radiation, the IR emission, and heat conduction through the dust mantle.

3 Results

The cometary matter is assumed to be composed of water ice and dust with, in volume, a proportion of 4:1. This corresponds to almost equal mass concentrations (see, e.g., [10]). The initial porosity is $\psi_0 = 0.5$, and the initial Hertz factor (the ratio between the grain-to-grain contact area and the cross-section of a grain) is $h_0 = 0.00$ [11], [5]. The radius of grains r_g, equal to the average radius of pores r_p, ranges from 0.5 mm to 0.1 mm. The initial temperature of the nucleus is $T = 20\,\mathrm{K}$ [10]. The porosity of the dust mantle is equal to 0.9, which is the value obtained in sticking experiments [6]. The thickness of the mantle L is $1\ \mathrm{cm} \leq L \leq 10\ \mathrm{cm}$. The thermal conductivity of the dust layer λ_d is $0.1\ \mathrm{mW\,m^{-1}\,K^{-1}} \leq \lambda_d \leq 0.12\ \mathrm{W\,m^{-1}\,K^{-1}}$ [12].

In Fig. 1 results are presented concerning the evolution of profiles of the Hertz factor. This parameter is a useful measure of cohesiveness of the material since it is proportional to the compressive strength of a granular medium. The profiles shown were obtained for $\lambda_d = 0.042\ \mathrm{W\,m^{-1}\,K^{-1}}$. The modeling concerns the period from the discovery of the comet until the present time.

It can be seen that already after one perihelion passage a significantly strengthened layer is formed. It has an easily distinguishable bottom which moves toward the centre of the nucleus, with the rate decreasing during few orbital periods. In [1], where the non-evolving orbit was considered, it was found that after several orbital periods the rate of this movement became equal to the rate of shrinking of the cometary radius, and the cohesive outer layer reached a stable thickness. The thickness of the modified layer depended on the size of the ice grains.

For $r_g = 0.5\,\mathrm{mm}$ the cohesive subsurface layer was less than 1 m thick, but for $r_g = 0.02\,\mathrm{mm}$ it exceeded 10 meters. In the present work we have found that the evolution of the strengthened layer depends also on the changes of the orbit. The initial increase of the thickness of the cohesive layer is followed by some decrease. This effect is especially prominent when the grains are big ($r_g = 0.5\,\mathrm{mm}$) or the dust layer is thin ($L = 2\,\mathrm{cm}$). The resulting present-day thicknesses of the cohesive outer layer of the nucleus are close to their values obtained in [1] for the non-evolving orbit.

It should be noted that because of sintering very small grains cannot last for a long time at the high temperatures characteristic of the cometary subsurface layer during approaches to the Sun. In real ice, where some distribution of grain sizes could be expected, smaller grains disappear and bigger ones grow. If the grains were initially very small, the subsurface layer should be characterized by an increased value of r_g. Thus, the assumption of grain size being constant with a very small value gives an upper estimate of the thickness of the cohesive subsurface layer of the nucleus.

The increase of density due to vapor migration was found to be confined to a layer about 1 m thick just below the dust mantle.

Concluding, it can be expected that sintering of ice grains leads to the formation of a cohesive subsurface layer much thicker than a layer of density increased due to migration of vapor in a thermal gradient.

References

[1] Kossacki K.J 1998. Comet P/Wirtanen: evolution of the subsurface layer. In *Proceedings of the IXt Rencontres de Blois* Editions Frontieres

[2] Kossacki K.J., N.I. Kömle, J. Leliwa-Kopystyński and G. Kargl 1997. Thermal and structural evolution of cometary subsurface layer: self-consistent model and experimental verification. *Icarus*, **128**, pp. 127–144

[3] Grün E., A. Bar-nun, J. Benkhoff, A. Bischoff, H. Düren, Hellmann, P. Hesselbarth, P. Hsiung, H.U. Keller, J. Klinger, J. Knölker, H. Kochan, H. Kohl, G. Kölzer, D. Krankowsky, P. Lammerzahl, K. Mauersberger, G. Neukum, A. Öhler, L. Ratke, K.Rössler, G. Schwehm, T. Spohn, D. Stöffler and K. Thiel 1991. Laboratory simulation of cometary processes: Results from KOSI experiments. In *Comets in the Post–Halley Era*, Vol. 1 (edited by R. L. Newburn, M. Neugebauer and J. Rahe), pp. 277–297. Kluwer, Dordrecht.

[4] Smoluchowski R., M. Marie, and A. McWilliam 1984. Evolution of density in solar system ices. *Earth, Moon, and Planets,* **30,** 281–288

[5] Kossacki K.J., N.I. Kömle, G. Kargl and G. Steiner 1994. The influence of grain sintering on the thermoconductivity of porous ice. *Planet. Space Sci.*, **42**, No. 5, pp. 383–389

[6] Hapke B. and H. Van Horn 1963. Photometric studies of complex surfaces with applications to the Moon. *J. Geophys. Res*, **68**, 4545

[7] Dominique D.L., G.W. Lockwood and D.T. Thompson 1995. Surface textural properties of icy satellites: a comparison between Europa and Rhea. *Icarus,* **115**, pp. 228–249

[8] Sekanina Z. 1984. Precession model for the nucleus of periodic comet Kopff. *Astron. Journal,* **89**, pp. 1573–1585

[9] Krolikowska M. Sitarski G. 1996 Evolution of the orbit of comet 46P/Wirtanen during 1947–2013. *Astron. Astrophys.*, **310**, pp. 992–998

[10] Podolak M. and D. Prialnik 1996. Models of the structure and evolution of the comet P/Wirtanen. *Planet. Space Sci.*, **44**, No. 7, pp. 655–664

[11] Wechsler A.E. and P.E. Glaser 1965. Pressure effects on postulated lunar materials. *Icarus,* **4**, pp. 335–352

[12] Kömle N.I. and G. Steiner 1992. Temperature evolution of porous ice samples covered by a dust mantle. *Icarus*, **96**, pp. 204–212

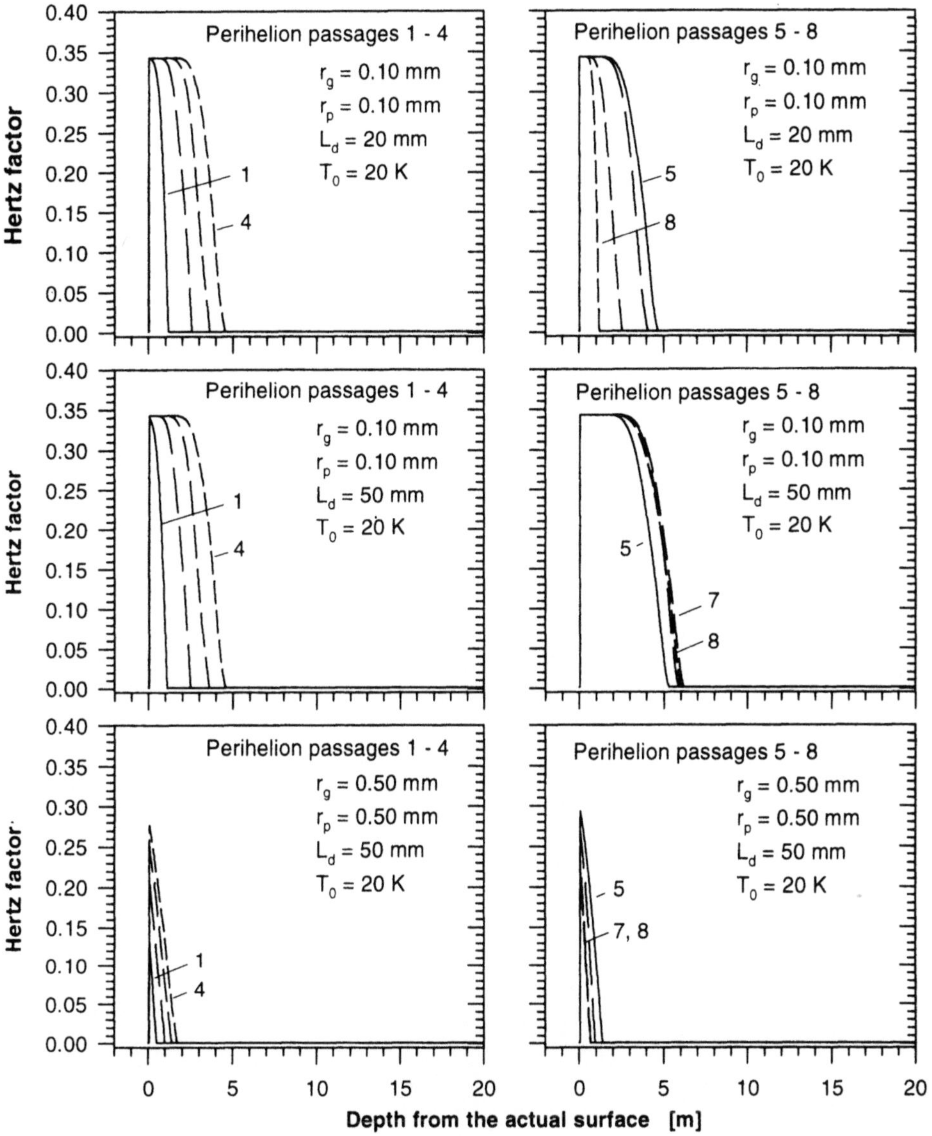

Fig. 1. Evolution of profiles of Hertz factor versus depth, calculated for the evolving orbit; see text. The curves are drawn for consecutive perihelion passages.

Water Ice as the Main Component of Icy Satellites

Jacek Leliwa-Kopystynski

Institute of Geophysics, University of Warsaw, ul. Pasteura 7, 02-093 Warszawa, Poland; and Space Research Center, Polish Academy of Sciences, ul. Bartycka 18A, 00-716 Warszawa, Poland
E-mail: jkopyst@fuw.edu.pl

Abstract. Water ice consists about a half of mass and therefore about 0.75 of volume of most of the icy satellites. Differentiated, with water ice forming outer shells, and undifferentiated models of internal structure of the icy satellites of the giant planets are mentioned. It is stressed that the modelling of the evolution of satellites' structure should be supported by laboratory experiments: (i) concerning rheology of compaction of icy/mineral granular porous media, and (ii) concerning kinetics of phase transitions of water ice in these media.

Keywords. Icy satellites, kinetics of phase transitions in ices, porous ices, rheology of ices

1 Introduction

The term "Solar System ice" denotes in general a solidified volatile and/or mixtures of solidified volatiles. The Solar System ices are mostly water ice H_2O, but also solidified CO_2, CO, NH_3, N_2, SO_2, CH_4 and many other simple molecules as well as the organics. On the surfaces of many of the satellites of the giant planets (Jupiter, Saturn, Uranus, and Neptune) water ice is the dominant component. Therefore, in the following we will adopt the common assumption that a satellite is composed of water ice, and silicates. The radii and the masses of all satellites but the smallest ones are well known. Therefore their densities are known as well, see Table 1. Taking into account that the densities of water ice are about 940, 1190, and 1360 kg m^{-3} at phases I, II, and VI, respectively, and that the density of the silicates is (3400 ± 400) kg m^{-3} the mass ratio C of silicates to total mass of the satellite can be estimated. This is rather simple for large satellites; however, the estimate can fail for smaller satellites because of the possible bulk primordial porosity left from an epoch of formation.

NATO ASI Series, Vol. I 56
Ice Physics and the Natural Environment
Edited by John S. Wettlaufer, J. Gregory Dash and Norbert Untersteiner

Table 1. Icy satellites and Pluto, ranged according to their radii (compiled according to [2], [4] and [5]). The table contains all currently known icy satellites having the size of the Neptunian satellite Nereid or larger. Only two satellites smaller than Nereid are also listed. They are irregular in shape, therefore their "radii" are related to the equivolumetric spheres. The Differentiated 2-, 3-, and 4-Layered models (D2L, D3L, D4L) and Non-Differentiated (ND) i.e., 1-Layered model are considered. If considerable global porosity can be or even must be taken into account a model is denote by letter ψ ("must" is for a case when the mean density of a satellite is less than density of ice and therefore existing of the pores is necessary for mass balance). The largest satellites from Ganymede to Triton, as well as Pluto are undoubtedly differentiated; moreover some of them e.g. Ganymede has metallic core [6]. The medium-size satellites from Rhea to Thetys are possibly differentiated. Differentiation of Enceladus and smaller satellites is doubtful. The smallest satellites, with densities lower than that of water ice, must be globally porous. Most probably they are undifferentiated bodies. The rough estimates for the maximal value of central temperature are related to post-accretional distributions of temperatures when infalling energy transforms into internal energy. The heating due to the decay of radionuclides distributed within minerals is not taken into account.

Satellite	Radius R km	Density ρ kg m^{-3}	Model	Mineral concentr. by mass C	Radius of a sphere corresponding to phase transition I ⇒ II km	II ⇒ VI km	Core radius r_{core} km	Central pressure p_{centr}	Rough estimate of central temp. K
J3 Ganymede	2631	1930	D4L	0.578	2484	2259	1811	6.93 Gpa	
S6 Titan	2575	1880	D4L	0.564	2420	2112	1743	6.35	
J4 Callisto	2400	1830	D4L	0.550	2230	1894	1596	5.35	
J2 Europa	1569	3040	D2L	0.953	----	----	1484	3.71	
N1 Triton	1360	2060	D3L	0.733	1108	----	1035	1.97	
Pluto	1195	1900	D3L	0.691	893	----	868	1.46	
U3 Titania	788.9	1700	D2L	0.620	----	----	536	0.55	
S5 Rhea	765	1330	D2L	0.407	----	----	416	0.38	
U4 Oberon	761.4	1640	D2L	0.593	----	----	503	0.50	
S8 Iapetus	730	1160	D2L	0.263	----	----	328	0.26	
P1 Charon	640	1100	D2L	0.202	----	----	258	0.17	

Table 1 (cont.)

Satellite	Radius	Density	Model	Mineral concentr. by mass	Radius of a sphere corresponding to phase transition		Core radius	Central pressure	Rough estimate of central temp.
	R	ρ		C	I ⇒ II	II ⇒ VI	r_{core}	p_{centr}	temp.
	km	kg m^{-3}			km	km	km		K
U2 Umbriel	584.7	1520	D2L	0.530	----	----	363	0.27	
U1 Ariel	578.9	1560	D2L	0.552	----	----	368	0.27	
S4 Dione	560	1430	D2L	0.476	----	----	329	0.22	
S3 Tethys	530	1210	D2L	0.310	----	----	255	0.15	
			ND	0.310	----	----	----	0.057	
S2 Enceladus	252	1200	D2L	0.301	----	----	120	40 Mpa	< 200
			ND	0.301	----	----	----	13	
			NDψ	Considerable porosity of the deep layers is possible					
U5 Miranda	235.8	1150	D2L	0.254	----	----	104	27	< 200
			ND	0.254	----	----	----	10	
			NDψ	Considerable porosity of the deep layers is possible					
1989 N1 Proteus	208 ± 5		The most likely range of bulk density is 700 to 2000 kg m^{-3} [4]; therefore:						
			D2L, ND, and NDψ models are possible						< 200
S1 Mimas	199.8	1137	D2L	0.241	----	----	86	18	< 200
			ND	0.241	----	----	----	7	
			NDψ	Considerable porosity of the deep layers is possible					
N2 Nereid	170± 25		The most likely range of bulk density is 700 to 2000 kg m^{-3} [4]; therefore:						
			D2L, ND, and NDψ models are possible						< 150
S10 Janus	89.5	670	NDψ	for any C ≥ 0 a global porosity must be ≥ 0.27				0.5	< 150
S11 Epimatheus	64.4	640	NDψ	for any C ≥ 0 a global porosity must be ≥ 0.33				0.24	< 150

2 Modelling of the Interiors of the Icy Satellites

Let as assume that a satellite was formed by homogeneous accretion from the circum-planetary disk and the mass ratio ice to rock was constant during the accretion. Therefore, as a result of accretion a body with uniform distribution of components was formed. The possibility of post-accretional segregation of components (gravitational differentiation) depends on many factors among which the size of the body is the most important. The smallest globes remained undifferentiated. For the bodies composed of different components of different densities (ice, silicates, iron) the radial distribution of components is the most important factor describing internal structure of the body. Two extreme cases of models are of special importance:

(1) Differentiated model, with denser components forming the central part with radius r_c (possibly differentiated into iron core at the center, surrounded by rocks) and the less dense component (ice) forming the envelope shells extending from r_c to R.

(2) Undifferentiated model with composition independent of central distance.

In both cases the density $\rho(r)$ depends on the local pressure $p(r)$ and temperature $T(r)$. In sufficiently large satellites the density of ice may change stepwise on certain levels corresponding to p, T conditions of phase transitions in ice. In particular, in water ice at about 200 K the phase equilibrium levels for I $\Rightarrow$ II and II $\Rightarrow$ VI transitions are about 0.2 GPa and 0.6 GPa, respectively. On the other hand, the pressure in the interiors of the smallest icy satellites combined with their low internal temperature allow for only very slow creep (slow rheology) of the material forming the satellite. So, if a small satellite have been formed as a porous body it is quite possible that the internal porosity has survived until the present time.

3 Laboratory Experiments Relevant to Ice Properties in Modelling of the Icy Satellites

From Table 1 it is seen that the pressure regimes inside different icy satellites differ by several orders of magnitude. Therefore two types of compaction experiments concerning icy/mineral granular mixtures are of special interest for the study of the evolution of small and of large icy satellites, respectively:

(1) Experiments concerning temperature- and pressure-depending rheology of loose (therefore porous) icy/mineral granular mixtures. The interesting p, T intervals are about 0 – 10 MPa and 80 – 270 K.

(2) Experimental studies of the kinetics of high pressure phase transitions within granular, non porous, icy/mineral mixtures.

Several series of experiments, about hundred individual runs in total, have been done [1]–[3]. The total duration of all runs was longer than a year. The piston-cylinder method was used with samples having volume 30 to 80 cm^3. The temporal changes of the volume of the sample were recorded continuously and therefore the compaction rate was derived. In both types of experiments the required output is compaction rate versus time. First of all it depends on actual temperature T and on rock-to-total mass ratio C. We also varied the granulation (average size of grains) of the samples; however, we have not found its influence on the compaction rate of a porous icy/mineral medium. It was found that in rheological experiments the compaction rate depends on pressure p. In such way the rate of decrease of porosity $d\psi/dt$ is given in the following form:

$$\frac{d\psi}{dt} = -\psi p^{k_1 \psi} \exp\left[k_2 + k_3 \psi + k_4 \psi T + \frac{k_5 + k_6 \ln(1 - C_V)}{T} \right] \quad \mathrm{s}^{-1} \qquad (1)$$

where p is in MPa, T in K, and C_V is the concentration of rock by volume related to nonporous sample. The coefficients are: $k_1 = 9.8$, $k_2 = 8.727$, $k_3 = 54.11$, $k_4 = -0.2625$ K^{-1}, and $k_6 = 6000$ K. Coefficient $k_5 = -4663$ K for water ice; however it is modified in presence of ammonia admixture in water ice. Experiments with ammonia-water ice $(NH_3)_x(H_2O)_{(1-x)}$ for $x \leq 0.28$ was conduced as well [3], since the presence of ammonia ice in satellites cannot be ruled out. Equation (1) was applied to modelling of the evolution of structure of the icy satellites during their accretion ([7], accretion time not longer than 10^6 y) and during post-accretional epoch ([5], up to 10^9 y).

Experiments concerning kinetics of phase transitions of water ice mixed with mineral grains showed that time-dependent concentration of the newly formed phase strongly depends on concentration of the mineral component. The presence of mineral grains makes the phase transition slower. On the other hand, the phase transformation is faster if the pressure difference ($p - p_{eq}$) is larger, where p is actual pressure and p_{eq} is the phase equilibrium pressure at given temperature T.

4 Conclusions

The laboratory study of icy/mineral mixtures in wide range of pressures and in low temperatures is of essential importance for modelling of the icy satellites. However, one must take into account that the duration of an experiment is of the order of days (10^{-2} y), accretion time is of the order of $10^2 - 10^6$ y, and postaccretional evolution can be as long as the age of the Planetary System, 4.6×10^9 y. The laboratory results must be extrapolated over a time interval more than 11 orders of magnitude longer. Therefore any of the experimentally found formulae must be considered with caution when applied to p, T regime which has not been studied. To accelerate rheological processes the p, T conditions of interest (expected within small icy satellites) have to be replaced by laboratory conditions $p_1 > p$ and/or $T_1 > T$. Similarly, to accelerate phase transitions the overpressure $\Delta p = (p - p_{eq})$ required from the point of view of physics of the satellites, has to be replaced in laboratory by considerably higher value $(p_1 - p_{eq})$.

Let us consider an example that underlines the importance of the studies of kinetics. One must realise that the transition of 1 km thick layer of ice I into ice II decreases the volume of ice by about 20% then "leaves empty" a layer of 200 m thick. For the largest satellites (Ganymede, Titan, Callisto) the hydrostatic relation $\Delta p = \rho g \Delta r$ gives pressure gradient $\Delta p/\Delta r \approx (1000 \text{ kg m}^{-3})(2 \text{m s}^{-2}) \approx 2 \text{ MPa km}^{-1}$ at the level of ice I – ice II phase equilibrium (about 150 km below the surface, see Table 1). However, 2 MPa is only 1% above the phase equilibrium pressure, so transition at low temperature may be very slow or metastable state can be expected. This may be related to tectonic activity of a satellite which could be active during a long time interval after its formation (e.g. Ganymede). Present-day activity of Europa confirmed by Galileo mission is still an enigma; undoubtedly this activity concerns the upper icy layers.

References

[1] Leliwa-Kopystynski, J., and N. Maeno. Icy/rocky porous mixtures: compaction experiments and interpretation. 1993, J. of Glaciology, v. 39, pp. 643–655.

[2] Leliwa-Kopystynski, J., L. Makkonen, O. Erikoinen, and K. J. Kossacki. Kinetics of pressure induced effects in water ice/rock granular mixtures and application to the physics of icy satellites. 1994, Planet. Space Sci., v. 42, pp. 545–555.

[3] Leliwa-Kopystynski, J. and K. J. Kossacki. Kinetics of compaction of granular ices H_2O, CO_2, and $(NH_3)_x(H_2O)_{(1-x)}$ at pressures 2 – 20 MPa and in temperatures 100 – 270 K. Application to the physics of the icy satellites. 1995. Planet. and Space Sci., v. 43, pp. 851–861.

[4] Thomas, P. C., J. Veverka and P. Helfenstein. Neptune's small satellites. 1995. In: Neptune and Triton. D. P. Cruishank (ed.). Univ. of Arizona Press, Tucson. Pp. 685– 699.

[5] Kossacki, K. J., and J. Leliwa-Kopystynski. Long term internal evolution of Mimas, Janus and Epimatheus. Submitted to Planet. Space Sci.

[6] Recent data of Galileo mission to Jovian System. 1997, JPL bulletins.

[7] Kossacki, K. J. and J. Leliwa-Kopystynski. Medium-size icy satellites: thermal and structural evolution during accretion. 1993, Planet. Space Sci., v. 41, pp. 729–741.

Surface Melting of Ice and Thunderstorm Electrification

B. L. Mason and J. G. Dash

Department of Physics, University of Washington, Seattle, WA 98195, USA

1 Introduction

In a typical thunderstorm the cloud is mainly electrically dipolar, with the principal positive charge in the upper reaches, and the main negative charge near the base. There is active ice formation due to the strong updraft and adiabatic cooling of humid air; temperatures in the negative zone range from about –5°C to –25°C. In this zone there are two main types of ice; small particles which are being carried upward in the updraft, and larger agglomerates, 'soft hail', which are falling. The principal charging occurs at collision sites between the two types: the small particles come away positively charged, and the hail negatively charged. Then the small particles continue upward, leaving a net negatively charged zone [1]. Lightning discharges occur between the two charge centers in the cloud and between the negative zone and the ground, so that the ground becomes negatively charged. Thunderstorms are most active during afternoons on seacoasts world wide, and hundreds of storms are active at all times; lightning discharges are so continuous that the Earth is more or less constantly negative with respect to the upper atmosphere. Without the storms, the Earth's charge would decay in a couple of hours. So much is known and accepted [2].

The principal mystery concerns the mechanism responsible for the charging during collisions: explanations have included frictional electricity, fracture, thermoelectricity, and induction. Baker and Dash have proposed that the mechanism involves thin liquid films on the colliding particles and ion segregation between the surface liquid and the ice interfaces, then the transfer of charged surface liquid between the colliding particles [3,4].

2 Experimental Setup and Procedure

To test the proposed Baker, Dash mechanism, we designed a setup of two quartz microbalances (QMB's), capable of measuring nanograms of mass, which are contained in a copper environmental chamber connected to a vacuum manifold. The QMB's are positioned facing each other, so that the ice layers which are grown on each QMB are separated by about one millimeter. The lower QMB is mounted on a simple audio speaker, so that by pulsing the speaker once, a single collision can be induced be-

NATO ASI Series, Vol. I 56
Ice Physics and the Natural Environment
Edited by John S. Wettlaufer, J. Gregory Dash and Norbert Untersteiner

tween the two ice layers. The resulting mass transfer, if it exists, may then be registered as discrete shifts of the frequencies of the two QMB's. In separate experiments under similar conditions charge transfer during collisions may be measured using an electrometer with a resolution of 10 fC. A video microscope provides imaging of the two ice layers in order to observe surface features on the ice with a resolution of approximately 10 μm.

The chamber can be cooled to –25°C, and each QMB is individually cooled by small thermoelectric coolers, so that the temperature difference between the ice layers can be controlled. The cooler layer grows by vapor deposition at the expense of the warmer sublimating layer. We measured charge transfer for a series of collisions during periods of such vapor growth, as a function of the differential growth rate of the layers (as measured by the QMB's). For each growth rate, we observed four collisions, separated by 2.5 minutes. Data were recorded for three temperatures: –6°C, –11°C and –16°C.

Under similar conditions at –11°C, we observed mass transfer for a series of collisions during periods of uniform differential growth rates. For each growth rate, the data from one QMB were averaged to reveal the average mass transfer per collision. Such series were performed for a range of growth rates, for both upper and lower QMBs, separately. Also, three series of 'near miss' collisions (without contact) were performed to distinguish whether the observed mass transfer was due to contact, versus enhanced vapor diffusion during the brief time that the layers were in close proximity before and after contact.

3 Results and Discussion

Figure 1 is a plot of charge transfer to the upper QMB vs. differential growth rate (upper - lower) at mean temperature –6°C. One μgm/cm^2 on the horizontal scale corresponds to 33 monolayers of ice. Each point is the average of 4 collisions. The data for –11°C and –16°C are similar, both in sign and magnitude. Experimental charge densities correspond to very dilute surface charges: one nC/cm^2 is roughly one electronic charge in 10^5 surface molecules. We estimated an upper bound for the contact area from the video data, so the charge densities shown in the figure are lower bounds and real densities are probably significantly larger than the scale indicates. Clearly when the upper layer is growing from the vapor it charges positively during collisions, and negatively when it is subliming. When there is no temperature difference between the layers and little or no growth, the charge transfer averages to null. These results are qualitatively consistent with observations by Baker *et al.* [5]. As the growth rate increases in magnitude, so does the magnitude of charge transfer. This does not persist, however, beyond differential growth rates of approximately 25 μgm/cm^{2}min, whereon the charging levels off and tends to diminish. For such extreme growth rates, the growing ice surface became visibly rougher with surface features of order 10 μm. While there is a clear variation of the charge transfer with differential growth rate, there does not appear to be any dependence on temperature for the three temperatures that were observed.

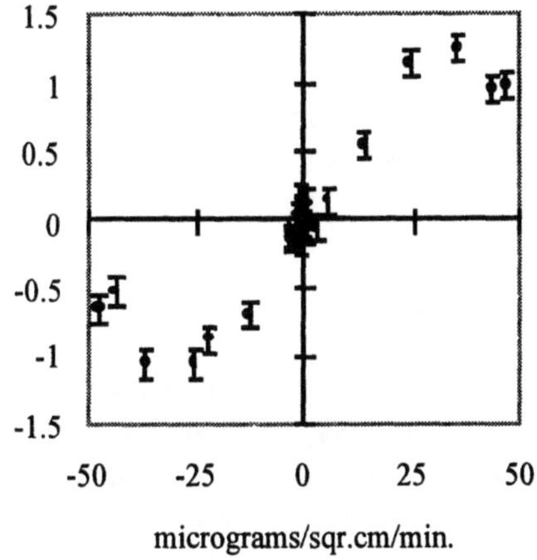

FIG. 1. Charge transfer vs. differential growth rate.

To date, we have observed mass transfer at –11°C. Figure 2 shows the average mass transfer to the upper QMB per collision, in equivalent monolayers, vs. differential growth rate. Although there is an as yet unexplained offset of the data from the origin, it is clear that the mass transfer to the upper QMB increases with increasing differential growth rate. When similar series of data runs were performed with 'near miss', no contact collisions, no mass transfer was observed, thus indicating that the observed mass transfer is due to actual contact, rather than enhanced vapor diffusion just before and after contact when the layers are in close proximity.

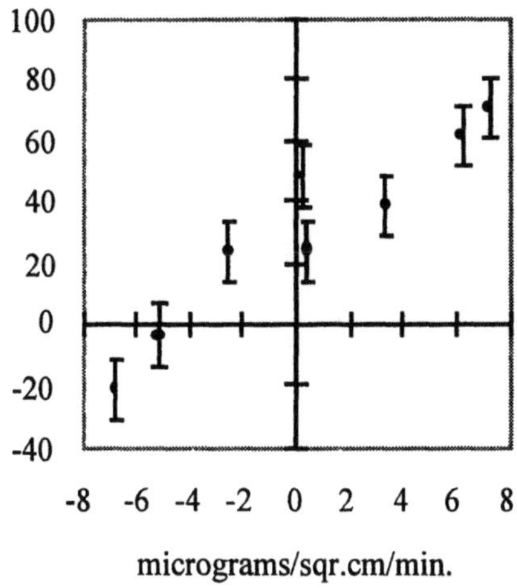

FIG. 2. Average mass transfer vs. differential growth rate

Acknowledgements

This work was supported by a grant from the Leonard X. Bosack and Bette M. Kruger Charitable Foundation and by the ARCS Foundation.

References

[1] Latham, J., *Q. J. R. Meteorol. Soc.* vol. 107, no. 452. pp. 277–298. Apr. 1981.
[2] Williams, E. R., *Sci. Am.*, vol. 259, no. 5. pp. 48–52, 61–5. Nov. 1988.
[3] Baker, M. B. and J. G. Dash, *J. Cryst. Growth*, vol. 97, pp. 770–776. 1989.
[4] Baker, M. B. and J. G. Dash, *J. Geoph. Res.*, vol. 99, no.D5. pp. 10,621–6. 1994.
[5] Baker, B., M. B. Baker, E. R. Jayaratne, J. Latham and C. P. R. Saunders, *Q. J. R. Meteorol. Soc.* vol. 113. pp. 1193–1215. 1987.

A Mathematical Model of Wide Subglacial Water Drainage channels

Felix S. L. Ng

Mathematical Institute, 24-29 St. Giles, Oxford OX1 3LB, ENGLAND

Keywords. Subglacial hydrology, wide drainage channels

1 Introduction

In the classical drainage theory proposed by Röthlisberger [1] and later extended by Nye [2], ice deformation closure of a subglacial channel is counteracted by melt-back of the channel walls, due to the heat dissipated by water flowing in the channel. The mathematical description of this is by means of an evolution equation for the channel cross-section, which has been formulated under the assumption that the cross-sectional shape is circular. For steady flow, the equation reduces to the equilibrium relation derived by Röthlisberger [1], linking water flux, water pressure and hydraulic gradient. For unsteady flow, it has been applied with success to models of glacier outburst floods, or *jökulhlaups*, in Iceland [e.g., 2] and elsewhere. However, it has recently been suggested that this theory can be generalized to the consideration of non-circular cross-sections. For instance, Hooke *et al.* [3] suggested broad and low channels as the cause of why measured subglacial water pressures are frequently underestimated by theory. Fowler & Ng [4] proposed a theory of Icelandic jökulhlaups and an improved model for simulating flood hydrographs, by considering a wide channel that has independently evolving depth and width. This configuration resembles the *canals* in Walder & Fowler's extension of the classical theory for deformable sediment beds [5]. In this paper, the first steps of analyzing the mathematics of *wide* subglacial channels of this type are presented. In particular, we outline the formulation of the evolution equations, for the case of 'hard bed' drainage where sediment processes can be excluded. We then derive the corresponding equilibrium results and discuss their implications.

2 Model Formulation

We consider turbulent water flow in a wide channel with depth $h(x,t)$ and half-width $l(t)$, as shown in Fig. 1a; x is the cross-channel coordinate and t is time. The basic depth evolution equation is

$$\frac{\partial h}{\partial t} = \hat{m} - \hat{w}, \tag{1}$$

where $\hat{m}$ and $\hat{w}$ respectively are the melt- and closure velocities at the roof.

NATO ASI Series, Vol. I 56
Ice Physics and the Natural Environment
Edited by John S. Wettlaufer, J. Gregory Dash and Norbert Untersteiner

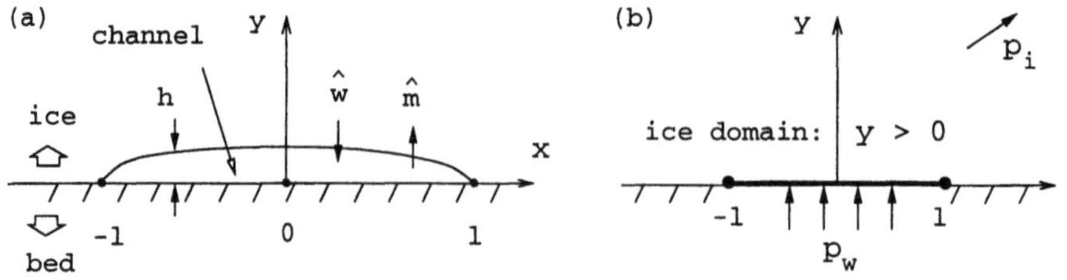

Fig. 1. (a) Definition diagram of the wide channel. (b) The deformation domain in the ice closure problem.

Ice melting. The problem of calculating the heat dissipated by the water flow is tackled by formulating the corresponding mass, momentum and energy equations [6], together with an appropriate turbulence parameterization. If this model is non-dimensionalized with length scales $[h]$ and $[l]$, then one can use a perturbation approximation based on $\delta = [h]/[l] \ll 1$, to show that channel water pressure p_w is essentially independent of x and uniform with y; also, determination of flow velocity and temperature involves solving a pair of elliptic p.d.e.'s with no slip and melting temperature boundary conditions.

Since flow in a wide channel is approximately parallel, a convenient turbulence parameterization to use is the eddy viscosity equation of Prandtl [6]. The melt-velocity, calculated to leading order from the temperature solution, is then

$$\hat{m}(x) \sim \frac{1}{\rho_i L}\sqrt{\frac{8}{\rho_w f}}(\Phi h)^{3/2}, \tag{2}$$

where Φ is the imposed hydraulic gradient, ρ_i and ρ_w are ice and water densities, L is latent heat, and f (≈ 0.01 to 0.1) is the channel friction coefficient [e.g., 6]. The expression in (2) is not valid close to the channel margins, because neglecting δ results in a *singular* approximation of the original p.d.e.'s, and also because the parallel approximation breaks down at the margins. However, crude approximation of the flow field there yields a melt-velocity estimate of the order of $10^{-2}\delta^{3/2}\hat{m}$, which is small. Lateral channel expansion by ice melting is therefore negligible.

Ice closure. Creep closure takes place when the far-field ice pressure p_i exceeds the channel pressure p_w. To calculate an approximate closure rate $\hat{w}(x)$ for the wide channel, one can solve the steady incompressible viscous flow equations in the domain shown in Fig. 1b. With a non-linear constitutive ice relation (e.g., Glen's law), solution is possible only by numerical methods. We circumvent this difficulty by first tackling the linear problem (where ice viscosity η_i is taken to be constant), then generalizing (heuristically) the results for non-linear rheology. The linear problem may be solved by using complex variable techniques developed for 2-D elastic crack problems [7], leading to the singular integral equation

$$N = -\frac{2\eta_i}{\pi}\int_{-l}^{l}\frac{d\hat{w}}{d\zeta}\frac{d\zeta}{\zeta - x}, \quad \text{for } |x| \leq l, \tag{3}$$

where $N = p_i - p_w$ is the effective channel pressure. With p_w (and hence N) constant across the channel, inversion of (3) gives $\hat{w}(x) = N\sqrt{l^2 - x^2}/2\eta_i$. The non-linear extension of this using Glen's law is achieved by putting $\eta_i^{-1} = N^2\eta_0^{-1}$, where the value of η_0 is about 10^{24} Pa^3 s. Lateral channel closure is also investigated, by evaluating $dl/dt = \lim_{x \to l}[\hat{w}/h_x]$ with the family of channel shapes $h = (l^2 - x^2)^\nu$, where $\nu > 0$. We find that stable margin retreat occurs when $\nu = 3/2$, with velocity $N^3 l/6\delta\eta_0$, but this is negligible for normal N values (1 to 10 bar).

3 Results and Discussions

To summarize, our channel evolution equations are

$$\frac{\partial h}{\partial t} = \frac{1}{\rho_i L}\sqrt{\frac{8}{\rho_w f}}(\Phi h)^{3/2} - \frac{N^3}{2\eta_0}\sqrt{l^2 - x^2}, \qquad \frac{dl}{dt} \approx 0. \tag{4}$$

At steady flow ($\partial h/\partial t = 0$), the channel shape $h \propto (l^2 - x^2)^{1/3}$ is predicted, with the constant of proportionality depending on Φ and N. A subsidiary result from the water flow problem is $Q = (8\Phi/\rho_w f)^{1/2}\int_{-l}^{l} h^{3/2}dx$, where Q is water flux; therefore, the equilibrium relation is $Q = \pi\rho_i L N^3 l^2/4\eta_0\Phi$. The width l does not appear to be able to vary in this theory. One way it could do so would be by conversion to a semi-circular Röthlisberger-channel (in a flood, for example) and subsequent widening. With l constant, we obtain $Q \propto \Phi^{-1}N^3$, thus a hard bed wide channel is of Röthlisberger-type (cf. $Q \propto \Phi^{-11/2}N^{12}$; [1]). For deformable beds however, l will be determined by other processes such as sediment erosion and deposition; then our model in (4) will have to be modified.

References

[1] Röthlisberger, H. 1972. Water pressure in intra- and subglacial channels. *J. Glaciol.*, **11**(62), 177–203.

[2] Nye, J. F. 1976. Water flow in glaciers: jökulhlaups, tunnels and veins. *J. Glaciol.*, **17**(76), 181–207.

[3] Hooke, R. LeB., T. Laumann and J. Kohler. 1990. Subglacial water pressures and the shape of subglacial conduits. *J. Glaciol.*, **36**(122), 67–71.

[4] Fowler, A. C. and Ng, F. S. L. 1996. The rôle of sediment transport in the mechanics of jökulhlaups. *Ann. Glaciol.*, **22**, 255–259.

[5] Walder, J. S. and A. Fowler. 1994. Channelized subglacial drainage over a deformable bed. *J. Glaciol.*, **40**(134), 3–15.

[6] Schlichting, H. 1960. Boundary layer theory. McGraw-Hill. New York.

[7] England, A. H. 1971. Complex variable methods in elasticity. Wiley.

High Uptake Efficiency and Conductivity of Polycrystalline Ice: Implication to UT/LS Clouds and Contrails

N.V. Persiantseva, O.B. Popovitcheva, and T.V. Rakhimova

Institute of Nuclear Physics, Moscow State University, Moscow, 119899, Russia

Abstract. The uptake of gases onto ice particles of upper tropospheric cirrus (UT) and lower polar stratosphere (LS) clouds depends on the efficiency and time of the reactive gas interaction that also determines the ice particle impact on heterogeneous chemistry. The HCl vapor uptake by polycrystalline ice produced by water freezing is investigated under laboratory conditions relevant to UT/LS. The uptake efficiency and ice conductivity were measured simultaneously in a Knudsen cell reactor. Regions with HCl water solutions in ice are likely to be formed during the HCl–ice interaction. This potentially defines a mechanism of effective and long HCl uptake by ice particles of cirrus clouds, polar stratospheric clouds (PSCs), and contrails due to fast freezing of submicron water drops.

Keywords. UT/LS clouds, polycrystalline ice, uptake efficiency, conductivity

1 Introduction

The upper troposphere/lower stratosphere (UT/LS) have been identified as altitude regions with high sensitivity to antropogeneous perturbations in connection with a possibly large impact on ozone depletion and climate. Predictions of antropogeneous influence are limited by uncertainties in our knowledge of trace gas effects on composition and heterogeneous activity of ice particles which were formed in the natural cirrus troposphere clouds and PSCs, and contrail clouds.

The structure of cirrus and PSCs particles is defined by the growth conditions and cooling velocity [1]. Observation of polycrystalline ice particles in the cirrus clouds at T ≅ −83°C and the identification of the conditions of PSC formation have indicated a common mechanism of ice production by submicron water drop freezing [2]. The ice particles in the aircraft contrails are formed by condensation of supersaturation water vapors and followed by fast freezing that can lead to a polycrystalline ice structure.

Interaction of acidic gases such as HCl with ice particles of PSCs is a key step in polar ozone hole chemistry [3]. The uptake efficiency of HCl, γ, on ice is defined by the fractional collision frequency that leads to the reactant loss on ice surface. Until recently, a single laboratory measurement $\gamma \approx 0.4$ has been performed at HCl vapor

NATO ASI Series, Vol. I 56
Ice Physics and the Natural Environment
Edited by John S. Wettlaufer, J. Gregory Dash and Norbert Untersteiner

pressure $p \cong 2 \cdot 10^{-4}$ Torr and T = 200K [4]. However, the high pressure of HCl imposes a problem of interpolation of this result to atmospheric conditions.

The present work is a laboratory investigation of HCl uptake by polycrystalline ice films which are produced by water freezing with different cooling rates. Measurements of HCl uptake efficiency were carried out at the vapor pressures and temperatures relevant to UT/LS. A phenomenon of the long, stationary and effective HCl uptake was observed simultaneously with an increase in ice conductivity. The formation of areas with high HCl concentration solutions is likely to occur during vapor uptake due to HCl molecule diffusion at the grain boundaries of polycrystalline ice. Measurements of water HCl solution conductivity at different HCl mole fractions and temperatures point to a mechanism of high uptake ability of polycrystalline ice.

2 Experimental Setup and Procedure

The experimental technique developed is a Knudsen cell. The Knudsen cell consists of two chambers separated by a valve. The upper chamber is coupled to a three dimensional quadrupole high frequency ion trap for mass spectrum analysis. The lower sample chamber is a stainless steel cup with the Teflon rod inside. Ar/HCl mixture used at the total pressure less 0.4 mTorr ensures the molecular flow regime and high accuracy for measurements of γ.

2.1 Preparation of Ice Film

In many experiments [4,6] the surface of PSCs are simulated by the water vapor deposition. Such films may be highly porous and have surface areas much larger than the geometrical area [7]. This may be one of the principal reasons why there is a great discrepancy between the different experimental data [6]. Solving this problem requires using special technology to prepare ice films with known surface area and structure.

In this work the ice films were produced by freezing distilled water. The water droplet was frozen on a cooled Teflon rod at T≈230K. The cooling rate is approximately 10K s^{-1}. Water is supercooled and rapidly crystallized to nonporous polycrystalline ice with fine grains [1]. Typical thicknesses of polycrystalline ice films produced were near 60 μm.

Then the ice films were heated to T≈260K when surface melting is likely to take place. This process of annealing has provided for the smooth surface of ice. After this the ice films were cooled to temperatures of interest T ≅ 150–240K. To prove that the surface area of ice film produced is near geometrical one, we changed the volume of initial water droplet and the surface of its spreading. But the uptake efficiency measured remained unchanged and depended only on T and p.

2.2 HCl Uptake Measurements

The measurements of HCl uptake efficiency, γ, by polycrystalline ice were carried out at initial pressure in the range $p \cong 10^{-7}-10^{-4}$ Torr and $T \cong 150$–240K. The two pressure regions of the different characteristic values of γ were observed at $T \cong 188$–208K [8]. At lower HCl pressure $p \cong 10^{-7}-10^{-6}$ Torr $\gamma \cong 0.1\pm0.02$. For the higher pressure $\gamma \cong 0.5\pm0.1$. Existence of the different values indicates the possible structure and phase changing of the ice surface as the HCl pressure increases. From comparison of our results with data of Fourier transform infrared studies [9] we have inferred that the low value $\gamma \cong 0.1$ corresponds to HCl interaction with pure ice at UT/LS conditions and the value $\gamma \cong 0.5$ indicates the formation of a new compound $HCl{\bullet}6H_2O$ during the HCl exposure of ice. So we see that the measurements [4] were carried out at HCl hexahydrate formation on the ice surface.

The typical duration of HCl exposure of ice in our experiments was near 10–60 min. But the important conclusion about long and stationary uptake without saturation was produced during exposure of 2.5 hours.

2.3 Electrical Measurements

To study a change in the ice sample during HCl exposure the method of ice electrical conductivity measurement was used. The electrodes constructed from platinum wire with a diameter≈ 400 µm were mounted on the Teflon rod surface at a distance of 1 cm from each other. The water droplet was located between electrodes. The thin glass wafer was used to achieve a good contact with electrodes and to restrict the ice film thickness about H≈ 400 µm. Electrical conductivity of sample, σ, was measured at 1 kHz using the LCR-meter over the temperature range $T \cong 150$–270K.

In one set of experiments the hydrochloric acid solutions were taken with HCl mole fraction, X_{HCl}, from $1.8\cdot10^{-3}$ % to 19.5% and frozen with cooling rates of $V_c \cong 0.5$–10 K/min. The σ dependence on X_{HCl}, shown on Fig. 1, indicates the change in phase of the solutions when X_{HCl} increases. At low concentrations HCl molecules are incorporated into the ice crystal, leading to a linear dependence of $\sigma(X_{HCl})$ due to the Bjerrum defect transport [10]. This effect is observed at $X_{HCl} \cong 5\cdot10^{-3}$ % and $T \cong 250$K. But as X_{HCl} increases a tendency of saturation by HCl molecules in a growing crystal of ice is observed [11] and HCl solution is located between the ice monocrystal grains [1,12]. So, σ is defined as a sum of effects from different areas of polycrystalline ice:

$$\sigma = \sigma_i h_i/H + \sigma_s h_s/H, \quad (1)$$

where h_i, h_s are the sizes of the ice monocrystals and HCl solution areas; σ_i, σ_s are their specific conductivities, respectively. The measured dependence $\sigma(X_{HCl})$ up to $X_{HCl} \geq 10^{-1}$% appears to be a square law. It indicates the influence of a second term in (1) due to growth both σ_s and h_s. However, when the value X_{HCl}* marked by a square for each T on Fig 1 is reached, a change in the dependence $\sigma(X_{HCl})$ to a linear law is observed. But the linear dependence $\sigma(X_{HCl})$ is characteristic for aqueous electrolyte

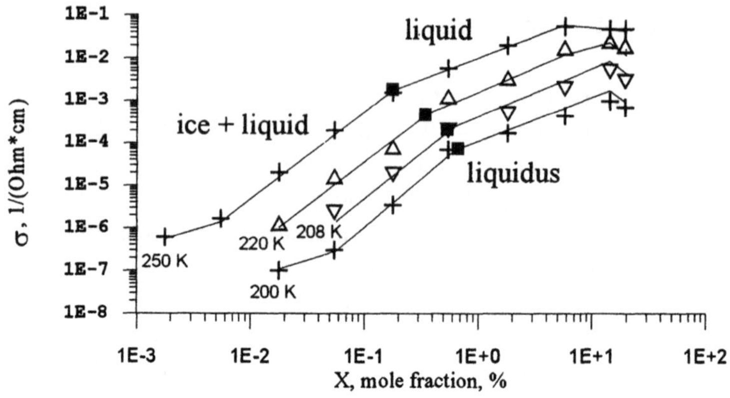

Fig. 1. Conductivity of HCl solutions vs the mole fraction of HCl at different temperatures. $V_c \cong 2$ K/min. Symbols ■ are marked the points of liquidus.

solutions. So, we can infer a transition across the liquidus of the phase-diagram to a liquid aqueous HCl solution.

At the highest X_{HCl} a decrease of σ is observed due to the increasing viscosity of the solutions. This effect is well known for highly concentrated electrolyte solutions.

3 Results and Discussion

A mass spectrometer and LCR-meter were employed for simultaneous analysis of the gas phase above ice surface and ice conductivity during HCl uptake. A temporal dependence of ice conductivity σ at $p = 6\cdot10^{-6}$ Torr and T = 208K is shown in Fig. 2. A continuous increase of ice conductivity is observed up to termination of HCl exposure. The value of γ appeared to be stationary $\cong 0.13\pm0.02$.

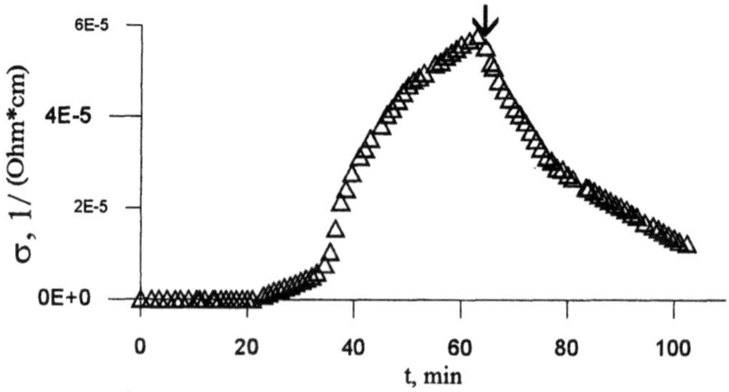

Fig. 2. Temporal dependence of conductivity during HCl uptake. $\gamma = 0.13$, $p \cong 6\cdot10^{-6}$ Torr, T = 208K. Arrow shows the moment of termination for HCl exposure.

In many works [5,6] the HCl–ice interaction is considered as a surface process due to low HCl diffusion coefficient D into the ice bulk. But large HCl uptake observed in our experiments can not be explained by only surface adsorption since HCl amount up taken during t ≥ 35 min is more then 10^3 ML. And the uniform bulk dissolution cannot be suggested because of the low value of HCl solubility in the ice monocrystal [13].

Hence, we infer the formation of areas with HCl solution at the ice grain boundaries during HCl uptake. During first time of HCl exposure the HCl concentration at the ice grain boundaries is low and HCl solution is solid. But enhancement in σ after $\Delta t \cong$ 35 min in Fig. 2 seems to be connected with formation of liquid HCl solution with $X_{HCl} > X_{HCl}$*. Therefore we can use the approximation for σ_s of liquid solution from Fig. 1 and use Eq. (1) to estimate σ of sample during HCl uptake. We found a good correlation with experimental values of conductivity until $\Delta t \cong 50$ min of exposure. Subsequent deviation of the linear law is likely to be a result of σ_s decreasing at the high HCl concentrations.

4 Conclusions

HCl impurities in ice have drastic effects on conduction properties at low frequency, and careful study of this effect has provided a clue to the mechanism of effective uptake ability of polycrystalline ice.

HCl molecules from gas phase are incorporated and distributed in the grain boundaries due to diffusion. The diffusion coefficient in the liquid solution is increased, facilitating the transference of HCl molecules from the ice surface into the ice bulk. This mechanism is consistent with a model of HCl–ice interaction suggested in [14] for the polycrystalline ice.

These investigations have important implications for the mechanisms of chemical uptake of polycrystalline ice in tropospheric cirrus and lower stratospheric clouds.

References

1. Water and aqueous solutions at subzero temperatures. Ed. by F. Franks. Plenum Press, N.Y., 1982.
2. Heymsfield A. J.Atmos. Sci. 1986. V.43. N8. P.851.
3. Solomon S., Garcia R.R. et al. Nature. 1986. V.321. P.755.
4. Leu M.T. Geophys. Res. Lett. 1988. V.15. N1. P.17.
5. Hanson D., Ravishankara A. J. Phys. Chem. 1992. V.96. P.2682.
6. Chu L.T., Leu M.T., Keyser L.F. Ibid. 1993. V.97. P. 7779.
7. Keyser L.F., M.T. Leu, S.B. Moore J. Phys. Chem., 1993. V. 97, P. 2800.
8. Persiantseva N.M., Popovitcheva O.B., Rakhimova T.V. Proc. International Colloquium 'Impact of aircraft emissions upon the atmosphere' Paris. 1996. V. II. P. 515.10563.
9. Koehler B.G., L.S.McNeill et al. J. Geophys. Res., 1993. V.98. P.10563.

11. Hobbs P. Ice Physics. Clarendon Press. Oxford. 1974.
11. Seidensticker R.G. J. Chem. Phys. 1972. V. 56. N. 6. P. 2853.
12. Wolff E., Mulvaney R., Oates K. Geophys. Res. Lett. 1989. V. 16. P. 487.
13. Domine F., Thibert E. et al. Geophys. Res. Lett. 1994. V. 21. N 7. P. 601
14. Popov A.M., Popovitcheva O.B., Rakhimova T.V. et al. Russian J. Phys. Chem. 1995. V. 69. P. 1823.

Methane Bubble Inclusions in Ice on High Latitude Lakes

Allan R. Phelps and Martin O. Jeffries

Geophysical Institute, University of Alaska Fairbanks, P.O. Box 757320, Fairbanks, Alaska 99775, U.S.A.

1 Introduction

Atmospheric concentrations of methane have nearly doubled in the last 200 years, and although concentrations are only a few ppm, a molecule of CH_4 has a 20 times greater greenhouse effect than a molecule of CO_2. Methane gas is included within bubbles in the seasonal ice cover on shallow high latitude lakes, and as a result, the ice and underlying water become a significant repository for methane. During the brief period of ice melt in the spring, a large pulse of the methane trapped in the ice and water is released to the atmosphere. During the break-up of 1997, for example, this pulse amounted to 2 g CH_4 m^{-2} during the 10 day period following ice-melt [*Phelps et al.*, In press].

The methane molecule is too large to be included within the crystal structure of ice at ambient temperatures and pressures, so included gasses are found in bubbles within the ice. Three types of macro bubbles can be observed in lake ice [*Gow and Langston*, 1977; *Jeffries et al.*, 1994]: ebullition bubbles evolved from the sediments arranged vertically above a point source; small spherical bubbles formed from the re-freezing of melted precipitation and/or overflow from the lake onto the ice surface ("white" ice); and bubbles forced out of solution into the ice/water interface from the supersaturation of dissolved gasses in the water. Ebullition bubbles are a spheroid shape, and can be virtually any size. Bubbles from gasses forced out of solution are vertically oriented cylinders, with a diameter of about 1 mm, and lengths varying from 1 to 20 cm, and can be interspersed with inverted tear-drop shaped bubbles. Initial freezing produces clear inclusion free "black" ice, but mid-way through the winter (~ January), distinct layers of high density tubular bubbles appear simultaneously across the lake, and these bubbles can contain high levels of methane. This paper describes methane concentrations in bubbles in ice on a shallow sub-arctic lake, and discusses possible nucleation mechanisms that can account for the observed bubble size and gas concentration.

NATO ASI Series, Vol. I 56
Ice Physics and the Natural Environment
Edited by John S. Wettlaufer, J. Gregory Dash and Norbert Untersteiner

2 Methods

During winter 1996–97, 7.5 cm diameter ice cores were taken from Mosquito Lake in Anchorage (61° 11.45′ N, 149° 48.67′ W), at two week intervals. Water samples and water chemistry data were obtained simultaneously. The cores were cut into 15 cm sections, weighed, placed in air-tight containers of pure nitrogen atmosphere fitted with serum stoppers, and melted. The containers were shaken vigorously for two minutes to equilibrate the headspace and the water, then samples of the headspace were withdrawn and analyzed for gas concentrations on a Shimadzu GC-8AIF Gas Chromatograph. Water samples were analyzed for gas content in a similar fashion. Water chemistry data were collected with a YSI 600XL Multi-Parameter Water Quality Probe *in situ*.

3 Results

A temporal plot of the CH_4 data (Fig. 1) shows a clear increase in CH_4 concentrations in the lowest sections of the ice during January, which coincided with the appearance of the cylindrical bubbles. A smaller increase in CH_4 concentrations can also be observed in the top layer of ice, due to periodic overflows of supersaturated water from below onto the ice surface. The middle layers, almost free of bubbles, contain negligible quantities of CH_4. There appears to be no relationship between ice growth rate and inclusion of cylindrical bubbles containing CH_4. A plot for CO_2 (Fig. 2), shows little change in concentrations within a layer over the course of the winter. Additionally, the gradient is reversed, with high CO_2 concentrations in the top layer containing re-frozen precipitation, and low CO_2 in the lower layer containing the cylindrical bubbles.

A plot of CH_4 concentrations in the water beneath the ice (Fig. 3) shows a pattern similar to the ice CH_4 concentrations (though an order of magnitude higher). The drop at the end of the season was probably due to venting of the dissolved CH_4 through thawed moats at the edges of the lake. Low oxidation/reduction potentials (Fig. 3) force the microbial metabolic pathway to methanogenesis, and the low levels of dissolved O_2 prohibit CH_4 consumption by methanotrophs. This enables dissolved CH_4 to reach much higher concentrations under the ice cover, compared to ice-free periods.

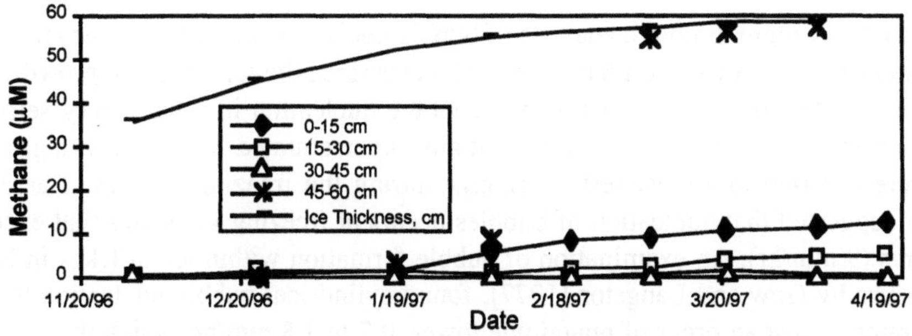

Fig. 1. CH_4 concentrations, four depths, Mosquito Lake ice.

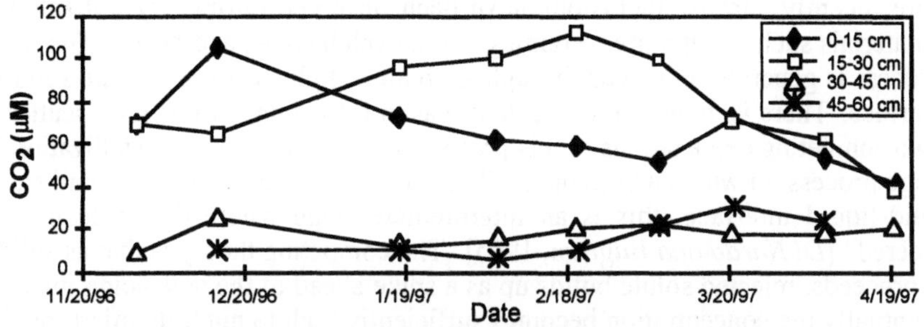

Fig. 2. CO_2 concentrations, four depths, Mosquito Lake Ice.

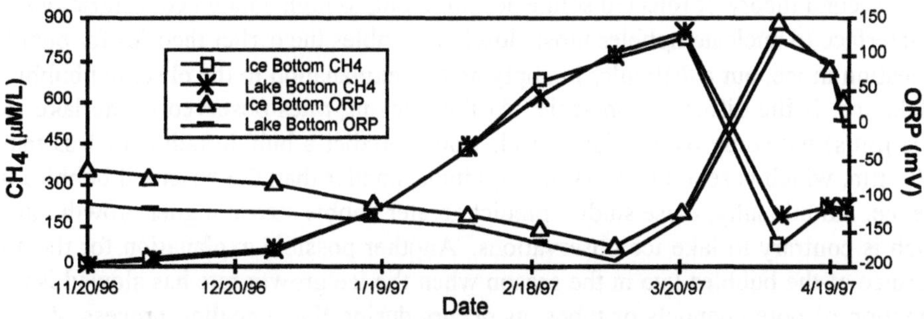

Fig. 3. Dissolved methane and oxidation reduction potential.

4 Discussion

The inclusion of bubbles within the ice fabric is a function of the rate of ice growth, the concentration and diffusion coefficients of dissolved gasses, and the concentration of dissolved particulates in the water which encourage nucleation of the bubbles. Carte [1961] applied a one-dimensional diffusion equation to the nucleation of bubbles at the ice/water interface, with bubble nucleation occurring when the concentration of dissolved gasses at the interface exceeded the equilibrium saturation by a factor of 30. A more thorough analysis of bubble nucleation at the ice-water interface was conducted by Bari and Hallett [1974] who examined the effects of ice growth rate and the introduction of particulates to encourage nucleation in a laboratory setting. However, their water samples were not initially supersaturated with dissolved gasses, and were exposed to atmospheric exchange during the freezing process. Bari and Hallett suggested that nucleation of bubbles occurs at freezing velocities that exceed 2.5 μm/s (9 mm/hr). An examination of bubble formation within ice of lakes in New Hampshire by Gow and Langston [1977], found cylindrical bubble nucleation to occur at rates almost an order of magnitude lower, 0.7 to 1.8 mm/hr, which they attributed to the higher impurity of the lake water compared to the distilled water used by Bari and Hallett.

More recently, similar inclusions have been observed during crystallization of other materials such as sapphire, ceramics, salol, cyclohexane, and xenon. The formation process is generally observed through anomalous light scattering at the solid liquid interface. There is, however, some debate as to the origins of this light scattering, with two competing explanations of this process. One theory is that crystallization is a two stage process, in which a "mesophase" boundary layer forms on the liquid side of the solid-liquid interface, this is an intermediate stage where the molecules are "preordered" [*Di Nardo and Bilgram*, 1995]. The competing theory is that as solidification proceeds, rejected solute builds up as a spike ahead of the advancing interface, and eventually the concentration becomes sufficiently high to nucleate micro-bubbles which produce the observed scattering [*Williams et al.*, 1990; *Mesquita and Cummins*, 1996].

The second theory of rejected solute accumulating to high enough concentrations at the interface to nucleate bubbles most closely resembles the earlier theories for bubble nucleation in ice, but a difficulty in applying this explanation to the observed bubbles in lake ice, is the difference in scale. At the freezing rates observed in the lake ice (0.04 μm/s) the study by Williams et al. would predict a bubble radius of less than 0.001 μm, which is several orders of magnitude smaller than the observed bubbles in lake ice. Additionally, these studies predict greater bubble size at higher growth rates, which is contrary to lake ice observations. Another possible explanation for the appearance of the bubbles late in the season when the ice growth rate has slowed is the formation of pore channels or tubes, as occurs during the annealing process of sapphire [*Kulinsky et al.*, 1996]. However, the lake thermal regime does not appear to support this explanation.

In this study, ice growth velocities do not appear to be a factor in cylindrical bubble nucleation. Early in the season, when maximum ice growth velocity occurs, the ice is free of cylindrical bubbles, which do not begin to nucleate until ice growth has slowed during late December and early January. The ice growth velocity during this period is 0.14 mm/hr, two orders of magnitude less than that found by Bari and Hallett [1974], and one order of magnitude less than Gow and Langstone [1977]. The increase in conductivity at the ice/water interface also observed during this nucleation period, implies that increasing concentrations of impurities excluded by the ice could enhance bubble nucleation.

The driving force in this study appears to be increasing concentrations of gasses that continue to be produced in the thawed sediments after the surface has iced over.

The determination of all dissolved gasses in the lake water was not possible, but during the nucleation period the concentration of dissolved methane in the water was 200 μM/L, which is about 50,000 times the equilibrium concentration with the atmosphere. A potential explanation for the appearance of bubbles as a uniform layer within the lake ice is that a pressure event could trigger the bubbles to come out of solution at the same time (as in opening a bottle of champagne). Changes in atmospheric pressure have been demonstrated to trigger ebullition from lake sediments [*Matson and Likens*, 1990]. Pressure changes could also be potentially triggered by release of lake water to the ice surface during episodes of overflow.

5 Acknowledgments

This work was made possible through the support of a number of organizations: the NASA Earth System Science Fellowship; the Alaska Space Grant Program, University of Alaska Fairbanks; the Center for Global Change and Arctic System Research, U.A.F.; and the Geophysical Institute, U.A.F.

6 References

Bari, S.A., and J. Hallett., Nucleation and growth of bubbles at an ice-water interface, *J. Glaciol.*, *13* (69), 489–520, 1974.

Carte, A.E., Air bubbles in ice, *Proceedings of the Physical Society (London)*, *77* (495), 757–768, 1961.

Di Nardo, S., and J.H. Bilgram, Fluctuations during freezing and melting at the solid-liquid interface of xenon, *Physical Review B*, *51* (13), 8012–8017, 1995.

Gow, A.J., and D. Langston, Growth history of lake ice in relation to its stratigraphic, crystalline and mechanical structure, U.S. Army CRREL 77-1, Hanover, N.H., 1977.

Jeffries, M.O., K. Morris, and W.F. Weeks, Structural-stratigraphic features and ERS-1 SAR backscatter characteristics of ice growing on lakes in N. W. Alaska, winter 1991–1992, *J. Geophys. Res*, *99* (C11), 22,459–22,471, 1994.

Kulinsky, L., J.D. Powers, and A.M. Glaeser, Morphological evolution of pre-perturbed pore channels in sapphire, *Acta Materialia*, *44* (10), 4115–4130, 1996.

Matson, M.D., and G.E. Likens, Air pressure and methane fluxes, *Nature*, *347*, 718–719, 1990.

Mesquita, O.N., and H.Z. Cummins, Comment on "Fluctuations during freezing and melting ... of xenon", *Phy. Rev. B*, *54* (2), 1410–1411, 1996.

Phelps, A.R., K.M. Peterson, and M.O. Jeffries, Methane efflux from high latitude lakes during spring ice-melt, *J. Geophys. Res. - Atmospheres*, In press.

Williams, L.M., M.R. Srinivasan, and H.Z. Cummins, Direct observation of micro-bubbles in directional solidification of salol, *Phy. Rev. Lett.*, *64* (13), 1526–1528, 1990.

Modelling Sea Ice Roughness in the Arctic

N. Steiner, M. Harder, P. Lemke

Institut für Meereskunde, Düsternbrooker Weg 20, D-24105 Kiel, Germany,
E-mail: nsteiner@ifm.uni-kiel.de

1 Introduction

The variability of sea ice in the polar regions is an important factor in the climate system, particularly because of its strong influence on heat and freshwater transports as well as momentum exchange between ocean and atmosphere. To describe these effects accurately, ice conditions need to be known over long time periods and wide regions. Models are able to produce such data but need to be verified by observations (Lemke et al., 1998). One classical model variable which can be validated quite well with remote sensing methods (SMMR, SSM/I) is the ice coverage. Modelled ice drift can be verified by comparing observed and simulated drift trajectories (Kreyscher et al., 1998). Ice thickness observations, however, are only rarely available from drillings, sonar measurements and laser altimeter recordings.

Another observable variable is sea ice roughness. Measurements of the geometrical shape of roughness elements like pressure ridge statistics are already available (e.g. Timco and Burden, 1997, Davis and Wadhams, 1995). Large-scale observations of ice roughness from satelliteborne radar sensors will be available when interpretation algorithms have been developed. Especially the Synthetic Aperture Radar (SAR) with its high spatial resolution (30 by 30 m) is sensitive to the surface roughness on the centimeter to decimeter scale. Under certain conditions even single ice ridges may be observed in the imagery (e.g. Carlström and Ulander, 1995). Harder (1996) introduced ice roughness as a new model variable which is defined as the accumulated deformation work which is acting on an ice floe. A fraction of the total deformation work is transformed into potential energy which is required to submerge pressure ridge keels and raise sails (Rothrock, 1975). In this study, a quantitative relationship between observed geometrical sea ice roughness and simulated large-scale deformation work is presented.

2 Model Description and Results

The applied dynamic-thermodynamic model (Harder, 1996, 1998) considers sea ice a two-dimensional continuum described by ice concentration A, horizontal ice drift velocity $\mathbf{u}$, ice thickness h, and ice roughness R. The ice rheology is based on the viscous-plastic model of Hibler (1979), and the energy balance is applied

NATO ASI Series, Vol. I 56
Ice Physics and the Natural Environment
Edited by John S. Wettlaufer, J. Gregory Dash and Norbert Untersteiner

according to Parkinson and Washington (1979). The model runs on a rotated spherical grid covering the whole Arctic. Time step is one day, spatial resolution is 1° (≈ 110 km). Daily atmospheric forcing data for wind, air temperature and humidity are taken from European Center for Midrange Weather Forecast analyses for 1986–1992. All other forcing data consist of climatological means. Roughness of sea ice is introduced as a scalar variable. Its spatial and temporal evolution is described by a prognostic budget equation:

$$\frac{\partial R}{\partial t} + \nabla \cdot (\mathbf{u}\, R) = \sigma \cdot \dot{\varepsilon} + \frac{R}{h} \min\left(0, \left.\frac{\partial h}{\partial t}\right|_{\text{thermo}}\right) \tag{1}$$

where the lefthand side describes the local rates of change and advection and the righthand side describes the source and sink terms for the roughness. R is derived from temporal integration of deformation work, given as an energy per area in Jm^{-2}. All deformation work which acts on an ice floe increases sea ice roughness. Sources of deformation work are external forces by surface winds and ocean currents which in turn cause internal forces during plastic deformation. These internal forces are derived as the product of the stress tensor σ and the strain rate $\dot{\varepsilon}$ by standard continuum mechanics (Rothrock, 1975). New ice is considered to have no initial roughness. As the ice volume is regarded as a carrier of ice roughness, it is assumed that R vanishes at the same rate as the ice volume melts. Typical values of R in the central Arctic are about 100 $\mathrm{KJ/m}^2$. Simulated sea ice roughness is transformed into observable geometrical sea ice roughness in terms of distributions of mean keel draft and mean ridge spacing. Fig.1 shows as an example mean values for February 1992 of the simulated number of pressure ridge keels exceeding 9 m draft. Winter measurements from submarine sonars (Bourke and McLaren, 1992, Fig. 7b) show similar results with values around 3 in the North Pole region and values up to 10 at the Canadian archipelago. Spring values of the simulated mean pressure ridge keel draft are about 9 m in the central Arctic, decreasing continually along the 0° meridian through the Fram Strait to values of about 5 m at 75°N. These values are in close agreement with the measurements from submarine sonars in Wadhams and Ono (1998) where a decrease of mean keel draft towards lower latitudes is described. The reported values of Bourke and McLaren (1992, Fig. 6a,b) are somewhat higher due to their higher cut-off value of 9 m in contrast to 5 m in Wadhams and Ono (1998). Simulated summer values of mean pressure ridge sail heights in the Laptev Sea are between 1.0 m and 1.5 m, which is in good agreement with laser-altimeter measurements of C.Haas (in Augstein et al.,1997) and Evers and Jochmann (1995).

To obtain geometrical sea ice roughness from model results, an idealized ridge geometry with Gaussian-shaped sail and keel profiles is assumed. We emphasize that we are interested in the statistics of a large number of ridging processes without considering individual ridge formations. This idea was already postulated by Thorndike (1992) who calculated draft distributions using a redistribution function. It has been shown that keel and sail height distributions fit a negative exponential (e.g. Davis and Wadhams, 1995). Therefore, an exponential distribution form for the number of keels per kilometer track per meter of draft increment, given by equations (1) to (3) in Davis and Wadhams (1995), is applied. Based on this, potential energy accumulated in pressure

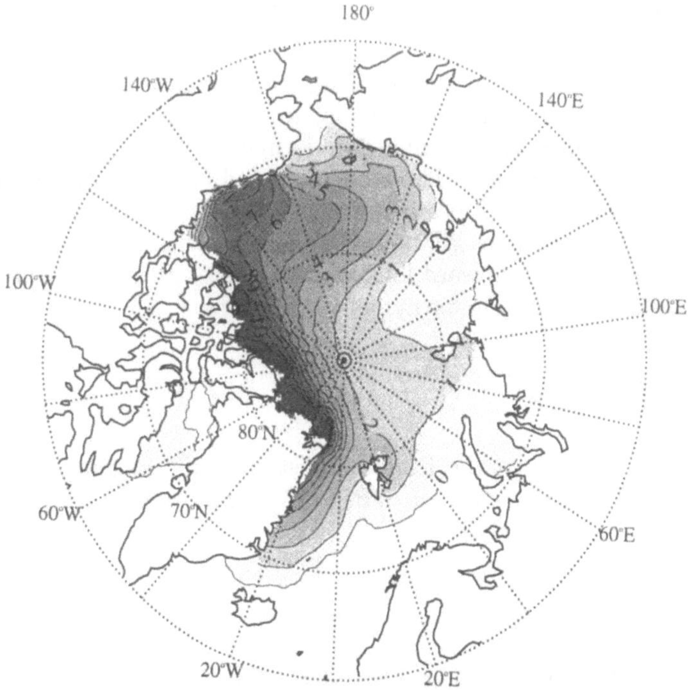

Figure 1: Simulated pressure ridge keel spatial frequency (number of keels exceeding a 9-m threshold per kilometer) for February 1992.

ridges is calculated. Only a fraction of the total deformation energy described by the model variable R is transformed into potential energy E_P. Therefore, we assume $E_P = c_R \cdot R$ with $c_R = 7.5\%$ being the proportionality factor. Small-scale numerical simulations of ridge growth yield values of $c_R \approx$(5–20)% (Hopkins et al. 1991, Hopkins and Hibler, 1991) and $c_R \approx$(5.9–10.4)% (Hopkins, 1994). The simulated total potential energy E_P per grid cell is equated to the potential energy accumulated in all pressure ridges in a grid cell. Mean keel draft h_m is obtained from this potential energy equation. The other simulated variables, deformed ice volume and area fraction, are calculated by integration over all pressure ridges in a grid cell. Compared to the calculations of Rothrock and Thorndike (1980) who related sea ice roughness to the roughness of mathematical functions, a profile of our simulated ice field would be a smoothed curve which retains the large features but loses detail where smoothing has little effect on the distribution of draft.

The ratio of sail height to keel depth as well as keel and sail slopes are taken from observations (e.g. Davis and Wadhams, 1995). Low-level cut-off draft is determined from simulated ice thickness h, and mean number of keels per unit distance is calculated as a function of the simulated ice roughness per volume R/h.

3 Summary

We have shown how the introduction of sea ice roughness as a variable in large-scale sea ice models is an additional tool to verify model results. Simulated ice roughness, described as accumulated deformation energy, is transformed into geometrical sea ice roughness. The model is based on the statistics of observed pressure ridge geometries. In particular when large-scale observations of sea ice roughness from radar sensors will be available in addition to the already existing observation methods (e.g upward looking sonar, laser altimeter), the roughness will be an important variable for model verifications.

Acknowledgments

This work is supported by the Bundesministerium für Bildung, Wissenschaft, Forschung und Technologie (grant 03PL018A). Atmospheric forcing data were derived from ECMWF analyses.

References

Augstein, E. and Cruise Participants. 1997. The expedition ARCTIC '96 of RV "Polarstern" (ARK XII) with the Arctic Climate System Study (ACSYS) *Berichte zur Polarforschung, 234, 30.*

Bourke, R.H. and A.S. McLaren. 1992. Contour mapping of Arctic Basin ice draft and roughness parameters, *J. Geophys. Res.*, 97(C11), 17.715–17.728.

Carlström, A. and L.M.H. Ulander. 1995. Validation of backscatter models for level and deformed sea-ice in ERS-1 SAR images, *Int. J. Remote Sensing*, 16(17), 3245–3266.

Davis, N.R. and P. Wadhams. 1995. A statistical analysis of Arctic pressure ridge morphology, *J. Geophys. Res.*, 100(C6), 10.915–10.925.

Evers, K.-U., P.Jochmann. 1995. Determination of the topography of pressure ice ridges in the Laptev Sea. *HSVA-Report*, E 258/95.

Harder, M. 1996. Dynamik, Rauhigkeit und Alter des Meereises in der Arktis — Numerische Untersuchungen mit einem großskaligen Modell, *Berichte zur Polarforschung, 203.*

Harder, M. 1998. Roughness, age and drift trajectories of sea ice in large-scale simulations and their use in model verifications, *Annals of Glaciology 25*, in press.

Hibler, W.D. 1979. A dynamic-thermodynamic sea ice model, *J.Phys.Oceanogr.*, 9, 815–845.

Hopkins, M.A. 1994. On the ridging of intact lead ice, *J. Geophys. Res.*, 99(C8), 16.351–16.360.

Hopkins, M.A. and W.D. Hibler. 1991. On the ridging of a thin sheet of lead ice, *Annals of Glaciology*, 15, 81–86.

Hopkins, M.A., W.D. Hibler and G.M. Flato. 1991. On the numerical simulation of the sea ice ridging process, *J. Geophys. Res.*, 96(C3), 4809–4820.

Kreyscher,M. and M. Harder. 1998. First results of the Sea Ice Model Intercomparison Project (SIMIP), *Annals of Glaciology 25*, in press.

Lemke, P., W.D. Hibler, G. Flato, M. Harder and M.Kreyscher. 1998. On the improvement of sea ice models for climate simulations: the sea ice model intercomparison project, *Annals of Glaciology 25*, in press.

Parkinson, C.L. and W.M. Washington. 1979. A large-scale numerical model of sea ice, *J. Geophys. Res.*, 84(C1), 6787–6803.

Rothrock, D.A. 1975. The energetics of plastic deformation of pack ice by ridging, *J. Geophys. Res.*, 80(33), 4514–4519.

Rothrock, D.A. and A.S. Thorndike. 1980. Geometric properties of the underside of sea ice, *J. Geophys. Res.*, 85(C7), 3955–3963.

Thorndike, A.S. 1992. Estimates of sea ice thickness distribution using observations and theory, *J. Geophys. Res.*, 97(C8), 12,601–12,605.

Timco, G.W. and R.P. Burden. 1997. An analysis of the shapes of sea ice ridges, *Cold Regions Science and Technology*, 25, 65–77.

Wadhams, P. and N. Ono. 1998. Downstream evolution of ice thickness in the Arctic Ocean — evidence for melting?, submitted to *J. Geophys. Res.*.

Dynamical Calculations for the Proton Ordered Ice II Structure

Y. Wang, S. L. Dong and J-C. Li

Department of Physics, UMIST, PO Box 88, Manchester, M60 1QD, UK

Abstract. The phonon density of states for ice II has been calculated using a lattice dynamics simulation. The calculated results show that the hydrogen bonds linked to the different six-molecule rings in the structure are significantly weaker in comparison with the interactions within the rings, which may result from the distortion of the hydrogen bonding in the structure.

Keywords. Lattice dynamics, neutron scattering, phonon density of states

1. Introduction

Neutron scattering is a powerful tool for studying solid materials due to its neutral charge characteristics. It has been frequently used to obtain structural and dynamical information by measuring elastic coherent[1,2] and inelastic incoherent neutron scattering[3,4] (IINS). IINS has a special advantage in studies of the hydrogen bond due to the large IINS cross section of hydrogen and its intensity is directly proportional to the weighted phonon density of states (PDOS) which can be compared to lattice dynamics (LD) simulations.

The inelastic neutron scattering (INS) spectra for ice II H_2O and D_2O at 15K[4] were measured using the TFXA[5] spectrometer at ISIS, Rutherford Appleton Laboratory, UK. The 'one-phonon' spectra were extracted from experimental data by subtracting the background scattering and multi-phonon parts.[6] The spectra show that there are sharp features in the translational (0-40 meV) and librational (50-120 meV) regions which may be due to the strong proton ordering in the structure according to diffraction[1,2] and dielectric studies[7]. LD simulations are a good way to interpret the features in the measured spectra for ice II.

Ice II is one of the high-pressure phases which partly resembles the structure of ice Ih, though ice II is proton-ordered. It exists at ~2.5 kbar (depending on the temperature). Its primitive unit cell is composed of two stacked six-molecule rings and the oxygens in these rings are labeled O(I) and O(II), respectively.[1,2] In ice Ih, they are in alignment looking along the c-axis, but in the ice II structure, these rings are twisted 15° relative to each other. This results in a decrease of the repeat distance along the c-axis of 7.36 $Å^2$ for ice Ih to 6.25 $Å^2$ for ice II. The six-molecule rings in one plane move relatively up and down parallel to the c-axis and enclose each other. Because of

NATO ASI Series, Vol. I 56
Ice Physics and the Natural Environment
Edited by John S. Wettlaufer, J. Gregory Dash and Norbert Untersteiner

this twist and vertical bending, O-O-O angles change to 83° and 99.6°, strongly diverging from the typical tetrahedral angle ~ 109.5°. These distortions cause changes not only in local configurations, but also in corresponding H-bond strengths.

2. Lattice Dynamical Calculations and Results

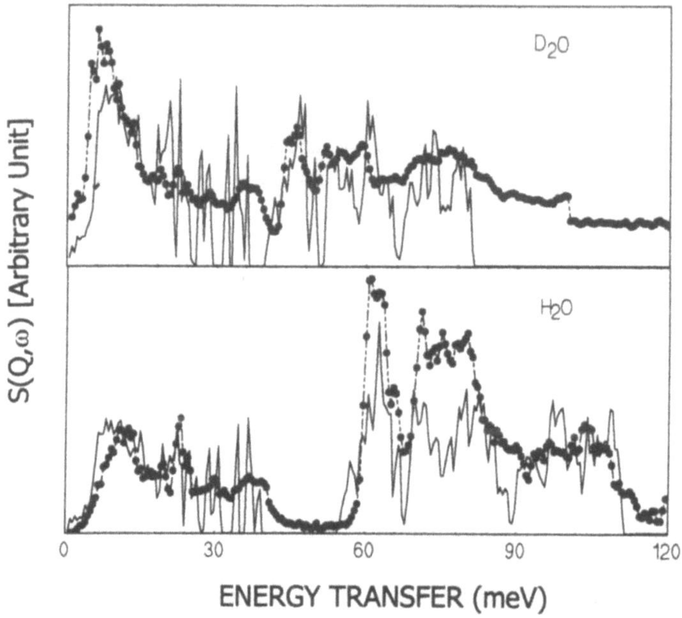

Fig 1. The calculated INS spectra, S(Q,ω), (continuous line) for the proton ordered ice II (H_2O and D_2O) compared with the measured INS spectra (dotted) in the range of up to 120 meV.

By diagonalization of the dynamical matrix (constructed by force constants), the eigenvalues and eigenvectors were obtained which represent the normal mode frequencies $\omega_j(\mathbf{q})$ and the zero-temperature displacements $\mathbf{e}_i(\mathbf{q},j)$ of atom in jth mode, where $\mathbf{q}$ is the phonon wave vector. The calculated PDOS is weighted by square displacements of atoms, atomic neutron scattering cross-section and multiplied by the Debye-Waller factor to give S(Q,ω) which can be directly compared to the one phonon INS spectrum.[8] The force constants were varied in order to obtain a best fit to the experimental spectrum. Each force constant used (see Table 1) is of different functionality. For instance, the O-O-O bending force constant mainly effects to the acoustic phonons at ~7 meV region and the H-O-H bending effects to the bending modes at about 200 meV (for H_2O). The calculation gives a unique set of force constants which are able to fit the main features of the measured spectra. The distortions discussed above will weaken the bond strengths, so small value of force constants for the distorted hydrogen bonds linking the six-fold rings should be introduced for these bonds in the LD calculation. However, the H-bonds within the rings are the same as in ice Ih.[4]

The calculated spectra for the ice II (H_2O and D_2O) up to 120 meV are shown in Figure 1. The energy positions of these calculated spectra (continuous lines) agree well with the measured neutron spectra (dotted data) though we note that the calculated intensities in the librational region are lower than the measured spectrum. In our calculations the resolution of instrument has not been considered so the data is sharper

Table 1. The force constants used in the LD calculations for ice II lattice compared to those for ice Ih (unit is eV/Å^2).

			Ice II	Ice Ih
O(I)-H	O(II)-H	Intramolecular stretching force constants within molecules in two different rings.	36.0	35.9
O(I)...H	O(II)...H	H-bonds within six-rings	2.2	2.1
O(I)...H		H-bonds between lateral rings	0.75	2.1
O(II)...H		H-bonds between vertical rings	0.75	1.1
H-O-H			3.1	3.2
H*-O(II)...H**		H* is H in O(II) ring, H** and H***	0.2	
H***-O(II)...H**		are H between vertical and lateral rings	0.88	
Other H-O...H			0.78	0.78
O...H-O				0.45
H-O...O			0.39	
O...O...O			0.16	0.16

H* is H(1), H** H(4) and H*** H(3) in Ref.[2] respectively.

than the experimental spectra. However, the sharpness in the translational optic region can not completely be explained by the resolution of the instrument, which is about 1.5% in whole energy region (from 2 meV to 500 meV) and the uncertainty in the energy is about 0.5 meV. Another factor could be due to the fact that the long distance interactions have not been included in our LD calculations.

In order to produce the sharp peak at 23 meV, the force constants of H-bond between the rings have to be weakened to 0.75 eV/Å^2 in comparison to those for Ih, 1.1 eV/Å^2 and 2.1 eV/Å^2. This may result from the distortion of the hydrogen bonds. Three-oxygen-atom angles are only 83.8° and 99.6° for these hydrogen bonds compared to the tetrahedral angle of 109.5°. More studies are needed to confirm it.

Acknowledgements

This work was funded by EPSRC. We would give many thanks to RAL for the use of neutron facilities.

References

1. E D Finch, et al, J. Chem. Phys., 49 (1968) 4361.
2. B Kamb, et al, J. Chem. Phys., **55** (1971) 1934
3. B Renker, Physics and Chemistry of Ice, Ed. E Whalley, et al, Ottawa,1973.
4. J-C Li, J. Chem. Phys., **105** (1996) 6733.
5. S F Parker, K E Horton and J Tomkinson, The TFXA User-Guide, Rutherford Appleton Laboratory, Technical Report No. RAL-TR-95-036,1995.
6. A I Kolesnikov and E F Sheka, Sov. Phys. Solid State, **25** (1983) 1303.
7. G J Wilson, et al, J. Chem. Phys., 43 (1965) 2384.
8. K Skold and D.L.Price, Neutron Scattering, Part A, Vol. **23**, (Academic Press, Inc., 1986) Chapt. 1.

Subject Index

The ASI Series Books Published as a Result of Activities of the Special Programme on Global Environmental Change

This book contains the proceedings of a NATO Advanced Research Workshop held within the activities of the NATO Special Programme on Global Environmental Change, which started in 1991 under the auspices of the NATO Science Committee.

The volumes published as a result of the activities of the Special Programme are:

Vol. 1: **Global Environmental Change.**
Edited by R. W. Corell and P. A. Anderson. 1991.
Vol. 2: **The Last Deglaciation: Absolute and Radiocarbon Chronologies.**
Edited by E. Bard and W. S. Broecker. 1992.
Vol. 3: **Start of a Glacial.** Edited by G. J. Kukla and E. Went. 1992.
Vol. 4: **Interactions of C, N, P and S Biogeochemical Cycles and Global Change.**
Edited by R. Wollast, F. T. Mackenzie and L. Chou. 1993.
Vol. 5: **Energy and Water Cycles in the Climate System.**
Edited by E. Raschke and D. Jacob. 1993.
Vol. 6: **Prediction of Interannual Climate Variations.**
Edited by J. Shukla. 1993.
Vol. 7: **The Tropospheric Chemistry of Ozone in the Polar Regions.**
Edited by H. Niki and K. H. Becker. 1993.
Vol. 8: **The Role of the Stratosphere in Global Change.**
Edited by M.-L. Chanin. 1993.
Vol. 9: **High Spectral Resolution Infrared Remote Sensing for Earth's Weather and Climate Studies.**
Edited by A. Chedin, M.T. Chahine and N.A. Scott. 1993.
Vol. 10: **Towards a Model of Ocean Biogeochemical Processes.**
Edited by G. T. Evans and M. J. R. Fasham. 1993.
Vol. 11: **Modelling Oceanic Climate Interactions.**
Edited by J. Willebrand and D.L.T. Anderson. 1993.
Vol. 12: **Ice in the Climate System.** Edited by W. Richard Peltier. 1993.
Vol. 13: **Atmospheric Methane: Sources, Sinks, and Role in Global Change.**
Edited by M. A. K. Khalil. 1993.
Vol. 14: **The Role of Regional Organizations in the Context of Climate Change.**
Edited by M. H. Glantz. 1993.
Vol. 15: **The Global Carbon Cycle.**
Edited by M. Heimann. 1993.
Vol. 16: **Interacting Stresses on Plants in a Changing Climate.**
Edited by M. B. Jackson and C. R. Black. 1993.
Vol. 17: **Carbon Cycling in the Glacial Ocean: Constraints on the Ocean's Role in Global Change.**
Edited by R. Zahn, T. F. Pedersen, M. A. Kaminski and L. Labeyrie. 1994.
Vol. 18: **Stratospheric Ozone Depletion/UV-B Radiation in the Biosphere.**
Edited by R. H. Biggs and M. E. B. Joyner. 1994.
Vol. 19: **Data Assimilation: Tools for Modelling the Ocean in a Global Change Perspective.**
Edited by P. O. Brasseur and J. Nihoul. 1994.

Vol. 20: **Biodiversity, Temperate Ecosystems, and Global Change.**
Edited by T. J. B. Boyle and C. E. B. Boyle. 1994.

Vol. 21: **Low-Temperature Chemistry of the Atmosphere.**
Edited by G. K. Moortgat, A. J. Barnes, G. Le Bras and J. R. Sodeau. 1994.

Vol. 22: **Long-Term Climatic Variations – Data and Modelling.**
Edited by J.-C. Duplessy and M.-T. Spyridakis. 1994.

Vol. 23: **Soil Responses to Climate Change.**
Edited by M. D. A. Rounsevell and P. J. Loveland. 1994.

Vol. 24: **Remote Sensing and Global Climate Change.**
Edited by R. A. Vaughan and A. P. Cracknell. 1994.

Vol. 25: **The Solar Engine and Its Influence on Terrestrial Atmosphere and Climate.**
Edited by E. Nesme-Ribes. 1994.

Vol. 26: **Global Precipitations and Climate Change.**
Edited by M. Desbois and F. Désalmand. 1994.

Vol. 27: **Cenozoic Plants and Climates of the Arctic.**
Edited by M. C. Boulter and H. C. Fisher. 1994.

Vol. 28: **Evaluating and Monitoring the Health of Large-Scale Ecosystems.**
Edited by D. J. Rapport, C. L. Gaudet and P. Calow. 1995.

Vol. 29: **Global Environmental Change Science: Education and Training.**
Edited by D. J. Waddington. 1995.

Vol. 30: **Ice Core Studies of Global Biogeochemical Cycles.**
Edited by R. J. Delmas. 1995.

Vol. 31: **The Role of Water and the Hydrological Cycle in Global Change.**
Edited by H. R. Oliver and S. A. Oliver. 1995.

Vol. 32: **Atmospheric Ozone as a Climate Gas.**
Edited by W.-C. Wang and I. S. A. Isaksen. 1995.

Vol. 33: **Carbon Sequestration in the Biosphere.**
Edited by M. A. Beran. 1995.

Vol. 34: **Climate Sensitivity to Radiative Perturbations: Physical Mechanisms and Their Validation.**
Edited by H. Le Treut. 1996

Vol. 35: **Clouds, Chemistry and Climate.**
Edited by P. J. Crutzen and V. Ramanathan. 1996.

Vol. 36: **Diachronic Climatic Impacts on Water Resources.**
Edited by A. N. Angelakis and A. S. Issar. 1996.

Vol. 37: **Climate Change and World Food Security.**
Edited by T. E. Downing. 1996.

Vol. 38: **Evaluation of Soil Organic Matter Models.**
Edited by D. S. Powlson, P. Smith and J. U. Smith. 1996.

Vol. 39: **Microbiology of Atmospheric Trace Gases.**
Edited by J. C. Murrell and D. P. Kelly. 1996.

Vol. 40: **Forest Ecosystems, Forest Management and the Global Carbon Cycle.**
Edited by M. J. Apps and D. T. Price. 1996.

Vol. 41: **Climate Variations and Forcing Mechanisms of the Last 2000 Years.**
Edited by P. D. Jones, R. S. Bradley and J. Jouzel. 1996.

Vol. 42: **The Mount Pinatubo Eruption: Effects on the Atmosphere and Climate.**
Edited by G. Fiocco, D. Fuà and G. Visconti. 1996.

Vol. 43: **Chemical Exchange Between the Atmosphere and Polar Snow.**
Edited by E. W. Wolff and R. C. Bales. 1996.

Vol. 44: **Decadal Climate Variability: Dynamics and Predictability.**
Edited by D. L. T. Anderson and J. Willebrand. 1996.

Vol. 45: **Radiation and Water in the Climate System: Remote Measurements.**
Edited by E. Raschke. 1996.

Vol. 46: **Land Surface Processes in Hydrology: Trials and Tribulations of Modeling and Measuring.** Edited by S. Sorooshian, H. V. Gupta, and J. C. Rodda. 1997.

Vol. 47: **Past and Future Rapid Environmental Changes: The Spatial and Evolutionary Responses of Terrestrial Biota.**
Edited by B. Huntley, W. Cramer, A. V. Morgan, H. C. Prentice, and J. R. M. Allen. 1997.

Vol. 48: **The Mathematics of Models for Climatology and Environment.**
Edited by J. I. Díaz. 1997.

Vol. 49: **Third Millennium BC Climate Change and Old World Collapse.**
Edited by H. N. Dalfes, G. Kukla, and R. Weiss. 1997.

Vol. 50: **Gravity Wave Processes: Their Parameterization in Global Climate Models.**
Edited by K. Hamilton. 1997.

Vol. 51: **Sediment Records of Biomass Burning and Global Change.**
Edited by J. S. Clark, H. Cachier, J. G. Goldammer, and Brian Stocks. 1997.

Vol. 52: **Solar Ultraviolet Radiation. Modelling, Measurements and Effects.**
Edited by C. S. Zerefos and A. F. Bais. 1997.

Vol. 53: **Atmospheric Ozone Dynamics. Observations in the Mediterranean Region.**
Edited by C. Varotsos. 1997.

Vol. 54: **The Stratosphere and Its Role in the Climate System**
Edited by G. P. Brasseur. 1997.

Vol. 55: **Modelling Soil Erosion by Water.**
Edited by J. Boardman and D. Favis-Mortlock. 1998.

Vol. 56: **Ice Physics and the Natural Environment.**
Edited by J. S. Wettlaufer, J. G. Dash, and N. Untersteiner. 1999.

MIX
Papier aus verantwortungsvollen Quellen
Paper from responsible sources
FSC® C105338

If you have any concerns about our products,
you can contact us on
ProductSafety@springernature.com

In case Publisher is established outside the EU,
the EU authorized representative is:
Springer Nature Customer Service Center GmbH
Europaplatz 3, 69115 Heidelberg, Germany

Printed by Libri Plureos GmbH
in Hamburg, Germany